AF389501

Energy Systems and Applications in Agriculture

Energy Systems and Applications in Agriculture

Editors

Muhammad Sultan
Md Shamim Ahamed
Redmond R. Shamshiri
Muhammad Wakil Shahzad

MDPI • Basel • Beijing • Wuhan • Barcelona • Belgrade • Manchester • Tokyo • Cluj • Tianjin

Editors

Muhammad Sultan
Department of Agricultural
Engineering
Bahauddin Zakariya
University
Multan
Pakistan

Md Shamim Ahamed
Biological and Agricultural
Engineering
University of California,
Davis
California
United States

Redmond R. Shamshiri
Engineering for crop
production
Leibniz-Institut für
Agrartechnik und
Bioökonomie e.V
Potsdam
Germany

Muhammad Wakil Shahzad
Mechanical and Construction
Engineering
Northumbria University
Newcastle upon Tyne
United Kingdom

Editorial Office
MDPI
St. Alban-Anlage 66
4052 Basel, Switzerland

This is a reprint of articles from the Special Issue published online in the open access journal *Energies* (ISSN 1996-1073) (available at: www.mdpi.com/journal/energies/special_issues/energy_agriculture).

For citation purposes, cite each article independently as indicated on the article page online and as indicated below:

LastName, A.A.; LastName, B.B.; LastName, C.C. Article Title. *Journal Name* **Year**, *Volume Number*, Page Range.

ISBN 978-3-0365-5008-4 (Hbk)
ISBN 978-3-0365-5007-7 (PDF)

Contents

About the Editors

Muhammad Sultan

Dr. Muhammad Sultan is an Associate Professor in the Department of Agricultural Engineering, Bahauddin Zakariya University, Multan (Pakistan). He holds B.Sc. (2008) and M.Sc. (2010) in Agricultural Engineering from the University of Agriculture, Faisalabad (Pakistan), and Ph.D. (2015) in Energy & Environmental Engineering from Kyushu University (Japan). He did postdoctoral research in Energy & Environmental Engineering (2017) at Kyushu University (Japan) and in Mechatronic Systems Engineering (2019) at Simon Fraser University (Canada). He has published more than 300 articles in international journals, conferences, books, and book chapters. He has been a reviewer for more than 100 SCIE journals and holds an editor role for 10 renowned journals with publishers like SAGE, MDPI, and Frontiers. His research focuses on developing energy-efficient temperature and humidity control systems for agricultural applications including greenhouse, fruits/vegetable storage, livestock, and poultry applications.

Md Shamim Ahamed

Dr. Shamim Ahamed is currently an Assistant Professor in the Department of Biological and Agricultural Engineering at the University of California, Davis (UC Davis). He completed his bachelor's in Agricultural Engineering and Masters's in Farm Power and Machinery from Bangladesh Agricultural University (BAU) and Ph.D. in Environmental Engineering from the University of Saskatchewan in Canada. He also worked as a faculty member at Bangladesh Agricultural University. Before joining UC Davis, he worked as a Postdoctoral fellow at the City University of Hong Kong and Concordia University in Montreal. Dr. is leading the Controlled Environment Engineering lab at UC Davis and has been working in energy systems optimization for Controlled Environments Agricultural (CEA) systems (greenhouses, indoor vertical farming, and livestock barns) since 2013. His research mainly focuses on thermal environment modeling, energy-efficient controls, and HVAC systems design for the next-generation CEA. The study aims to integrate renewable energy to operate the heating, ventilation, air conditioning, and dehumidification system to achieve the net-zero energy status for CEA facilities. His research also spans automated fault detection and diagnosis of automatic control of HVAC systems and other control systems like nutrient management and CO2 enrichment, and lighting.

Redmond R. Shamshiri

Dr. Redmond R. Shamshiri holds a Ph.D. in agricultural automation with a focus on control systems and dynamics. He is a scientist at the Leibniz-Institut für Agrartechnik und Bioökonomie and a lecturer at the Technische Universität Berlin working toward the digitization of agriculture for food security. His main research fields include simulation and modeling for closed-field plant production systems, LPWAN sensors, wireless control, and autonomous navigation. His works have appeared in over 110 publications, including peer-reviewed journal papers, book chapters, and conference proceedings. He is the founder of Adaptive AgroTech Consultancy Network and serves as a section editor and reviewer for various peer-reviewed journals in the field of smart farming.

Muhammad Wakil Shahzad

Dr. Muhammad Wakil Shahzad is a senior lecturer in the Mechanical and Construction Engineering Department, Northumbria University (NU), Newcastle Upon Tyne, United Kingdom.

He is an expert in renewable energy storage and its applications for water treatment, hybrid desalination processes and life cycle cost analysis. He has won many international awards for his innovative desalination cycle, including the National Energy Globe Award Saudi Arabia 2021, Sustainability Medal 2020, Global Innovation Award 2020, National Energy Globe Award Saudi Arabia 2020 and 2019, Excellence and Leadership Award 2019, and IDA Environmental & Sustainability Award 2019. His research has been highlighted at Yahoo Business, Nature Middle East, Arab News, and many other national and international platforms.

Dr. Shahzad has a PhD in Mechanical Engineering from the National University of Singapore and research training from KAUST Saudi Arabia. He has extensive experience in intellectual property development and the commercialization of innovative technologies. He holds eleven international patents. To date, he has published four books, twenty book chapters, more than eighty peer-reviewed journal papers, and more than 110 conference papers. He also received three best paper awards in international conferences. He is an editor of International Communications in Heat and Mass Transfer, an editorial board member of SN Applied Sciences, and a guest editor for topical collections. He is a Chartered Engineer and a mentor for International Desalination Association's Young Leader Program (IDA-YLP). He is also a member of many professional organizations, including the Institute of Mechanical Engineers, International Desalination Association (IDA), International Water Association (IWA), and American Society of Mechanical Engineers (ASME).

Article

Potential Investigation of Membrane Energy Recovery Ventilators for the Management of Building Air-Conditioning Loads

Hadeed Ashraf [1,†], Muhammad Sultan [1,*,†], Uzair Sajjad [2], Muhammad Wakil Shahzad [3], Muhammad Farooq [4], Sobhy M. Ibrahim [5], Muhammad Usman Khan [6] and Muhammad Ahmad Jamil [3]

1　Department of Agricultural Engineering, Faculty of Agricultural Sciences & Technology, Bahauddin Zakariya University, Multan 60800, Pakistan; hadeedashraf15@gmail.com
2　Department of Energy and Refrigerating Air-Conditioning Engineering, National Taipei University of Technology, Taipei 10608, Taiwan; energyengineer01@gmail.com
3　Department of Mechanical and Construction Engineering, Northumbria University, Newcastle Upon Tyne NE1 8ST, UK; muhammad.w.shahzad@northumbria.ac.uk (M.W.S.); muhammad2.ahmad@northumbria.ac.uk (M.A.J.)
4　Department of Mechanical Engineering, University of Engineering and Technology, Lahore 39161, Pakistan; engr.farooq@uet.edu.pk
5　Department of Biochemistry, College of Science, King Saud University, P.O. Box 2455, Riyadh 11451, Saudi Arabia; syakout@ksu.edu.sa
6　Department of Energy Systems Engineering, Faculty of Agricultural Engineering and Technology, University of Agriculture, Faisalabad 38040, Pakistan; usman.khan@uaf.edu.pk
*　Correspondence: muhammadsultan@bzu.edu.pk; Tel.: +92-333-610-8888
†　These authors contributed equally to this work.

Citation: Ashraf, H.; Sultan, M.; Sajjad, U.; Shahzad, M.W.; Farooq, M.; Ibrahim, S.M.; Khan, M.U.; Jamil, M.A. Potential Investigation of Membrane Energy Recovery Ventilators for the Management of Building Air-Conditioning Loads. *Energies* 2022, 15, 2139. https://doi.org/10.3390/en15062139

Academic Editors: Antonio Gagliano and Hom Bahadur Rijal

Received: 19 January 2022
Accepted: 10 March 2022
Published: 15 March 2022

Publisher's Note: MDPI stays neutral with regard to jurisdictional claims in published maps and institutional affiliations.

Abstract: The present study provides insights into the energy-saving potential of a membrane energy recovery ventilator (ERV) for the management of building air-conditioning loads. This study explores direct (DEC), Maisotsenko cycle (MEC) evaporative cooling, and vapor compression (VAC) systems with ERV. Therefore, this study aims to explore possible air-conditioning options in terms of temperature, relative humidity, human thermal comfort, wet bulb effectiveness, energy saving potential, and CO_2 emissions. Eight different combinations of the above-mentioned systems are proposed in this study i.e., DEC, MEC, VAC, MEC-VAC, and their possible combinations with and without ERVs. A building was modeled in DesignBuilder and simulated in EnergyPlus. The MEC-VAC system with ERV achieved the highest temperature gradient, wet bulb effectiveness, energy-saving potential, optimum relative humidity, and relatively lower CO_2 emissions i.e., 19.7 °C, 2.2, 49%, 48%, and 499.2 kgCO_2/kWh, respectively. Thus, this study concludes the hybrid MEC-VAC system with ERV the optimum system for the management of building air-conditioning loads.

Keywords: membrane energy recovery ventilator; energy recovery potential; Maisotsenko cycle evaporative cooling; building air-conditioning; human thermal comfort; Pakistan

1. Introduction

1.1. Background

Energy consumption for domestic use/the building sector has forever been soaring, gradually increasing up to 36% (for 2021) of the total global energy consumption [1]. Pakistan is among the countries with high energy consumption in the building sector (including residential/commercial and public services), leading up to 65% (for 2021) of its total energy consumption [2], whereas the energy consumption in the industry and agriculture/forestry sectors in Pakistan was 26% and 9% (for 2021) of the total energy consumption [2]. In Pakistan, for 2020, more than half (i.e., ~55%) of the total energy consumption in the residential/commercial and public services sectors was consumed in space heating/cooling [3]. This is due to the geophysical location of the country. Most of Pakistan's area lies in the hot desert climate classification of the Köppen climate classification, which leads to increased need of space heating/cooling in the winter/summer [4]. Figure 1 shows the sector-wise energy consumption in Pakistan and energy consumption in buildings. According to Figure 1, residential/commercial and public service buildings consume 264,734 TJ (around 65% of the total sector-wise energy consumption), of which more than half (~55%) is consumed in space heating/cooling. In Pakistan, buildings are not built concerning energy savings or utilization. Generally, standalone vapor compression (VAC) systems are considered suitable for building air-conditioning in Pakistan that contributes to greenhouse gas emissions, ultimately leading to global warming and increased CO_2 in the atmosphere. Therefore, alternate building air-conditioning systems are fundamentally needed. Evaporative cooling systems could potentially prove to be an environmentally friendly building air-conditioning option. Moreover, energy in the form of conditioned air is wasted from the exhausts of buildings. This waste energy can be recovered using membrane energy recovery ventilators (ERVs). The ERVs can recover heat, as well as humidity, from the stale exhaust and exchange them with the outside fresh air.

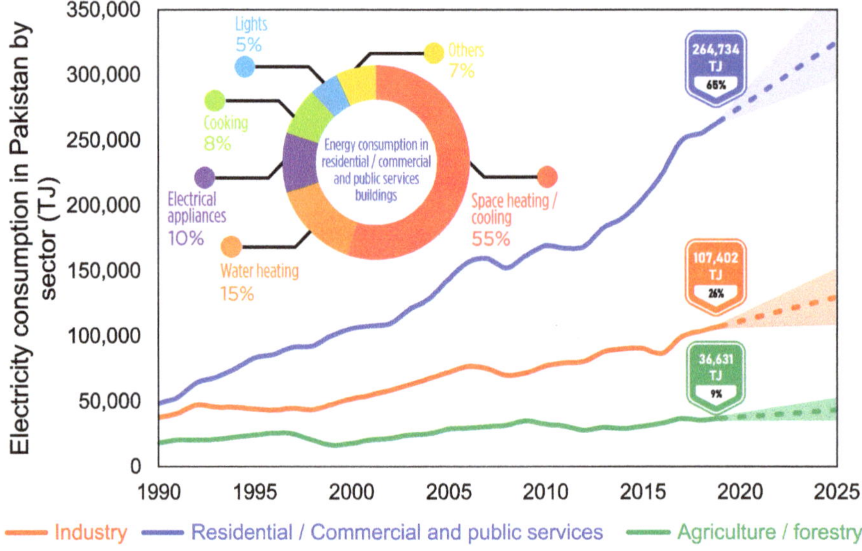

Figure 1. Energy consumption in Pakistan by sector and energy consumption in buildings, reproduced from [5].

1.2. Multi-Stage Air-Conditioning Systems

Evaporative cooling systems coupled with standalone conventional vapor compression systems for the management of building air-conditioning loads and associated applications have been extensively studied in the literature [6–15].

Chun et al. [16] thermodynamically analyzed a MEC system coupled with membrane based dehumidification. The authors concluded that the system performance directly correlated with the outside temperature and humidity ratio. Kowalski et al. [17] studied the indirect evaporative cooling and the DEC systems in Poland. The authors concluded that the indirect evaporative cooling system outperformed the DEC. Additionally, the proposed system performed better in the dedicated outside air scenario compared to the recirculation air scenario. Da Veiga et al. [18] modeled an evaporative cooling system for building roofs and experimentally validated the results. The authors found a strong correlation between the performance of the proposed system and irradiation, outside temperature, relative humidity, and windspeed. In a follow-up study, da Veiga et al. [19] further studied the

DEC system on a global level for building roofs. The results indicated that the DEC system is more likely to perform better in a cold semi-arid climate (BSk) and warm desert climate class (BWh) of the Köppen climate classification areas [19]. Tewari et al. [20] experimentally studied the DEC system for office building air-conditioning. According to the results, the optimum relative humidity, temperature, and windspeed for human thermal comfort were 35–85%, 28.8 °C, and 0.75–1.5 m/s, respectively. Kim et al. [21] studied the integration of DEC and indirect evaporative cooling systems and compared them with a conventional VAC system in an outdoor air-conditioning system. According to the results, the multi-staging of evaporative cooling systems resulted in a total energy consumption of 1.1 kWh compared to the conventional VAC system (i.e., 2.1 kWh). Cui et al. [22] theoretically investigated a MEC system for air-conditioning applications from the viewpoint of wet bulb (WBE) and dewpoint (DPE) effectiveness. According to the results, the proposed system achieved 1.2 to 1.3 WBE and 0.8 to 0.9 DPE under varying inlet temperature and relative humidity conditions. Gómez et al. [23] experimentally investigated two modes of air flow in a prototype polycarbonate indirect evaporative cooling system. According to the results, heat transfer, cooling capacity, and thermal effectiveness were improved in the case of the polycarbonate evaporative cooling system. Heidarinejad et al. [24] theoretically investigated an indirect evaporative cooling system and a MEC system. Numerical modeling was used to predict the performances of both the systems. According to the results, the WBE performance of the MEC system comparatively increased by 60%. Cui et al. [25] used computation fluid dynamics for the performance prediction of the MEC system. The results from the study were within the ±10% range of the actual experimental data. Moreover, the authors summarized that the MEC system could potentially achieve higher WBE and DPE for a larger length-to-height ratio (at least 200 times), lower air velocity (<1.5 m/s), and smaller channel height (<10 mm). Moshari et al. [26] studied the performance of a multi-stage evaporative cooling system for a variety of ambient conditions. According to the results, the WBE of the multi-stage evaporative cooling system peaked at 0.9, whereas it was 0.6 in the case of a standalone indirect evaporative cooling system. According to the results, the multi-staging of evaporative cooling systems can improve the performance at the expense of a higher energy input. Cui et al. [27] investigated the performance of a multi-stage VAC and indirect evaporative cooling system. According to the results, the EC component pretreated the outside air, resulting in a ~47% drop in the cooling load on the compressor at the VAC stage, which ultimately leads to an appreciable increase in the energy-saving potential. The numerical results from the study were within a 9.7% range of actual experimental data. Campisi et al. [28] studied the energy efficiency in a case study of Italy. The authors concluded that the best energy-saving technique was the installation of solar thermal panels with a heat pump for optimum energy savings and lesser CO_2 emissions. Obando et al. [29] investigated the effect of the temperature of the water on the DEC system for livestock building air-conditioning. According to the experimental results, the supply air temperature dropped 0.7 °C per drop in temperature, and the supply air relative humidity increased by ~0.9%. Badiei et al. [30] numerically predicted the MEC performance. The authors concluded that the energy building model accurately predicted the performance of the MEC system as compared to the numerical model. Nada et al. [31] analyzed the DEC with cellulose pads for the climatic conditions of Egypt. According to the results, the DEC system achieved a maximum WBE of 0.85 at different ambient conditions. He et al. [32] experimentally studied a solar-operated dehumidification unit to pretreat the outside air before entering an evaporative cooling unit. According to the results, the proposed system could achieve a supply temperature and relative humidity of 28 °C and 70%, which satisfied the required human thermal comfort levels. Boukhanouf et al. [33] designed and developed a MEC system for the management of building air-conditioning loads in an arid climate. The study proposed porous ceramic and heat pipes for heat exchange purposes. According to the results, the proposed system achieved a peak WBE, COP, and temperature gradient (i.e., referred to as the subtraction of the outside and system outlet temperatures) of 0.8, 11.43, and 14 °C, respectively. Zanchini et al. [34]

investigated the MEC-VAC system in terms of energy-saving potential for the climatic conditions of North Italy. According to the results, the proposed system yielded 38% energy and electricity savings as compared to the conventional VAC system. Khandelwal et al. [35] studied the energy-saving potential of the DEC and MEC systems coupled with a water chiller. According to the results, the proposed systems yielded peak energy savings i.e., 12% and ~15.7%, respectively, whereas the predicted mean vote (PMV) was between −1 and +1.

The results from these studies led to general summarizations that (i) multi-stage EC systems can potentially improve the performance from the viewpoints of the WBE, DPE, and system outlet humidity and temperature, and (ii) the MEC system can result in a relatively higher performance for hot and dry climatic conditions. Figure 2 represents the Köppen's climate classification of the world and Pakistan. Pakistan is represented by the warm desert climate (referred to as BWh), which imposes a need for air conditioning.

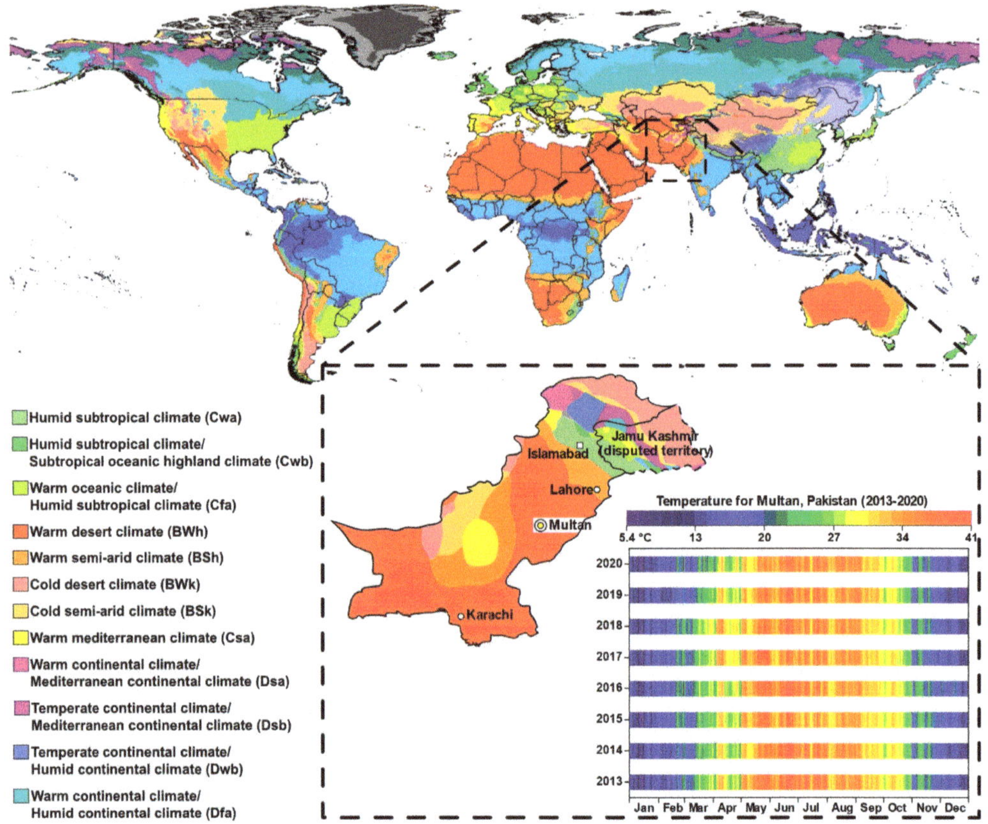

Figure 2. Köppen climate conditions of Pakistan and a temperature profile for Multan, Pakistan, reproduced from [36,37].

1.3. Membrane Energy Recovery Ventilators

Gao et al. [38] studied the membrane energy recovery ventilator (ERV) from the viewpoints of latent heat, sensible heat, and total energy exchange efficiency. According to the results, the developed model accurately predicted the performance of the ERV within ±7% of the actual experimental data for both the summer and winter conditions. Abadi et al. [39] studied the condensation phenomenon in the ERV core. According to the results, the exhaust sensible effectiveness was increased due to condensation. However, in countries like Pakistan where most of the region lies in a hot and dry area, condensation is not a problematic factor. Huang et al. [40] experimentally investigated an indirect evaporative cooling system coupled with a heat recovery ventilator in different regions of China. According to the results, the performance of the proposed system increased in hot and humid regions (i.e., Guangzhou). Additionally, the authors concluded that the energy-saving potential of the vertical arrangement of the indirect evaporative cooling system was relatively more (~almost twice) compared to the horizontal arrangement. Chen et al. [41] investigated the performance of a MEC system coupled with the ERV system for air conditioning in wet markets of Hong Kong. The authors concluded that a MEC-based ERV system resulted in a maximum energy-saving potential (specifically in the summer, up to 45% more) compared to the standalone total energy or heat recovery. Qiu et al. [42] developed a regression model for the performance prediction of an ERV system. The authors declared that thermal properties of the membrane in the ERV core were not a prerequisite. The developed numerical polynomial model accurately predicted the results within ±8% of the experimental data. Zhong et al. [43] modeled the performance of an ERV system for building air conditioning. The performance of the ERV was influenced by effectiveness, occupancy of the building, and outside air conditions. Moreover, the peak yearly energy efficiency of the ERV in hybrid mode was 43%. Rasouli et al. [44] studied an ERV for ten-story building in four different climates of the USA. The authors concluded that the peak energy savings during cooling mode in the summer season was 20% under specified control conditions, whereas it was 40% in the heating mode in the winter. Moreover, the authors declared that the proposed system yielded a higher performance in humid climates with higher latent loads. Zhou et al. [45] modeled the performance of the ERV system in Shanghai and Beijing. According to the results, the ERV performed better in Shanghai under winter conditions, resulting in peak energy savings as compared to Beijing.

From the above discussion, to the best of the authors' knowledge, it can be stated that membrane energy recovery ventilators are not widely used in Pakistan. Therefore, this study provides a comprehensive assessment on the potential investigation of ERV systems for the management of building air-conditioning loads in Pakistan. According to Figure 2, Pakistan is represented by BWh class of Köppen's climate classification, which enforces a need for air conditioning. In this regard, this study explores combinations of direct (DEC), Maisotsenko cycle (MEC) evaporative cooling, and typical mechanical vapor compression (VAC) systems with a membrane energy recovery ventilator in Pakistan.

2. Materials and Methods

2.1. Building Energy Simulation

A nonresidential, office building (reported in the authors' previous work [6]) was used as a baseline for this study. The total area for the building was 3251.74 m^2, of which 847.5 m^2 was conditioned and 2404.24 m^2 was unconditioned of the building area. Figure 3 shows the proposed systems/configurations for the management of building air-conditioning loads. All the proposed systems/configurations presented in Figure 3 were designed in the AHU module of DesignBuilder [46]. All the thermophysical properties of the baseline building were incorporated in the design in DesignBuilder and simulated using the EnergyPlus simulation engine [47]. Table 1 shows the details of the design specifications of the building construction and air-conditioning systems. In the present study, standalone direct evaporative cooling (DEC), Maisotsenko cycle evaporative cooling (MEC), conventional vapor compression (VAC), and hybrid MEC-VAC systems are explored as possible alternative options for building air conditioning coupled with two configurations i.e., with and without ERV and recirculation. Figure 3 shows the proposed configurations and air-conditioning systems for energy recovery and building air conditioning. The working phenomenon of each system/configuration studied in this manuscript is given below:

- DEC system (Figure 3a). This system humidifies and simultaneously cools the outside air, increasing the humidity of the supply air in an isenthalpic manner.

- MEC system (Figure 3b). This system provides two streams of air, i.e., sensibly cooled air and humidified hot air. Ideally, this system can achieve below the wet bulb temperature, i.e., dewpoint temperature.
- VAC system (Figure 3c). This system is the conventional mechanical vapor compression system based on refrigerant cooling.
- MEC-VAC system (Figure 3d). This system essentially behaves like a hybrid of the MEC and the VAC systems. The MEC system pretreats the outside air (reducing its temperature) to reduce the cooling coil loads in the VAC system.
- Air-conditioning system with ERV and recirculation configuration (Figure 3). This configuration allows the energy from the stale exhaust air to be recovered and exchanged into the supply air before exhaust. Additionally, the recirculation mode is used in this configuration to divert/mix a portion of the outside air depending upon the thermal comfort requirements of the inside conditions.
- Air-conditioning system without ERV and recirculation configuration (Figure 3). This configuration does not allow any sensible/latent energy recovery from the stale air stream, essentially behaving like a simple air handling unit (AHU) consisting of the four proposed air-conditioning systems.

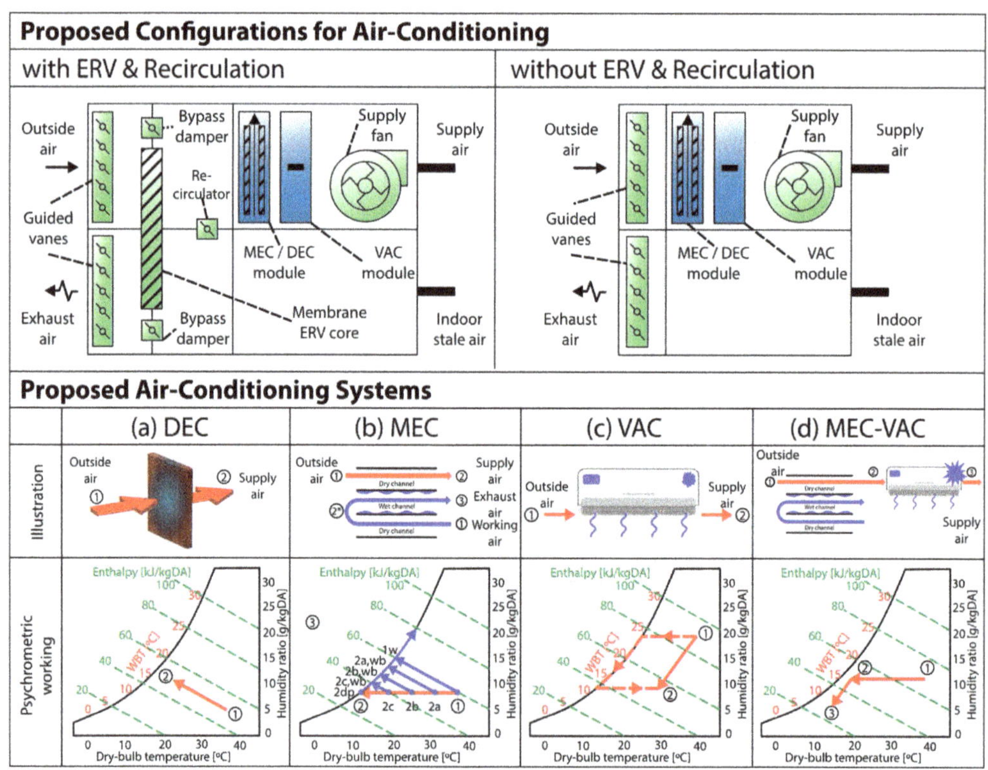

Figure 3. Proposed configurations and air-conditioning systems for energy recovery and building air conditioning.

Table 1. Design specifications of the building construction and air-conditioning systems.

Building Construction Information		
Type	**Details**	**Properties**
Wall 1st layer	Brickwork outer	4 in
Wall 2nd layer	Cement	0.5 in
Wall 3rd layer	Brickwork inner	4 in
Wall 4th layer	Cement	0.5 in
Flat roof Layer 1	Concrete at R 0.0625/in	0.5 in
Flat roof Layer 2	Cement	0.5 in
Floor 1st layer	Concrete aggregate at R 0.0625/in	2 in
Floor 2nd layer	Cast concrete (dense)	5 in
Window	Clear	0.118 in
	W	59.84 in
Dimensions	H	72.04 in
Sill height		48.03 in
	Panels	3
	Frame: Aluminum (no insulation break)	
	Frame width	1.5 in
Area of the building		3251.74 m^2
Area of conditioned space		847.5 m^2
Zones in conditioned space	24/47 (excluded zones include corridors, kitchen, and washrooms.)	
Lighting power density fractions	Visible	0.2
	Radiant	0.7
Population	Person population	0.295 persons/conditioned area
	Metabolic factor	0.9
Schedule	Workday schedule	On at: 08:00 Off at: 16:00
	Workdays/week	5
Clothing	Winter	1.145 clo
	Summer	0.775 clo
	Typical Pakistani wear (Shalwar Kameez) [48,49]	
Fan	Total efficiency	75%
	Motor efficiency	85%
	Fan pressure	600 Pa
	Motor in air stream (*draw through system*)	Yes
AHU information		
ERV	Type	Counterflow
	Design Sensile effectiveness	0.75
	Design Latent effectiveness	0.65
	Silica-PE substrate	105 μm
	Silica-PE substrate permeability (at 50 °C, 50% RH)	6.2×10^{-10} mol·m/m^2/s/Pa
	Silica-PE substrate density	600 kg/m^3
	Silica-PE substrate thickness	105×10^{-6} m
	Silica-PE thermal conductivity	0.44 W/m/k
	PU-PEO coating film	
	PU-PEO coating permeability (at 50 °C, 50% RH)	1.8×10^{-11} mol·m/m^2/s/Pa
	PU-PEO coating density	1210 kg/m^3
	PU-PEO coating thickness	2×10^{-6} m
	PU-PEO thermal conductivity	0.159 W/m/k

Table 1. *Cont.*

Building Construction Information		
Type	**Details**	**Properties**
VAC	VAC rated COP Coil type Condenser type Rated evaporator fan power per volume flow rate Operation	3 Single speed Air cooled 773.3 W/m^3/s On sensible load
DEC	*Direct research special* module Design wet bulb effectiveness Water pump power sizing factor	 0.90 90.0 W/m^3/s
MEC	*Indirect research special* module Design wet bulb effectiveness Design dewpoint effectiveness Water pump power sizing factor	 1.06 0.75 90.0 W/m^3/s
Recirculation Thermostat Humidistat	Varies in zones based on cooling load requirement	0.64 (average) 24 °C 60%

2.2. Performance Assessment Indicators

The temperature gradient, relative humidity, predicted mean vote (PMV), psychrometric profile, carbon dioxide (CO_2) emissions, predicted percentage of dissatisfied (PPD), wet bulb effectiveness, electricity consumption, and energy-saving potential were used as the performance indicators. The wet bulb effectiveness of the proposed system was calculated using Equation (1):

$$\varepsilon_{wb} = \frac{T_{OA,DBT} - T_{SA,DBT}}{T_{OA,DBT} - T_{OA,WBT}} \tag{1}$$

where ε_{wb} represents the wet bulb effectiveness; T represents the dry bulb temperature (°C); and the subscripts SA, OA, WBT, and DBT denote the supply air, outside air, wet bulb, and dry bulb temperatures, respectively. Equation (2) was used to calculate the cooling capacity of the proposed systems:

$$Q = mC_p(\Delta T) \tag{2}$$

where Q represents cooling capacity of the system (kW), m represents the mass flow rate of air in the system (i.e., ~14.5 kg/s for this study), C_p represents the specific heat capacity and ΔT denotes the difference in the system outlet temperatures between OA and SA (°C). Equations (3)–(6) were used to calculate the cooling capacity of the VAC or hybrid VAC systems, with the latent heat transfer taken into account:

$$m = \rho_{air}u \tag{3}$$

$$Q_{net} = m(h_{SA} - h_{OA}) \tag{4}$$

$$Q_{sensible} = m(h_{SA} - h_{OA})X_{min} \tag{5}$$

$$Q_{latent} = Q_{net} - Q_{sensible} \tag{6}$$

where ρ_{air}, u, h, X_{min}, Q_{net}, $Q_{sensible}$, and Q_{latent} represent the density of the air (kg/m^3) volumetric flow rate (m^3/s), total enthalpy of air (J/kg), minimum of the two humidity ratios at OA and SA (kg/kg), the total cooling capacity, the total sensible cooling capacity and the total latent cooling capacity (kW). Equation (7) was used to calculate the average area of a human body to further facilitate the estimation of the total heat and mass transfer from the human body:

$$A_{hb} = 0.202 M^{0.425} H^{0.725} \tag{7}$$

where A, M, and H represent the average area, mass, and height of a typical human body. Subscript hb represents the human body. Equations (8)–(17) were used to calculate the heat gained from the human body (Q_{hb}), sol–air temperature (T_{sa}), heat gained from exterior surfaces (Q_{ext}), heat gained from fans and motors (Q_m), heat gained from walls (Q_{wall}), heat gained from windows ($Q_{glazing}$), heat gained from the floor (Q_{sp}), heat gains/losses from infiltration (ACH), latent load gains/losses through infiltration ($Q_{l,i}$), and sensible load gains/losses through infiltration ($Q_{s,i}$), respectively. The details of the parameters can be found from the cited literature [6].

$$Q_{hb} = (Q_{sensible} + Q_{latent})_{skin} + (Q_{sensible} + Q_{latent})_{lungs} \tag{8}$$

$$T_{sa} = T_a + \frac{\alpha_{solar} q_{solar}}{h_o} - \frac{E\varphi(T_a^4 - T_{sur}^4)}{h_o} \tag{9}$$

$$Q_{ext} = h_o A_s(T_a - T_s) + \alpha_{solar} A_s q_{solar} - E A_s \varphi\left(T_a^4 - T_{sur}^4\right) \tag{10}$$

$$Q_m = \frac{P_r L_f U_f}{\eta_m} \tag{11}$$

$$Q_{wall} = U_{wall} A_s(T_{s-a} - T_{in}) \tag{12}$$

$$Q_{glazing} = Q_c + Q_e + Q_f = U_g A_g(T_{in} - T_{out}) \tag{13}$$

$$Q_{sp} = U_{floor} P_{floor}(T_{in} - T_{out}) \tag{14}$$

$$ACH = \frac{m_{in}}{V} \tag{15}$$

$$Q_{l,i} = \rho_{air} h_{fg} ACHV(X_{in} - X_{out}) \tag{16}$$

$$Q_{s,i} = \rho_{air} C_p ACHV(T_{in} - T_{out}) \tag{17}$$

2.3. Human Thermal Comfort Indices

Comfort for human beings was assessed from the viewpoints of Fanger's predicted mean vote (PMV) and predicted percentage of dissatisfied (PPD) [50,51]. Equations (18)–(22) were used to calculate the PMV and PPD. Details of the parameters can be found from the cited literature [6].

$$
\begin{aligned}
\text{PMV} = {}& [0.303\, e^{(-0.036 MBR)} + 0.028](MBR - HL) \\
& -3.05[5.73 - 0.007(MBR - HL) - P_{OA}] \\
& -0.42[(MBR - HL) - 58.15] - 0.0173\, MBR(5.87 - P_{OA}) \\
& -0.0014\, MBR(34 - T_{OA}) \\
& -3.96\, e^{-8}\, RCNB[(CT + 273)^4 - (MRT + 273)^4] \\
& -RCNB\, h_{conv}(CT - T_{OA})
\end{aligned}
\tag{18}
$$

$$
\begin{aligned}
CT = {}& 35.7 - 0.0275(MBR - HL) \\
& - R_c\{(MBR - HL) - 3.05[5.73 - 0.007(MBR - HL) - P_{OA}\} \\
& -0.42[(MBR - HL) - 58.15] - 0.0173\, MBR(5.87 - P_{OA}) \\
& -0.0014\, MBR(34 - T_{OA})
\end{aligned}
\tag{19}
$$

$$
h_{conv} = \begin{cases} 2.38(CT - T_{OA})^{0.25} & \text{for} \quad 2.38\,(CT - T_{OA})^{0.25} > 12.1\,\sqrt{u_{OA}} \\ 12.1\,\sqrt{u_{OA}} & \text{for} \quad 2.38\,(CT - T_{OA})^{0.25} < 12.1\,\sqrt{u_{OA}} \end{cases}
\tag{20}
$$

$$
r_{cb} = \begin{cases} 1.0 + 0.2\, CI & \text{for} \quad CI < 0.5\, clo \\ 1.05 + 0.1 CI & \text{for} \quad CI > 0.5\, clo \end{cases}
\tag{21}
$$

$$PPD = 100 - 95\,e^{\left(-0.03353\text{PMV}^4 - 0.2179\text{PMV}^2\right)} \tag{22}$$

where MBR denotes the metabolic rate in W m^{-2}, HL denotes the loss of heat from the body in W m^{-2}, T_{OA} denotes the room's temperature in °C, P_{OA} denotes the vapor pressure of the outside air, $RCNB$ denotes the cloth–naked body ratio, CI denotes clothes' insulation in clo, CT denotes clothes' temperature in °C, h_{conv} denotes the convective heat transfer coefficient in W m^{-2} °C^{-1}, MRT denotes the mean radiant temperature in °C, and u_{OA} denotes the air flow velocity in m s^{-1}.

3. Results and Discussion

3.1. Design Parameters

Effectiveness of a high-efficiency energy recovery ventilator (ERV) used as design effectiveness in DesignBuilder at varied design flow rates are presented in Figure 4. Experimental data for the effectiveness was obtained from the literature [52]. This specific membrane ERV was selected for the study due to its low-pressure drop at the supply air side, high effectiveness towards blocking of a variety of gases, prevention of recirculation of odors into the supply air, and smaller lead time. According to Figure 4, design effectiveness alpha represents the effectiveness at 57.7 ft/min active face velocity, whereas design effectiveness beta represents the effectiveness at 107.5 ft/min active face velocity. The sensible design effectiveness at a 75% heating and cooling flow rate was 0.75, whereas it was 0.70 at a 100% heating and cooling flow rate. The latent design effectiveness at a 75% heating and cooling flow rate was 0.65, whereas it was 0.60 at a 100% heating and cooling flow rate. Figure 5 shows the numerical values of both wet bulb and dewpoint effectiveness of the experimental lab-scale prototypes of the DEC and MEC systems, respectively, for the summer months, obtained from the authors' previous work [53]. The design values were used in simulation models developed in DesignBuilder. The design values used in DesignBuilder are presented in Table 1. The developed model was simulated for Multan Pakistan in EnergyPlus.

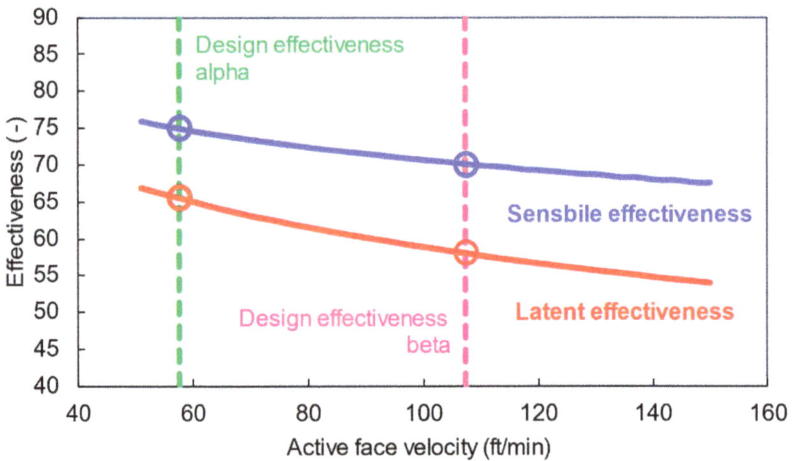

Figure 4. Membrane energy recovery ventilator design effectiveness for DesignBuilder at varied flow rates, reproduced from the literature [52].

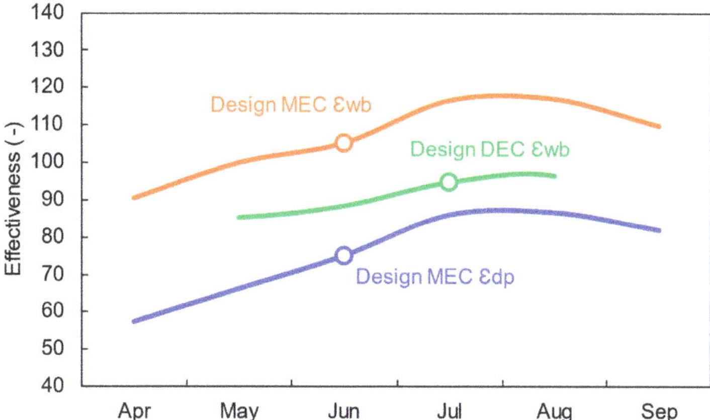

Figure 5. Design effectiveness of lab-scale evaporative cooling systems (i.e., MEC εwb, MEC εdp, and DEC εwb) for DesignBuilder for the summer in Multan [53].

3.2. Performance Profiles of the Systems

3.2.1. Temperature/Humidity

The results from the simulation are presented in Figures 6 and 7. Figure 6 shows the marginal distribution curves of the temperature gradient of the proposed system configurations against the ambient temperature for the summer months of the study area. According to Figure 6, the MEC-VAC system with the ERV and recirculation configuration achieved a maximum temperature gradient of 19.8 °C (on 16 June), with an average 13.2 °C in the summer months. MEC-VAC without the ERV system achieved a maximum temperature gradient of 16.6 °C, with an average of 10.7 °C. It is worth mentioning that the distribution curves in Figure 6 represent the data height in terms of count rather than the relative frequency or density of the data. The DEC and MEC systems could only achieve a maximum temperature gradient of 9.2 °C and 10.8 °C, respectively, without ERV and recirculation configuration, with an average temperature gradient of 4.7 °C and 5.5 °C, respectively. It is worth mentioning that the temperature gradient refers to the subtraction of outside/ambient air and the system outlet temperature. Meanwhile, both the standalone evaporative cooling systems achieved a maximum temperature gradient of 5.9 °C and 10.4 °C, respectively, with the ERV and recirculation configuration in the summer months having an average of 2.3 °C and 5.4 °C. Contrarily, the standalone VAC system achieved a maximum temperature gradient of 13.4 °C and 19.7 °C without and with the ERV and recirculation configurations, with an average temperature gradient of 9.3 °C and 13.2 °C, respectively. From these results, it can be deduced that the standalone VAC system with ERV configuration and the hybrid MEC-VAC system with ERV configuration performed best as compared to the other systems. It could be argued that the VAC with ERV performed nearly equal to the hybrid MEC-VAC with ERV; however, the VAC with ERV system consumed more electricity as compared to the MEC-VAC with ERV system. According to the marginal distribution curves, the MEC-VAC with ERV system yielded an overall smooth performance from the point of view of the supply air temperature. The MEC-VAC with ERV system maintained an overall temperature of ~20–22 °C throughout the summer months. Similarly, the VAC with ERV system also maintained a similar profile of temperature throughout the summer months. The marginal distribution curves of the humidity of the proposed system configurations against the ambient relative humidity for the summer months of the study area are presented in Figure 7. According to Figure 7, the maximum ambient air relative humidity for the summer months was 81%, with an average relative humidity of 48%. It is worth mentioning that the distribution curves in Figure 7 represent the data height in terms of count rather than relative frequency or density of the data. Consequently, the standalone DEC system without and with ERV configuration achieved a maximum relative humidity of 92% and 95%, with an average relative humidity of 76% and 85%, respectively. Similarly, the standalone MEC system without and with ERV configuration achieved a maximum relative humidity of 89% and 90%, with an average relative humidity of 65% and 66%, respectively. The comprehensive results of the study are presented in Table 2.

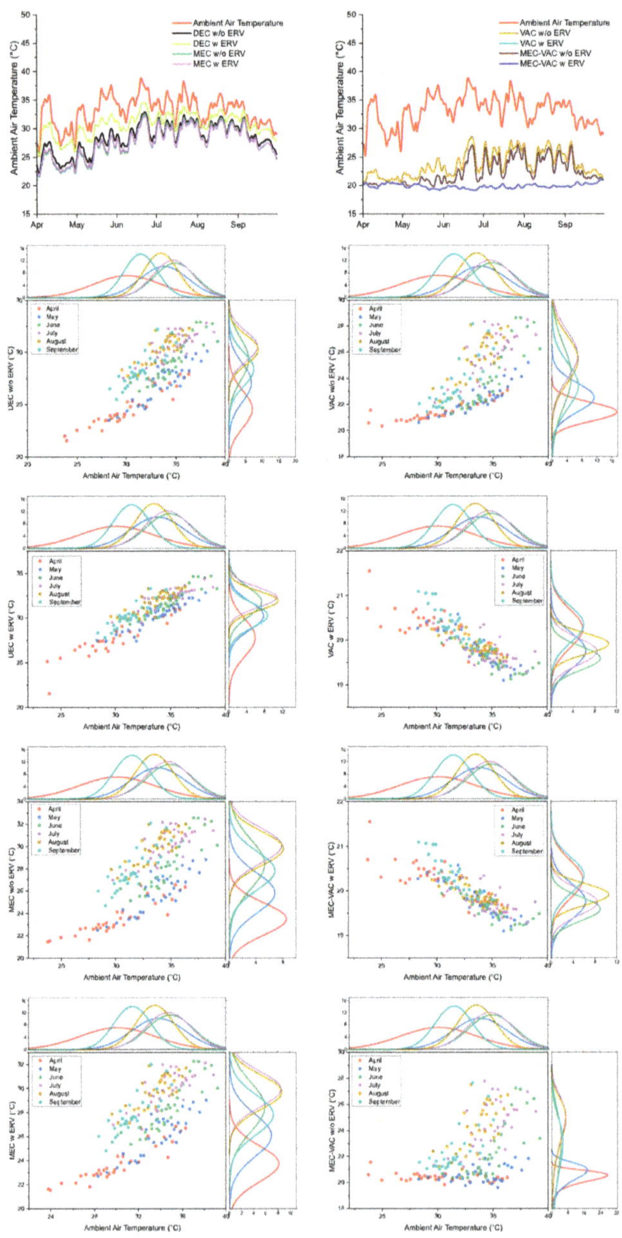

Figure 6. Marginal distribution curves of the proposed system configurations' temperature against the ambient temperature for the summer months of Multan (Pakistan).

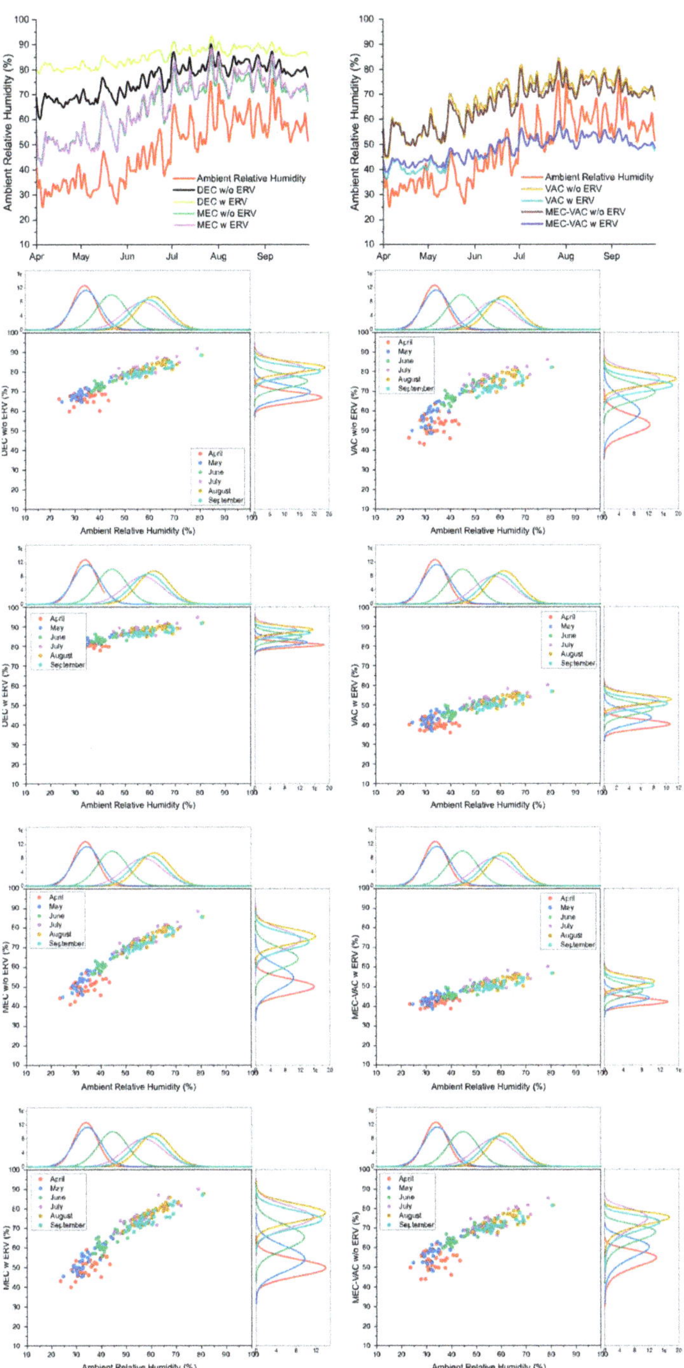

Figure 7. Marginal distribution curves of the proposed system configurations' relative humidity against the ambient relative humidity for the summer months of Multan (Pakistan).

Table 2. Comprehensive results of the performances of the proposed systems.

	Temperature Gradient (°C)		Relative Humidity (%)		MRT (°C)		PMV (-)	PPD (%)		CO$_2$ Emissions (kgCO$_2$/kWh)	WBE (-)	Electricity Consumption (kWh/day)	
	Max.	Avg.	Max.	Avg.	Max.	Avg.		Range	Avg.		Max.	Max.	Avg.
DEC w/o ERV	9.2	4.7	92	76	36.1	32.3	−2.43–3.43	17.6–96.7	50	134.5	0.55	281.6	266.5
DEC w/ERV	5.9	2.3	95	85	36.9	33.4	−0.89–3.0	17.6–96.7	50	135.6	0.31	281.5	268.6
MEC w/o ERV	10.8	5.5	89	65	36.0	32.0	−2.84–3.0	13.4–94.5	46.3	159.2	0.65	354.6	315.3
MEC w/ERV	10.4	5.4	90	66	35.8	32.5	−2.77–3.0	14.4–94.0	45.7	160.2	0.64	354.6	317.4
VAC w/o ERV	13.4	9.3	86	68	34.2	30.2	−3.0–1.39	14.4–98.8	54.0	878.5	1.27	3018.6	1739.7
VAC w/ERV	19.7	13.2	60	48	29.6	28.2	−2.28−−3.0	84.6–99.5	98.7	538.1	2.2	2176.6	1065.7
MEC-VAC w/o ERV	16.6	10.7	86	68	33.5	29.5	0.71−−3.0	9.4–99.3	67.2	749.5	1.46	3106.7	1484.2
MEC-VAC w/ERV	19.8	13.2	60	48	31.9	29.6	−2.2−−3.0	83.2–99.5	98.7	499.2	2.2	2176.6	988.5

On the other hand, the standalone VAC system without and with ERV configuration achieved a maximum relative humidity of 86% and 60%, with an average relative humidity of 68% and 48%, respectively. Contrarily, the hybrid MEC-VAC system without and with ERV configuration achieved a maximum relative humidity of 85% and 60%, with an average relative humidity of 66% and 49%, respectively. According to these results, the standalone VAC system with the ERV and recirculation configuration and the hybrid MEC-VAC system with the ERV and recirculation configuration achieved the desired average relative humidity level (i.e., 40–60% [54]) prescribed for human thermal comfort during the summer months of the study area. From Figures 6 and 7, it can be concluded that the standalone VAC system with the ERV and recirculation configuration, and the hybrid MEC-VAC system with the ERV and recirculation configuration achieved the desired performance level in terms of the humidity and temperature required for optimum human thermal comfort.

3.2.2. Mean Radiant Temperature

Figure 8 shows the psychrometric performance profile of the proposed system configurations correlating with the mean radiant temperature and human thermal comfort. According to Figure 8, the standalone DEC system without and with the ERV and recirculation configuration achieved a maximum mean radiant temperature of 36.1 °C and 36.9 °C and average MRT of 32.3 °C and 33.4 °C, respectively. However, according to the color-coding in Figure 8, the average MRT of this system should be <26 °C to achieve the optimum human thermal comfort.

Similarly, the standalone MEC system without and with the ERV and recirculation configuration achieved a maximum mean radiant temperature of 36.0 °C and 35.8 °C with an average mean radiant temperature of 32.0 °C and 32.5 °C, respectively. However, according to the color-coding in Figure 8, the average MRT of this system should also be <26 °C to achieve the optimum human thermal comfort. Contrarily, the standalone VAC system without and with the ERV and recirculation configuration achieved a maximum mean radiant temperature of 34.2 °C and 29.6, with an average mean radiant temperature of 30.2 °C and 28.2 °C, respectively. However, according to the color-coding in Figure 8, the average MRT of the VAC without the ERV system should be <26 °C, whereas, in the case of the VAC with the ERV system, it should be 30.5 °C–31.9 °C. Although the VAC system with the ERV and recirculation configuration achieved the desired temperature and relative humidity levels (as per Figures 6 and 7), the system failed to achieve the required mean radiant temperature (as per Figure 8). On the other hand, the hybrid MEC-VAC system without and with the ERV and recirculation configuration achieved a maximum mean radiant temperature of 33.5 °C and 31.9 °C, with an average 29.5 °C and 29.6 °C MRT. According to the color-coding in Figure 8, the average MRT for the MEC-VAC without the ERV system should be 29.6 °C–30.5 °C, whereas the MRT for the MEC-VAC system with the ERV system should be 29.6 °C–31.9 °C, respectively. According to these results, the hybrid MEC-VAC system with the ERV and recirculation configuration achieved the desired MRT system outlet humidity and temperature.

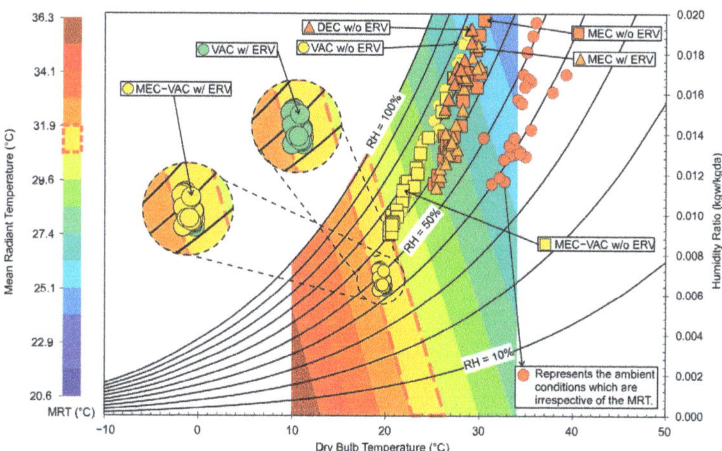

Figure 8. Performance profile of the proposed system configurations correlating with the mean radiant temperature and thermal comfort.

3.2.3. Human Thermal Comfort

The profile of the proposed system/configurations in terms of the predicted percentage dissatisfied (PPD) and predicted mean vote (PMV) for the summer months of the study area is presented in Figure 9. According to Figure 9, the standalone DEC system without the ERV and recirculation configuration achieved a PMV of −2.43 to 3.43. It is worth mentioning that the distribution curves in Figure 9 represent the data height in terms of count rather than the relative frequency or density of the data. Similarly, the standalone DEC system with the ERV and recirculation configuration achieved a PMV of −0.89 to 3.00. Similarly, the standalone MEC system without the ERV and recirculation configuration achieved a PMV of −2.84 to 3.00. In addition, the standalone MEC system with the ERV and recirculation configuration achieved a PMV of −2.77 to 3.00. Additionally, the standalone VAC system without the ERV and recirculation configuration achieved a PMV of −3.00 to 1.39. On the other hand, the standalone VAC system with the ERV and recirculation configuration achieved a PMV of −2.28 to −3.00. Contrarily, the hybrid MEC-VAC system without the ERV and recirculation configuration achieved a PMV of 0.71 to −3.00. In addition, the hybrid MEC-VAC system with the ERV and recirculation configuration achieved a PMV of −2.20 to −3.00. From the predicted percentage of dissatisfied (PPD) point of view, according to Figure 9, both the standalone DEC system without and with the ERV and recirculation configuration achieved a PPD of 17.6–96.7%, with an average of 50% PPD during the summer months of the study area. Additionally, the standalone MEC system without and with the ERV and recirculation configuration achieved a PPD of 13.4–94.5% and 14.4–4.0%, with an average PPD of 46.3% and 45.7%, respectively. Moreover, the standalone VAC system without and with the ERV and recirculation configuration achieved a PPD of 14.4–98.8% and 84.6–99.5%, with an average PPD of 54.0% and 98.7%, respectively.

In the case of the hybrid systems, the MEC-VAC system without and with the ERV and recirculation configuration achieved a PPD of 9.4–99.3% and 83.2–99.5%, with an average PPD of 67.2% and 98.7%, respectively. According to the results, the standalone VAC system and the hybrid MEC-VAC system with the ERV and Recirculation configuration achieved more than 98% PPD throughout the summer months, which caused dissatisfaction among the human subjects. In the case of MEC-VAC with the ERV and recirculation configuration, although high PMV and PPD correlate to discomfort under certain conditions, it could be tackled by using an air-conditioning schedule, economizer, and time-to-time operation of the outside air mixing/recirculation.

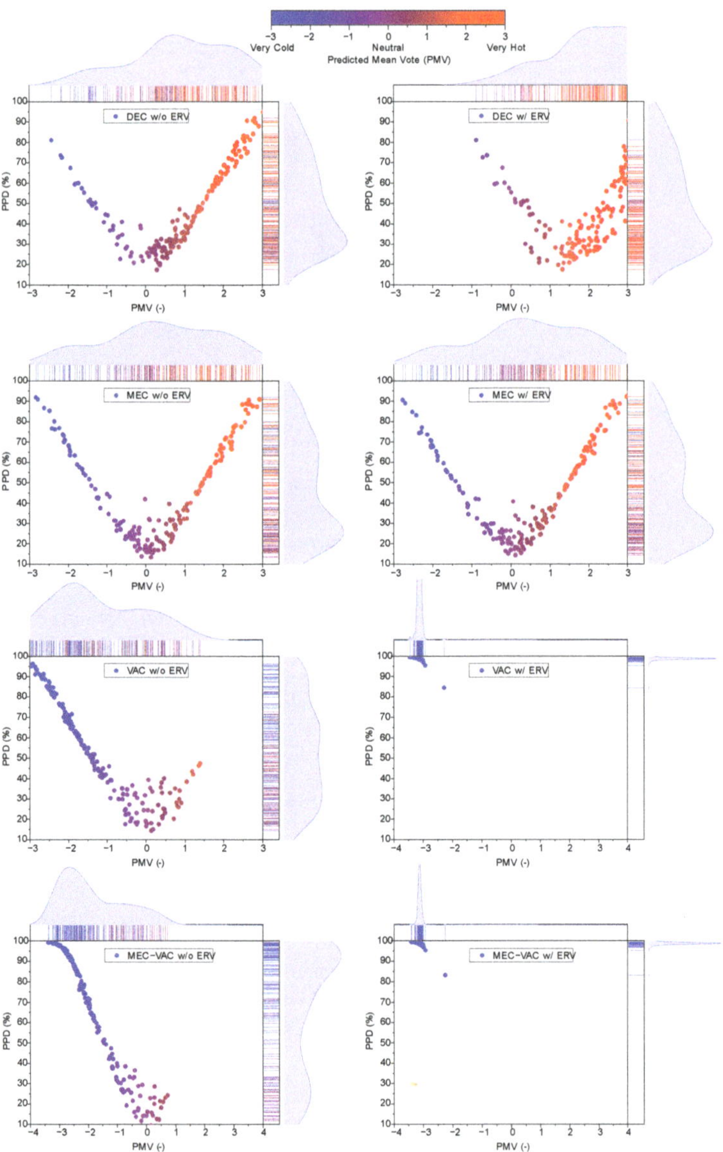

Figure 9. Variation in thermal comfort indices (i.e., PMV and PPD) of the proposed system configurations for the summer months.

3.2.4. Carbon Dioxide Emissions

Figure 10 shows the annual carbon dioxide (CO_2) emissions per kWh electricity usage of the proposed system configurations. A CO_2 emission factor of 0.56 $kgCO_2$/kWh of electricity was used in Pakistan [55]. According to Figure 10, the standalone DEC system without the ERV and recirculation configuration emitted 134.5 $kgCO_2$/kWh, closely followed by the standalone DEC system with the ERV and recirculation configuration, which emitted 135.6 $kgCO_2$/kWh. Similarly, the standalone MEC system without the ERV and recirculation configuration emitted 159.2 $kgCO_2$/kWh. Similarly, the standalone MEC system with the ERV and recirculation configuration emitted 160.2 $kgCO_2$/kWh. In the case of the standalone VAC system without the ERV and recirculation configuration, the CO_2 emissions were 878.5 $kgCO_2$/kWh, whereas it was 538.1 kg/CO_2/kWh in the case of the VAC system with the ERV and recirculation configuration. On the other hand, the hybrid MEC-VAC system without the ERV and recirculation configuration emitted 749.5 $kgCO_2$/kWh, whereas the MEC-VAC system with the ERV and recirculation configuration emitted 499.2 $kgCO_2$/kWh.

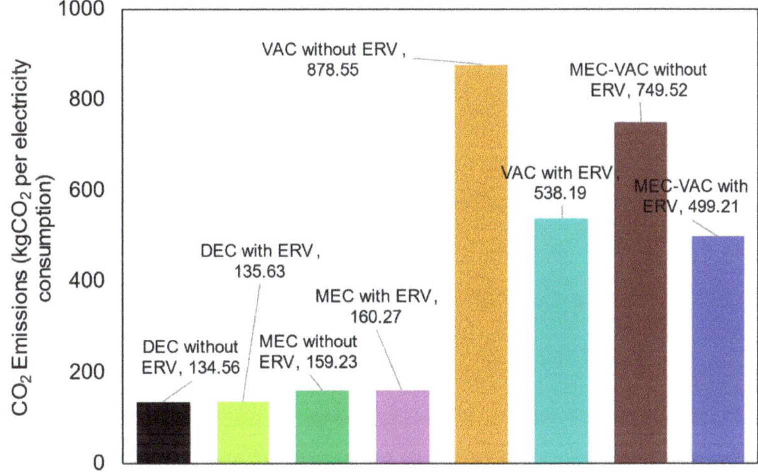

Figure 10. CO_2 emissions per kWh electricity usage of the proposed system configurations.

From these results, it can be safely concluded that, although the CO_2 emissions of the standalone EC systems are relatively low, these systems fail to achieve the desired human thermal comfort and fail to provide the optimum management of building air-conditioning loads. On the other hand, the standalone VAC system emits higher CO_2 but consumes high energy and fails to achieve the desired human thermal comfort and to provide optimum management of building air-conditioning loads. In contrast, the hybrid MEC-VAC system with the ERV and recirculation configuration emits relatively lower CO_2 per kWh and promises to deliver optimum thermal comfort, as well as optimum management of building air-conditioning loads.

3.2.5. Wet Bulb Effectiveness

The profile of wet bulb effectiveness (WBE) of the proposed system configurations for the summer months of the study area is presented in Figure 11. According to Figure 11, the hybrid MEC-VAC system with the ERV and recirculation configuration achieved a maximum wet bulb effectiveness of 2.2 in August. However, the MEC-VAC system without the ERV and recirculation configuration was the only system that achieved a maximum WBE of 1.46 in August. Similarly, the standalone VAC system with the ERV and recirculation configuration also achieved a maximum wet bulb effectiveness of 2.2 in August. However, the VAC system without the ERV and recirculation configuration only achieved a highest WBE of 1.27 in August. Contrarily, the standalone MEC system with the ERV and recirculation configuration achieved the highest WBE of 0.64 in August. However, the MEC without the ERV and recirculation configuration achieved the highest WBE of 0.65 in May. Similarly, the standalone DEC system without the ERV and recirculation configuration achieved a maximum wet bulb effectiveness of 0.55 in June. On the other hand, the DEC system with the ERV and recirculation configuration achieved a maximum wet bulb effectiveness of 0.31 in July. According to these results, the standalone VAC system and the hybrid MEC-VAC system with the ERV and recirculation configuration achieved the maximum wet bulb effectiveness; however, the standalone VAC system with ERV and recirculation was unable to achieve the desired mean radiant temperature (as per Figure 8) and was therefore unable to optimally manage the building air-conditioning loads. The hybrid MEC-VAC system, however, produced the best wet bulb effectiveness and was able to achieve the desired mean radiant temperature (as per Figure 8), thereby optimally managing the building air-conditioning loads.

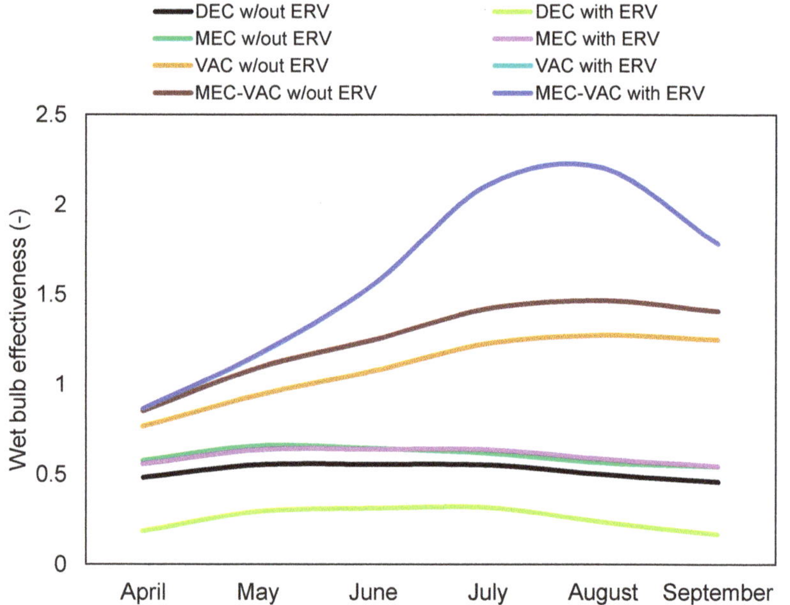

Figure 11. Monthly averaged profile of wet bulb effectiveness of the proposed system configurations for the summer months.

Evidently, from Figure 11, the evaporative cooling systems were thermodynamically limited to the outside conditions of the study area; therefore, their performance was limited. However, for the hybrid air-conditioning system, the system's output was beyond the psychrometric limits of EC.

3.2.6. Electricity Consumption

The annual electricity usage of the proposed systems is presented in Figure 12. According to the results, the standalone DEC system without the ERV and recirculation configuration consumed a maximum and minimum annual electricity of 281.6 kWh/day and 231.3 kWh/day, with an average of 266.5 kWh/day annual electricity. Similarly, the DEC system with the ERV and recirculation configuration consumed a maximum and minimum annual electricity of 281.5 kWh/day and 231.7 kWh/day, with an average of 268.6 kWh/day annual electricity. On the other hand, the standalone MEC system without the ERV and recirculation configuration consumed a maximum and minimum an nual electricity of 354.6 kWh/day and 231.3 kWh/day, with an average of 315.3 kWh/day annual electricity. Similarly, the MEC system with the ERV and recirculation configuration consumed a maximum and minimum annual electricity of 354.6 kWh/day and 232.4 kWh/day, with an average of 317.4 kWh/day annual electricity. However, the standalone VAC system without the ERV and recirculation configuration consumed a maximum and minimum annual electricity of 3018.6 kWh/day and 231.3 kWh/day, with an average of 1739.7 kWh/day annual electricity. Similarly, the VAC system with the ERV and recirculation configuration consumed a maximum and minimum annual electricity of 2176.6 kWh/day and 240.0 kWh/day, with an average of 1065.7 kWh/day annual electricity. In contrast, the hybrid MEC-VAC without the ERV and recirculation configuration consumed a maximum and minimum annual electricity of 3106.7 kWh/day and 231.3 kWh/day, with an average of 1484.2 kWh/day annual electricity. However, the MEC-VAC system with the ERV and recirculation configuration consumed a maximum and minimum annual electricity of 2176.6 kWh/day and 232.4 kWh/day, with an average of 988.5 kWh/day annual electricity. Evidently, from the results, the standalone evaporative cooling systems were not feasible from the viewpoint of energy savings. However, the standalone VAC and the hybrid MEC-VAC systems with the ERV and recirculation configuration proved to have energy savings throughout the year.

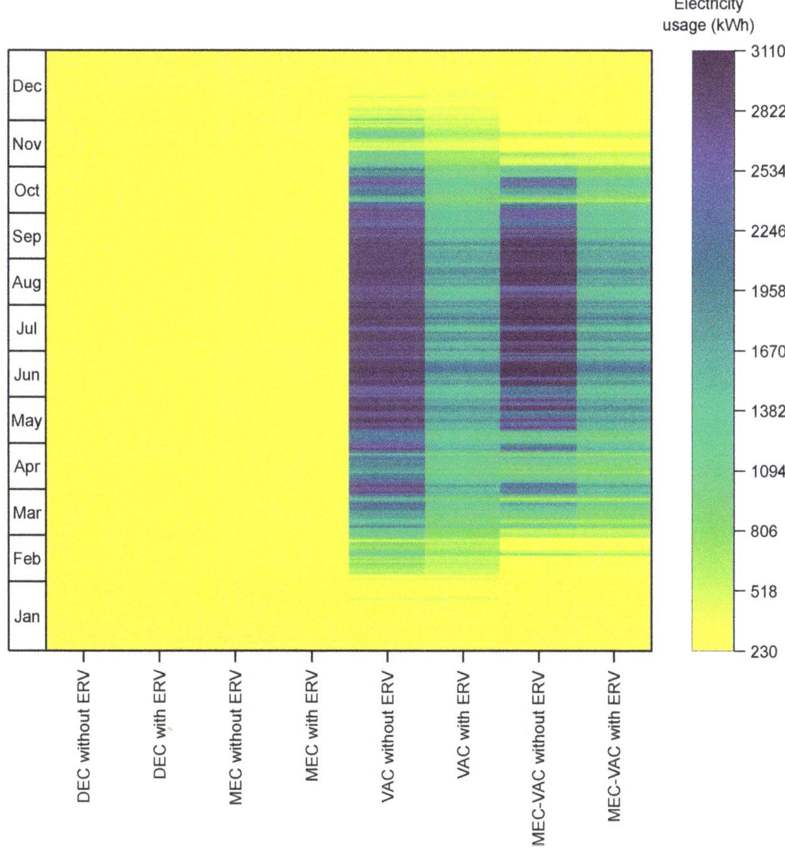

Figure 12. Annual electricity usage of the proposed system configurations.

3.2.7. Energy Saving Potential

Correlation contours of the energy-saving potential (%) of the membrane energy recovery ventilator and recirculation configuration of the DEC system in terms of the outside humidity and temperature for the summer months are presented in Figure 13. According to Figure 13, the energy-saving potential of the standalone DEC system with the ERV and recirculation configuration increased with the increasing ambient temperature and relative humidity. According to the results, the standalone DEC system with the ERV and recirculation configuration achieved a maximum and minimum energy-saving potential of 47.8% and 29.1%, with an average energy-saving potential of 39.4%. Similarly, other systems with the ERV and recirculation configuration also saved energy as compared to their respective counterpart systems without the ERV and recirculation configuration.

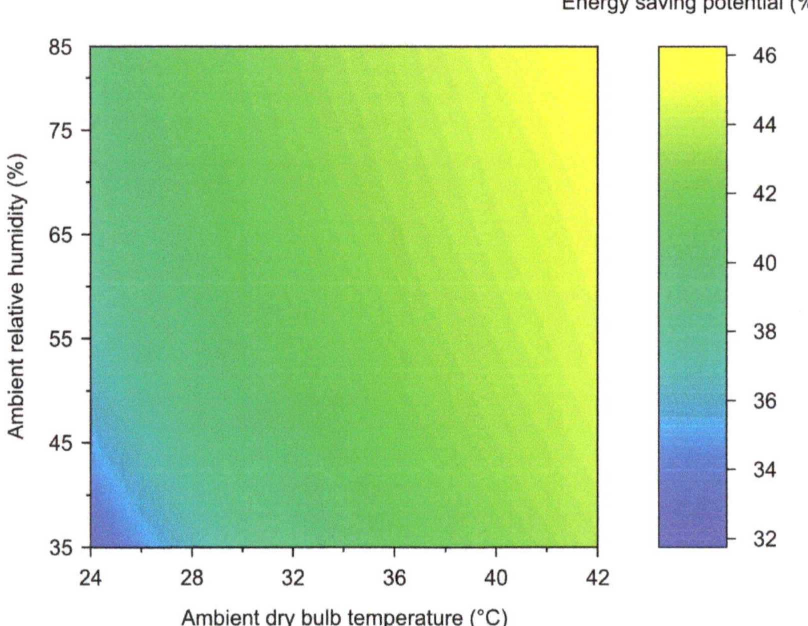

Figure 13. Energy-saving potential (%) of the membrane energy recovery ventilator and recirculation configuration of the DEC system in terms of the outside humidity and temperature for the summer months.

Being a developing country, it is extremely challenging for Pakistan to stop expanding energy interests. Therefore, it is crucial to investigate potential energy technologies in the air-conditioning sector. The present research study is limited to numerical simulations that can only provide an approximate estimation of the energy-saving potential of ERV systems. However, in future research directions, the ERV systems could be experimentally investigated in commercial buildings of Pakistan. Moreover, the experimental energy saving potential of such systems could be validated with the numerically simulated results presented in the current study. Being a cheaper source of air-conditioning, conventional air-conditioning techniques coupled with ERV and recirculation renders incredible possibilities for accomplishing energy security, independence to address energy issues, natural assurance, and supportable monetary development. Moreover, energy conservation, efficiency improvement of the current air-conditioning systems, and mitigation measures can also be undertaken to mitigate CO_2 emissions.

4. Conclusions

The current study intended to explore possible alternative options, i.e., standalone direct (DEC), Maisotsenko cycle (MEC) evaporative cooling systems, typical mechanical vapor compression (VAC), and hybrid MEC-VAC systems coupled with two configurations, i.e., with and without membrane energy recovery ventilator (ERV) and recirculation, for the management of building air-conditioning loads in Multan (Pakistan). Commonly, the DEC and the VAC systems are used for building air conditioning without any regard to the waste of energy through building exhaust. Therefore, eight possible combinations of the above-mentioned systems are proposed in this study. The building model was developed and simulated in DesignBuilder and EnergyPlus. The essential conclusions of the study are:

- The MEC-VAC system with the ERV and recirculation configuration provided a temperature gradient of 19.7 °C with average relative humidity of 49%, whereas other systems failed to compete in terms of the temperature gradient and relative humidity.
- All the systems except the MEC-VAC with the ERV and recirculation configuration failed to achieve the required mean radiant temperature, matching with their respective psychrometric performances.
- All the systems except the standalone evaporative cooling systems achieved slightly cool to extremely cool PMV with 20–98% PPD.
- The VAC with ERV and recirculation and MEC-VAC with ERV and recirculation systems achieved a maximum wet bulb effectiveness of 2.2 out of all the proposed systems.
- Among the studied systems, the standalone VAC system without the ERV and recirculation configuration resulted in the highest CO_2 emissions, i.e., 878.5 $kgCO_2/kWh$, whereas the MEC-VAC system with the ERV and recirculation configuration resulted in a relatively lower CO_2 emission, i.e., 499.2 $kgCO_2/kWh$.

The hybrid MEC-VAC system with the ERV and recirculation configuration could achieve the desired results from the viewpoints of the temperature gradient, relative humidity, wet bulb effectiveness, mean radiant temperature, and CO_2 emissions. Therefore, this study concludes the hybrid MEC-VAC system with the ERV and recirculation configuration is the optimum feasible option for the management of building air-conditioning loads.

Author Contributions: Conceptualization, H.A. and M.S.; Data curation, H.A., M.S. and U.S.; Formal analysis, H.A., M.S., U.S., M.W.S. and M.F.; Funding acquisition, M.S. and S.M.I.; Investigation, H.A., M.S., M.W.S., M.U.K. and M.A.J.; Methodology, H.A., M.S., U.S., M.W.S., M.F. and M.A.J.; Project administration, M.S. and S.M.I.; Resources, M.S.; Software, H.A., U.S. and M.W.S.; Supervision, M.S.; Validation, H.A., M.S. and U.S.; Visualization, H.A., M.S., S.M.I. and M.U.K.; Writing—original draft, H.A. and M.S.; and Writing—review and editing, M.W.S., M.F., S.M.I., M.U.K. and M.A.J. All authors have read and agreed to the published version of the manuscript.

Funding: The authors acknowledge the support from the Researchers Supporting Project number (RSP-2021/100), King Saud University, Riyadh, Saudi Arabia.

Institutional Review Board Statement: Not applicable.

Informed Consent Statement: Not applicable.

Data Availability Statement: Data are contained within the article.

Acknowledgments: The authors acknowledge the support from the Researchers Supporting Project number (RSP-2021/100), King Saud University, Riyadh, Saudi Arabia. This research was carried out in the Department of Agricultural Engineering, Bahauddin Zakariya University, Multan, Pakistan with the support of BZU Director Research/ORIC grants awarded to Principal Investigator Muhammad Sultan.

Conflicts of Interest: The authors declare no conflict of interest.

References

1. United Nations Environment Programme. *2021 Global Status Report for Buildings and Construction: Towards a Zero-Emission*; United Nations Environment Programme: Nairobi, Kenya, 2021.
2. International Energy Agency. *World Energy Balances*; International Energy Agency: Paris, France, 2021.
3. European Commission. *In Focus: Energy Efficiency in Buildings*; European Commission: Brussels, Belgium, 2020.
4. Beck, H.E.; Zimmermann, N.E.; McVicar, T.R.; Vergopolan, N.; Berg, A.; Wood, E.F. Present and Future Köppen-Geiger Climate Classification Maps at 1-Km Resolution. *Sci. Data* **2018**, *5*, 180214. [CrossRef] [PubMed]
5. International Energy Agency Data and Statistics. Available online: https://www.iea.org/data-and-statistics (accessed on 12 December 2021).
6. Noor, S.; Ashraf, H.; Sultan, M.; Khan, Z.M. Evaporative Cooling Options for Building Air-Conditioning: A Comprehensive Study for Climatic Conditions of Multan (Pakistan). *Energies* **2020**, *13*, 3061. [CrossRef]
7. Harby, K.; Gebaly, D.R.; Koura, N.S.; Hassan, M.S. Performance Improvement of Vapor Compression Cooling Systems Using Evaporative Condenser: An Overview. *Renew. Sustain. Energy Rev.* **2016**, *58*, 347–360. [CrossRef]
8. Ibrahim, N.I.; Al-Farayedhi, A.A.; Gandhidasan, P. Experimental Investigation of a Vapor Compression System with Condenser Air Pre-Cooling by Condensate. *Appl. Therm. Eng.* **2017**, *110*, 1255–1263. [CrossRef]
9. Kojok, F.; Fardoun, F.; Younes, R.; Outbib, R. Hybrid Cooling Systems: A Review and an Optimized Selection Scheme. *Renew. Sustain. Energy Rev.* **2016**, *65*, 57–80. [CrossRef]
10. Liu, Y.; Yang, X.; Li, J.; Zhao, X. Energy Savings of Hybrid Dew-Point Evaporative Cooler and Micro-Channel Separated Heat Pipe Cooling Systems for Computer Data Centers. *Energy* **2018**, *163*, 629–640. [CrossRef]
11. Raza, H.M.U.; Ashraf, H.; Shahzad, K.; Sultan, M.; Miyazaki, T.; Usman, M.; Shamshiri, R.R.; Zhou, Y.; Ahmad, R. Investigating Applicability of Evaporative Cooling Systems for Thermal Comfort of Poultry Birds in Pakistan. *Appl. Sci.* **2020**, *10*, 4445. [CrossRef]
12. Sultan, M.; Ashraf, H.; Miyazaki, T.; Shamshiri, R.R.; Hameed, I.A. Temperature and Humidity Control for the Next Generation Greenhouses: Overview of Desiccant and Evaporative Cooling Systems. In *Next-Generation Greenhouses for Food Security*; IntechOpen: London, UK, 2021.
13. Noor, S.; Ashraf, H.; Sultan, M.; Miyazaki, T.; Mahmood, M.H.; Khan, Z.M. Investigation of Direct and Indirect Evaporative Cooling Options for Greenhouse Air Conditioning in Multan (Pakistan). In Proceedings of the International Exchange and Innovation Conference on Engineering & Sciences (IEICES), Fukuoka City, Japan, 22–23 October 2020; pp. 110–115.
14. Noor, S.; Ashraf, H.; Hussain, G.; Sultan, M.; Miyazaki, T.; Shakoor, A.; Mahmood, M.H.; Riaz, M. Spatiotemporal Investigation of Evaporative Cooling Options for Greenhouse Air-Conditioning Application in Pakistan. *Fresenius Environ. Bull.* **2021**, *30*, 21.
15. Ashraf, H.; Noor, S.; Sultan, M.; Khan, Z.M. Energy Saving Potential of Evaporative Cooling Systems Compared to Traditional Air Conditioners. In Proceedings of the International Conference on Mechanical Engineering—2020 (ICME-20), Lahore, Pakistan, 29–30 January 2020; UET: Lahore, Pakistan, 2020; pp. 75–81.
16. Chun, L.; Gong, G.; Peng, P.; Wan, Y.; Chua, K.J.; Fang, X.; Li, W. Research on Thermodynamic Performance of a Novel Building Cooling System Integrating Dew Point Evaporative Cooling, Air-Carrying Energy Radiant Air Conditioning and Vacuum Membrane-Based Dehumidification (DAV-Cooling System). *Energy Convers. Manag.* **2021**, *245*, 114551. [CrossRef]
17. Kowalski, P.; Kwiecień, D. Evaluation of Simple Evaporative Cooling Systems in an Industrial Building in Poland. *J. Build. Eng.* **2020**, *32*, 101555. [CrossRef]
18. da Veiga, A.P.; Güths, S.; da Silva, A.K. Evaporative Cooling in Building Roofs: Theoretical Modeling and Experimental Validation (Part-1). *Sol. Energy* **2020**, *207*, 1122–1131. [CrossRef]
19. da Veiga, A.P.; Güths, S.; da Silva, A.K. Evaporative Cooling in Building Roofs: Local Parametric and Global Analyses (Part-2). *Sol. Energy* **2020**, *207*, 1009–1020. [CrossRef]
20. Tewari, P.; Mathur, S.; Mathur, J.; Kumar, S.; Loftness, V. Field Study on Indoor Thermal Comfort of Office Buildings Using Evaporative Cooling in the Composite Climate of India. *Energy Build.* **2019**, *199*, 145–163. [CrossRef]
21. Kim, M.-H.; Jeong, J.-W. Cooling Performance of a 100% Outdoor Air System Integrated with Indirect and Direct Evaporative Coolers. *Energy* **2013**, *52*, 245–257. [CrossRef]
22. Cui, X.; Chua, K.J.; Yang, W.M.; Ng, K.C.; Thu, K.; Nguyen, V.T. Studying the Performance of an Improved Dew-Point Evaporative Design for Cooling Application. *Appl. Therm. Eng.* **2014**, *63*, 624–633. [CrossRef]
23. Velasco Gómez, E.; Tejero González, A.; Rey Martínez, F.J. Experimental Characterisation of an Indirect Evaporative Cooling Prototype in Two Operating Modes. *Appl. Energy* **2012**, *97*, 340–346. [CrossRef]
24. Heidarinejad, G.; Moshari, S. Novel Modeling of an Indirect Evaporative Cooling System with Cross-Flow Configuration. *Energy Build.* **2015**, *92*, 351–362. [CrossRef]
25. Cui, X.; Chua, K.J.; Yang, W.M. Numerical Simulation of a Novel Energy-Efficient Dew-Point Evaporative Air Cooler. *Appl. Energy* **2014**, *136*, 979–988. [CrossRef]
26. Moshari, S.; Heidarinejad, G.; Fathipour, A. Numerical Investigation of Wet-Bulb Effectiveness and Water Consumption in One-and Two-Stage Indirect Evaporative Coolers. *Energy Convers. Manag.* **2016**, *108*, 309–321. [CrossRef]
27. Cui, X.; Chua, K.J.; Islam, M.R.; Ng, K.C. Performance Evaluation of an Indirect Pre-Cooling Evaporative Heat Exchanger Operating in Hot and Humid Climate. *Energy Convers. Manag.* **2015**, *102*, 140–150. [CrossRef]

28. Campisi, D.; Gitto, S.; Morea, D. An Evaluation of Energy and Economic Efficiency in Residential Buildings Sector: A Multi-Criteria Analisys on an Italian Case Study. *Int. J. Energy Econ. Policy* **2018**, *8*, 185–196.

29. Obando, F.A.; Montoya, A.P.; Osorio, J.A.; Damasceno, F.A.; Norton, T. Evaporative Pad Cooling Model Validation in a Closed Dairy Cattle Building. *Biosyst. Eng.* **2020**, *198*, 147–162. [CrossRef]

30. Badiei, A.; Akhlaghi, Y.G.; Zhao, X.; Li, J.; Yi, F.; Wang, Z. Can Whole Building Energy Models Outperform Numerical Models, When Forecasting Performance of Indirect Evaporative Cooling Systems? *Energy Convers. Manag.* **2020**, *213*, 112886. [CrossRef]

31. Nada, S.A.; Elattar, H.F.; Mahmoud, M.A.; Fouda, A. Performance Enhancement and Heat and Mass Transfer Characteristics of Direct Evaporative Building Free Cooling Using Corrugated Cellulose Papers. *Energy* **2020**, *211*, 118678. [CrossRef]

32. He, W.; Xilian, L.; Yuhui, S.; Min, Z.; Zhaolin, G. Research of Evaporative Cooling Experiment in Summer of Residential Buildings in Xi'an. *Energy Procedia* **2018**, *152*, 928–934. [CrossRef]

33. Boukhanouf, R.; Amer, O.; Ibrahim, H.; Calautit, J. Design and Performance Analysis of a Regenerative Evaporative Cooler for Cooling of Buildings in Arid Climates. *Build. Environ.* **2018**, *142*, 1–10. [CrossRef]

34. Zanchini, E.; Naldi, C. Energy Saving Obtainable by Applying a Commercially Available M-Cycle Evaporative Cooling System to the Air Conditioning of an Office Building in North Italy. *Energy* **2019**, *179*, 975–988. [CrossRef]

35. Khandelwal, A.; Talukdar, P.; Jain, S. Energy Savings in a Building Using Regenerative Evaporative Cooling. *Energy Build.* **2011**, *43*, 581–591. [CrossRef]

36. Ashraf, H.; Sultan, M.; Shamshiri, R.R.; Abbas, F.; Farooq, M.; Sajjad, U.; Md-Tahir, H.; Mahmood, M.H.; Ahmad, F.; Taseer, Y.R.; et al. Dynamic Evaluation of Desiccant Dehumidification Evaporative Cooling Options for Greenhouse Air-Conditioning Application in Multan (Pakistan). *Energies* **2021**, *14*, 1097. [CrossRef]

37. Kottek, M.; Grieser, J.; Beck, C.; Rudolf, B.; Rubel, F. World Map of the Köppen-Geiger Climate Classification Updated. *Meteorol. Zeitschrift* **2006**, *15*, 259–263. [CrossRef]

38. Gao, H.; Li, Z.; Qiu, S.; Yang, B.; Li, S.; Wen, Y. Energy Exchange Efficiency Prediction from Non-Linear Regression for Membrane-Based Energy-Recovery Ventilator Cores. *Appl. Therm. Eng.* **2021**, *197*, 117353. [CrossRef]

39. Abadi, I.R.; Aminian, B.; Nasr, M.R.; Huizing, R.; Green, S.; Rogak, S. Experimental Investigation of Condensation in Energy Recovery Ventilators. *Energy Build.* **2022**, *256*, 111732. [CrossRef]

40. Huang, S.; Li, W.; Lu, J.; Li, Y. Experimental Study on Two Type of Indirect Evaporative Cooling Heat Recovery Ventilator. *Procedia Eng.* **2017**, *205*, 4105–4110. [CrossRef]

41. Chen, Y.; Luo, Y.; Yang, H. Fresh Air Pre-Cooling and Energy Recovery by Using Indirect Evaporative Cooling in Hot and Humid Region—A Case Study in Hong Kong. *Energy Procedia* **2014**, *61*, 126–130. [CrossRef]

42. Qiu, S.; Li, S.; Wang, F.; Wen, Y.; Li, Z.; Li, Z.; Guo, J. An Energy Exchange Efficiency Prediction Approach Based on Multivariate Polynomial Regression for Membrane-Based Air-to-Air Energy Recovery Ventilator Core. *Build. Environ.* **2019**, *149*, 490–500. [CrossRef]

43. Zhong, X.; Wu, W.; Ridley, I.A. Assessing the Energy and Indoor-PM2.5-Exposure Impacts of Control Strategies for Residential Energy Recovery Ventilators. *J. Build. Eng.* **2020**, *29*, 101137. [CrossRef]

44. Rasouli, M.; Simonson, C.J.; Besant, R.W. Applicability and Optimum Control Strategy of Energy Recovery Ventilators in Different Climatic Conditions. *Energy Build.* **2010**, *42*, 1376–1385. [CrossRef]

45. Zhou, Y.P.; Wu, J.Y.; Wang, R.Z. Performance of Energy Recovery Ventilator with Various Weathers and Temperature Set-Points. *Energy Build.* **2007**, *39*, 1202–1210. [CrossRef]

46. Designbuilder Software Ltd. *DesignBuilder 6.1*; DesignBuilder: Gloucestershire, UK, 2021.

47. U.S. Department of Energy (DOE); National Renewable Energy Laboratory (NREL). *EnergyPlus 9.3*; National Renewable Energy Laboratory (NREL): Golden, CO, USA, 2020.

48. Sheldon, R.A. Fundamentals of Green Chemistry: Efficiency in Reaction Design. *Chem. Soc. Rev.* **2012**, *41*, 1437–1451. [CrossRef]

49. Akbari, H.; Havenith, G.; Al-Sahhaf, A. A Database of Static Clothing Thermal Insulation and Vapor Permeability Values of Non-Western Ensembles for Use in ASHRAE Standard 55, ISO 7730, and ISO 9920: Discussion. *ASHRAE Conf.* **2015**, *121*, 215.

50. Fanger, P.O. *Thermal Comfort: Analysis and Applications in Environmental Engineering*, 1st ed.; Danish Technical Press: Copenhagen, Denmark, 1970; ISBN 9780070199156.

51. *ISO 7730*; Ergonomics of the Thermal Environment—Analytical Determination and Interpretation of Thermal Comfort Using Calculation of the PMV and PPD Indices and Local Thermal Comfort Criteria. International Standardization Organization: Geneva, Switzerland, 2005.

52. CORE. Energy Recovery Solutions Core M-ERV250. Available online: https://core.life/en/resources/spec/m-erv250/ (accessed on 10 October 2021).

53. Raza, H.M.U. Investigation of Evaporative Cooling Based Low-Cost Air-Conditioning Technologies for Pakistan. Ph.D. Thesis, Agricultural Engineering, Bahauddin Zakariya University, Multan, Pakistan, 2018.

54. ASHRAE. *ASHRAE Handbook—Fundamentals (SI)*; American Society of Heating, Refrigerating and Air-Conditioning Engineers: Atlanta, GA, USA, 2017.

55. Yousuf, I.; Ghumman, A.R.; Hashmi, H.N.; Kamal, M.A. Carbon Emissions from Power Sector in Pakistan and Opportunities to Mitigate Those. *Renew. Sustain. Energy Rev.* **2014**, *34*, 71–77. [CrossRef]

Article

Investigation of Energy Consumption and Associated CO_2 Emissions for Wheat–Rice Crop Rotation Farming

Muhammad N. Ashraf [1], Muhammad H. Mahmood [1,*], Muhammad Sultan [1,*], Redmond R. Shamshiri [2,*] and Sobhy M. Ibrahim [3]

1   Department of Agricultural Engineering, Bahauddin Zakariya University, Multan 60800, Pakistan; m.noumanashraf436@gmail.com

2   Department of Engineering for Crop Production, Leibniz Institute for Agricultural Engineering and Bioeconomy, 14469 Potsdam, Germany

3   Department of Biochemistry, College of Science, King Saud University, Riyadh 11451, Saudi Arabia; syakout@ksu.edu.sa

*   Correspondence: hamidmahmood@bzu.edu.pk (M.H.M.); muhammadsultan@bzu.edu.pk (M.S.); rshamshiri@atb-potsdam.de (R.R.S.); Tel.: +92-333-610-8888 (M.S.)

Abstract: This study investigates the input–output energy-flow patterns and CO_2 emissions from the wheat–rice crop rotation system. In this regard, an arid region of Punjab, Pakistan was selected as the study area, comprising 4150 km^2. Farmers were interviewed to collect data and information on input/output sources during the 2020 work season. The total energy from these sources was calculated using appropriate energy equivalents. Three energy indices, including energy use efficiency (η_e), energy productivity (η_p), and net energy (ρ), were defined and calculated to investigate overall energy efficiency. Moreover, the data envelopment analysis (DEA) technique was used to optimize the input energy in wheat and rice production. Finally, CO_2 emissions was calculated using emissions equivalents from peer-reviewed published literature. Results showed that the average total energy consumption in rice production was twice the energy consumed in wheat production. However, the values of η_e, η_p, and ρ were higher in wheat production and calculated as 5.68, 202.3 kg/GJ, and 100.12 GJ/ha, respectively. The DEA showed the highest reduction potential in machinery energy for both crops, calculated as -42.97% in rice production and -17.48% in wheat production. The highest CO_2 emissions were found in rice production and calculated as 1762.5 kg-CO_2/ha. Our conclusion indicates that energy consumption and CO_2 emissions from wheat–rice cropping systems can be minimized using optimized energy inputs.

Keywords: cereal production; crop rotation; energy analysis; DEA; CO_2 emissions

Citation: Ashraf, M.N.; Mahmood, M.H.; Sultan, M.; Shamshiri, R.R.; Ibrahim, S.M. Investigation of Energy Consumption and Associated CO_2 Emissions for Wheat–Rice Crop Rotation Farming. *Energies* **2021**, *14*, 5094. https://doi.org/10.3390/en14165094

Academic Editor: Mariusz J. Stolarski

Received: 9 July 2021
Accepted: 13 August 2021
Published: 18 August 2021

Publisher's Note: MDPI stays neutral with regard to jurisdictional claims in published maps and institutional affiliations.

1. Introduction

Wheat (Triticum aestivum) and rice (Oryza sativa) are the main cereal crops grown worldwide, constituting 54% of the total cereal production [1]. The mean annual global wheat and rice production are recorded as 646.9 and 654.8 million metric tonnes, respectively [2]. These grains are the staple food of 85% of the world population [3,4]. Pakistan is an agricultural country, having a share of 3.32% and 1.3% in the world's wheat and rice production, respectively [2]. The demand for these grains is increasing tremendously, due to increasing growth in the world's population. Forecasts showed an increase in world wheat demand up to 60% of the current production—by 2050 [5]. Moreover, arable land is decreasing, due to housing societies and other domestic/commercial purposes. This necessitates efficient energy use in agricultural production to enhance yields and sustain food security. Moreover, food storage and processing is also an important element of food security [6–8]. In this regard, energy-efficient and environment-friendly storage systems can play an important role in avoiding off-season food shortages [9–11].

Agriculture is an energy conversion process, in which solar energy is converted into food and fiber through photosynthesis [12]. It becomes more energy-intensive due to the use of fossil fuels, fertilizers, machinery, and electricity to enhance overall production [13]. Agricultural energy flow is classified as direct energy input and indirect energy input. Direct energy input includes energy consumed on the farm during various operations e.g. labor, fuel, machinery, water, and electricity. Indirect energy input is the energy consumed during the production process of different input sources e.g., fertilizers, machinery, seeds and biocides [14]. These energy inputs can also be divided into renewable and non renewable energy inputs. Renewable energy includes the energy input from machinery and labor while non-renewable energy constitutes all other input sources [15]. The amount and type of energy input in agricultural production mainly depend on the socioeconomic characteristics of the farm and farmers [16–18]. These characteristics include farmer's experience and education, farm size, source of irrigation, local climate, soil and crop type and farmer's landholdings [15,19,20]. However, the intensive use of energy in agricultural production creates health and environmental concerns [13]. For example, the combustion of fossil fuels emits a large amount of carbon dioxide (CO_2) and other greenhouse gasses (GHGs) into the atmosphere. Similarly, the production processes of fertilizers, machinery, electricity, and chemicals emit a huge amount of GHGs. Higher concentrations of these gasses in the atmosphere creates an alarming situation for the atmospheric chemistry of the globe [21]. For instance, the concentration of CO_2 in the atmosphere has reached 419.05 ppm and is increasing further rapidly [22]. It leads to global warming and climate change, the most challenging issues of the current century [23], to which end the United Nations Framework Convention on Climate Change has aimed to limit global warming up to 2 °C in the current century. In this regard, the present decarbonization rate, of 1.6 percent per year, has to be increased to 6.4 percent per year, otherwise global temperatures may increase by four degrees Celsius by the end of this century [24]. Furthermore, though Pakistan is ranked seventh among countries vulnerable to climate change, per capita emissions of the country are among the lowest in the world [25]. This said, the majority GHGs emissions in Pakistan come from the energy and agriculture sectors, accounting for 46% and 41% of total emissions, respectively [26]. Therefore, efficient, and optimized energy use in agricultural production is necessary in agricultural countries like Pakistan if emissions are to be reduced further.

Several studies have been conducted worldwide on energy-use patterns and CO_2 emissions from agricultural production. These studies include energy consumption analysis in wheat [27,28], maize [29], rice [30,31], and soybean [32] production in India; energy and economic analyses in the production of wheat [33–35], rice [13], sugar beet [36], and fruits [37–39] in Iran; energy modeling and CO_2 emissions assessment for wheat production in New Zealand [18,40–42]; comparative input–output energy analysis in agricultural production in Indonesia and Thailand [43,44]; the assessment of CO_2 emissions and energy flow in cotton [17,45], fruits [20,46,47], and vegetables [48] in Turkey; the effects of different fertilizer management practices on CO_2 emissions from different crop production systems, energy and water footprint assessment in grain crops in Australia [16,49,50]; energy budget evaluation in wheat production in China [51]. Finally, in Pakistan, Ashraf et. al. [15] and Khan et. al. [52] investigated energy consumption patterns in wheat and rice production respectively. However, no such study was found for the wheat–rice crop rotation system in an arid region of Punjab, Pakistan.

The main objective of this study was to investigate and compare input–output energy flow and CO_2 emissions in wheat–rice crop rotation systems in an arid region of Punjab, Pakistan. This study provides insights into energy-use optimization in wheat and rice production, using the data envelopment analysis (DEA) technique. Moreover, the study reveals the relationship between grain yield and energy input, using mathematical models developed from the data and information collected from farmers in the study area. The relationship between total input energy and CO_2 emission is also developed to analyze the impact of energy consumption on carbon dioxide emissions. This study contributes solutions to optimizing energy use for wheat and rice production.

2. Research Methodology

This study was conducted in Kabirwala (30°24′ N, 71°52′ E), a Tehsil of Khanewal District, Punjab, Pakistan. The map of the study area is shown in Figure 1. Kabirwala is situated at the bank of the river Chenab and has fertile lands for agricultural cultivations. The study area is comprised was 4150 km^{2} of which more than 80% was cultivatable land. Most of the population in the study area is engaged with the agriculture sector, indicating the economic importance of agriculture in this area. Wheat–rice is the prominent crop rotation in Tehsil. According to the Pakistan Bureau of Statistics [53], the Khanewal district has a significant share (3.9%) of Pakistan's total wheat production, while 1.6% of total rice production was in Punjab. The mean annual minimum and maximum temperatures in the study area are 11 °C and 46 °C, respectively. The study area lies in an arid region with a mean annual rainfall of 177.4 mm. Most of the rainfall (41.5%) in the study area occurs in the monsoon season i.e., July and August. However, canal water and groundwater (pumped by tube-wells) are used for irrigation to meet crop's water requirements throughout the year. In this regard, data on irrigation water and other energetic aspects i.e., seeds, labor, fuel, fertilizer, machinery, chemicals, and electricity were collected from farmers through face-to-face interviews and questionnaires during the 2020 work season. A total of 23 farmers were selected and visited randomly from the study area to collect data and information. The sample size was calculated using the Cochran formula given in Equation (1) [54].

$$N = \frac{nq^2E^2}{ne^2 + q^2E^2} \tag{1}$$

where N is the required sample size, n is the number of landholdings, q is a constant taken as 1.96 for 95% reliability, E^2 is the variance, and e is the margin of error.

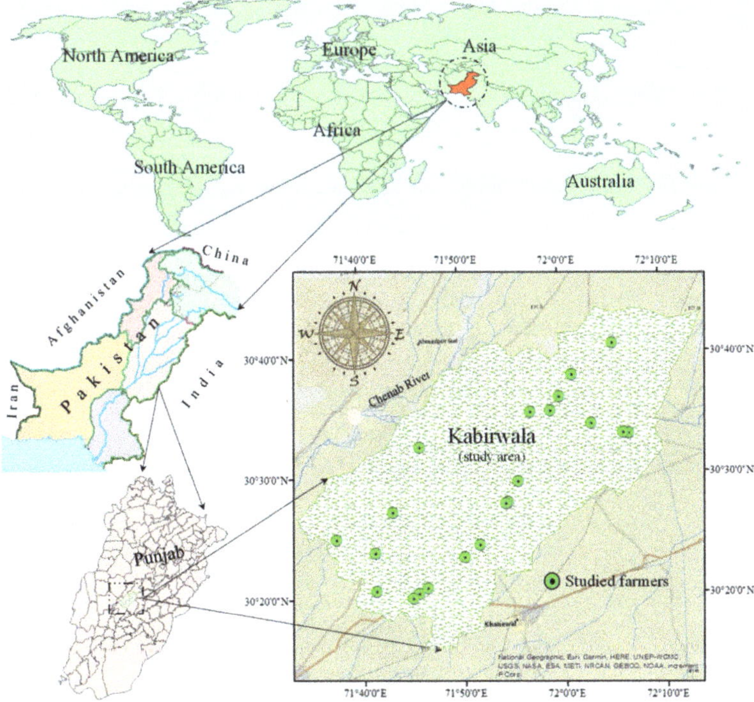

Figure 1. Map of the study area and distribution of the studied farmers inside the study area.

Preliminary data evaluation revealed that there were variations in energy inputs/outputs among the studied farmers. These variations were mainly due to differences in the socioeconomic characteristics of a specific farm and/or farmer. It is evident that variations in these properties of the farms/farmers affect the management practices and eventual crop yield [42]. The socioeconomic structures of the farms and farmers are given in Table 1. The table depicts minimum, maximum, and mean values of seven socioeconomic characteristics of the farm/farmers i.e., farmers' education, farmers' experience, farm size, number of farm laborers, number of tractors, source of irrigation, and canal-to-tube-well irrigation ratio. We reported that a farmer's education and farm size are directly related to their energy consumption; farmers' education and farm size vary positively with energy use [42] In this regard, the average farmer's education and farm size were found to be 8 years of schooling and 8.78 ha, respectively. Similarly, the number of a farm's laborers and tractors also increases with its size. The highest number of farm laborers and tractors observed were seven and two, respectively, corresponding to the largest farm at 31.03 ha. Moreover the farmers' mean experience was 18.35 years, indicating the higher technical knowledge of farmers in the study area. The farmers' greater experience also suggests that the studied farmers have been intergenerationally involved in agriculture. Additionally, sources of irrigation are also an important factor in the overall energy consumption of agricultural production [15]. Therefore, the consumption of canal water is ideal for irrigation, as it eliminates pumping energy and hence decreases overall energy usage. However, the main source of irrigation in the study area was tube-well water, due to the limited availability of canal water throughout the year. The canal-to-tube-well irrigation ratio was observed as 31:61. These socioeconomic factors were considered in the present study.

Table 1. Socioeconomic properties of the farms and farmers in the study area.

Property	Min	Max	Mean
Farmer's education (yrs.)	2	14	8.00
Farmer's experience (yrs.)	6	30	18.35
Farm size (ha)	2.07	31.03	8.78
Farm labor (No.)	1	7	2.30
No. of tractors	0	2	0.87
Source of irrigation		C + T	
Canal to tube-well ratio		31:61	

C and T stand for canal and tube-well, respectively.

The present study evaluates input–output energy patterns for wheat–rice crop rotation systems from sowing until harvesting. The system boundaries for this study are shown in Figure 2. The figure depicts this study's evaluation of the energy flow inside a farm and does not consider transporting outputs from the farm gate to the end consumer Quantitative data on input sources were used to calculate total energy consumption, using energy equivalents. Energy equivalents from different studies were reviewed technically and the most appropriate equivalents were used in the present study; those used in this study are given in Table 2. However, energy consumption from farm equipment was calculated using Equation (2) [55]. Furthermore, energy efficiency in wheat–rice cropping patterns was investigated based on three indices, energy-use efficiency, energy productivity and net energy (Equations (3)–(5)). The term energy efficiency is used to denote the overall performance of energy inputs to generate energy outputs. This performance was investigated based on the above-mentioned energy indices.

$$E_e = \frac{W\,B}{L} \tag{2}$$

$$\eta_p = \frac{G_a}{E_i} \tag{3}$$

$$\eta_e = \frac{E_o}{E_i} \tag{4}$$

$$\rho = E_o - E_i \tag{5}$$

where E_e is the energy equivalent for farm implement/tractor (MJ/hr), B is a constant (MJ/kg), W refers to the weight of a tractor/implement (kg), and L refers to the economic life span of a tractor/implement (hr). The appropriate values for B, W, and L were taken from the literature [55–57]. η_p is energy productivity (kg/GJ), G_a is the weight of grains (kg/ha), E_i is the total input energy (GJ/ha), η_e is energy-use efficiency, E_o is the total output energy (GJ/ha), and ρ is net energy (GJ/ha).

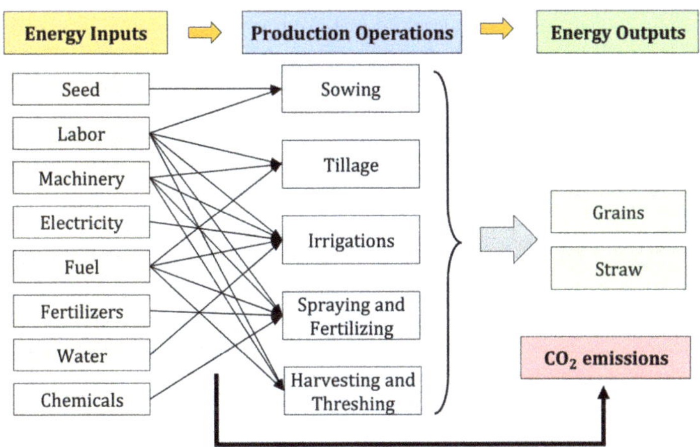

Figure 2. System boundaries and energy flow pattern for input–output energy analyses in a wheat–rice crop rotation system.

Table 2. Energy equivalents used to calculate total input–output energy in wheat and rice production.

Inputs/Outputs	Equivalent (MJ/Unit)	Reference
Inputs		
wheat seed (kg)	20.1	[51]
rice seed (kg)	14.7	[13]
diesel (l)	56.31	[28]
labor (h)	1.96	[28]
nitrogen (kg)	60.6	[28]
phosphorous (kg)	11.1	[28]
potassium (kg)	6.7	[28]
water (m^3)	1.02	[28]
tractor operation (h)	138	[55]
disc harrow operation (h)	149	[55]
plough operation (h)	180	[55]
combine harvester operation (h)	116	[55]
rotavator operation (h)	148	[55]
chemicals (ml)	0.102	[28]
fungicides (kg)	216	[20]
herbicides (kg)	238	[20]
insecticides (kg)	101.2	[20]
electricity (kWh)	11.93	[51]
Outputs		
wheat grains (kg)	15.7	[28]
wheat straw (kg)	12.5	[28]
rice grains (kg)	17	[55]

Data envelopment analysis (DEA) was also conducted to calculate the technical efficiencies of the farmers in the study area. Technical efficiency is the performance of an individual farmer based on the ratio of weighted outputs to weighted inputs. Data envelopment analysis (DEA) is an analytical technique that compares different decision-making units (DMUs) against their relative performances in a specific operation [58]. DMUs are the specific units that must be compared in DEA e.g., farmers, in the present study. The technical efficiencies of the DMUs were calculated by an input-oriented CCR (Charnes-Cooper-Rhodes) model [59]. In this regard, three inputs i.e., diesel, seed, and fertilizer energies were taken as inputs, and grains (kg/ha) as output, in the studied CCR model. The input-oriented CCR model works on the principle of achieving a constant output while minimizing inputs. Technically efficient and inefficient DMUs were critically studied to investigate potential improvements in inefficient DMUs (farmers). Frontier Analyst 4 was used to conduct DEA in this study. However, the mathematical interpretation of the technical efficiency of DEA is given in Equation (6) [60]. Furthermore, the grain yield (kg/ha) was also estimated by the multiple regression technique (Equation (7) and (8)). These equations were developed using grain yields as a response, while energy inputs from fertilizer, seed, and diesel as continuous predictors.

$$TE_a = \frac{\sum_{i=1}^{n} u_i y_{ia}}{\sum_{j=1}^{m} v_j x_{ja}} \tag{6}$$

Applying linear programming to solve Equation (6):

$$\text{Maximize } TE = \sum_{i=1}^{n} u_i y_{ia}$$

$$\text{Subject to } \sum_{i=1}^{n} u_i y_{ia} - \sum_{j=1}^{m} v_j x_{ja} \leq 0$$

$$\text{Df } \sum_{j=1}^{m} v_j x_{ja} = 1, \ u_i \geq 0, \ v_j \geq 0, \ a = 1, 2, 3, \ldots\ldots, b$$

$$G_w = 306.5583(F) + 1180.8498(S) - 142.3587(D) - 362.6101 \tag{7}$$

$$G_r = 293.8132(F) + 4172.6152(S) + 47.31394(W) - 356.6136 \tag{8}$$

where TE_a is the technical efficiency of ath DMU, y_i refers to quantity of nth output, u_i and v_j refer to the weights of nth output and mth input, respectively. i and j denote the number of outputs (i = 1, 2, 3, ... , n) and inputs (j = 1, 2, 3, ... , m), respectively. x_j is the quantity of mth input, and a refers to the number of DMUs (a = 1, 2, 3, ... , b). G_w and G_r are the grains yield (kg) of wheat and rice crops, respectively. F represents energy from fertilizers (GJ), S is the energy input from seed (GJ), W and D are the energy inputs (GJ) from water and diesel, respectively. The study uses the Origin software developed by the OriginLab Corporation for the regression analyses and for the development of the graphs.

In addition to the energy analyses, the carbon footprint from wheat and rice production in the study area was also assessed, by investigating carbon dioxide (CO_2) emissions. In this regard, energy units were converted into kg-CO_2 units using the emission equivalents given in Table 3 while considering the guidelines given by [61]. These emissions were further converted to kg-CO_2/ha units using Equation (9). Other GHGs emissions, like N_2O and CH_4, were not considered in this study due to their minor contribution to the total emissions from agricultural production [36]. The CO_2 emissions from the production processes of fertilizers, electricity, machinery, chemicals, and fuel were considered to calculate the total carbon footprint. Organic fertilizers were not considered due to their limited or absent use in wheat and rice cultivations in the study area. In the case of emissions from electricity generation, only 64% of the total electrical energy input was converted to emissions equivalence. In Pakistan, total electricity generation is generated by hydropower plants (36%) and thermal power plants (64%).

$$CE_{ha} = \frac{CE}{E_i} \tag{9}$$

where CE_{ha} refers to CO_2 emissions in kg-CO_2/ha, CE is CO_2 emissions in kg-CO_2/MJ and E_i is the total input energy (MJ/ha).

Table 3. Input energy and associated equivalent CO_2 emissions used in this study to evaluate carbon footprint while considering the guidelines given by [61].

Input Energy	Equivalent (kg CO_2/MJ)
nitrogen	0.05
phosphorous	0.06
potassium	0.06
biocides	0.06
diesel	0.0687
electricity	0.0192
machinery	0.10

3. Results and Discussion

Data and information on energy sources were collected from the farmers and are given in Table 4. The table depicts the averages of consumption of different energy sources in wheat and rice production. The average labor consumption in rice production was observed as 242.698 h/ha, which was four times higher than that used in wheat production. This was due to the higher energy consumption in transplanting rice plants and the higher water requirement of this crop. Similarly, the usage of diesel, electricity, water, and machinery in rice production was much higher than their usage in wheat production. However, other inputs, i.e., fertilizers and chemicals, were used in a similar pattern to produce both crops. These inputs were converted into energy units (GJ/ha) and are exhibited in Figures 3 and 4. The energy input from different sources in wheat and rice production are shown in Figures 3 and 4, respectively. It is evident from the figure (Figure 3) that there are variations among the studied farmers in diesel and electrical energy inputs. The main reason for these discrepancies was the type of alternative source of irrigation. For example, some farmers used tractor-operated tube-wells for irrigation water pumping, which increased diesel energy input, while others used electricity-operated tube-wells, which increased electrical energy input. However, no major variations were found in the use of other energy inputs, hence, this study mainly focuses on the average energy consumption. The energy classifications for wheat and rice production are shown in Figures 5 and 6. The average total energy input in wheat production was calculated as 21.36 GJ/ha which was further divided into direct and indirect energy inputs (Figure 5). The indirect energy input was calculated as 11.96 GJ/ha and dominated the direct energy use following the previous study [15]. The total energy was also classified as renewable and non-renewable energy inputs. The average renewable energy was found to be 1.14 GJ/ha which was lesser than non-renewable energy input i.e., 20.21 GJ/ha. Soltani et. al. [34] also reported a similar trend of energy consumption in their study on energy analysis in wheat production. However, fertilizer was the top input energy source in wheat production and accounts for 39% of the total energy consumption. Similarly, diesel was the second-highest energy source, accounting for 22.6% of total energy consumption, while energy from chemical energy was found to be the lowest (0.42% of total energy) among all the energy sources. These results are in accordance with the results of previous studies [27,41]. Singh et. al. [27] conducted a study on energy-use patterns in wheat cultivation in Punjab, India. The present study in Punjab, Pakistan, showed similar results due to comparable climatic and socioeconomic conditions. However, Gündogmus and Bayramoglu [62] reported diesel as the highest energy input source in their study on organic farming. The contradiction in these results is due to the minimum use of fertilizers in organic farming as reported in their study. On the other hand, Yuan et. al. [51] found electricity to be the highest energy input source in their study on energy flow assessment in wheat production. This is due to the use of an electricity-operated water-pumping system in their study, while tractor-operated tube-wells were mostly used for irrigation in the present study. Finally, fertilizer was found to be the main source of energy in wheat production and thus an important input source in controlling total energy consumption.

Table 4. Description of average use of output and input energy sources for wheat and rice production in the study area.

Production Factors	Units/ha	Wheat	Rice
inputs			
seed	kg	125.053	23.0824
diesel	L	85.5408	252.154
labor	h	59.4992	242.698
fertilizer	kg		
N	-	126.525	127.134
P2O5	-	59.7945	62.2378
K2O	-	3.67804	9.06375
water	m3	3197.99	13137.7
machinery	h		
tractor	-	19.1878	72.1281
plough	-	3.58482	4.37146
rotary hoe	-	2.50457	1.89331
combine harvester	-	0.97132	1.66165
chemicals	mL	867.493	375.844
chemicals	kg	0	4.87158
electricity	kWh	100.568	354.091
outputs			
grain	kg	4325.38	4441.24
straw	kg	4287.55	

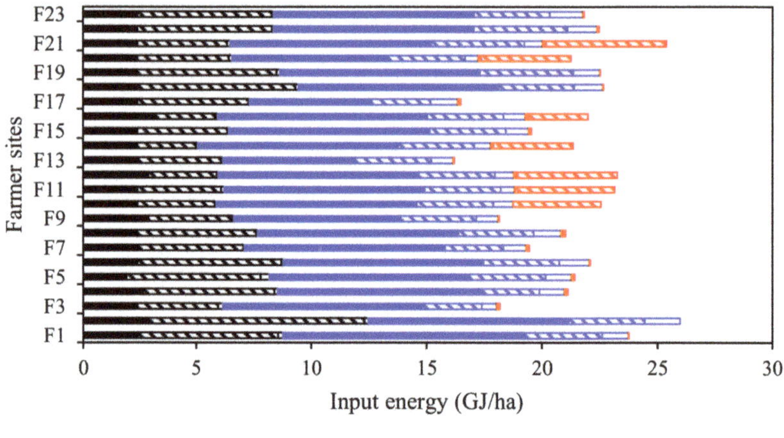

Figure 3. Energy input (GJ/ha) from various sources in wheat cultivation. The labels F1–F23 represent different farmers in the study area.

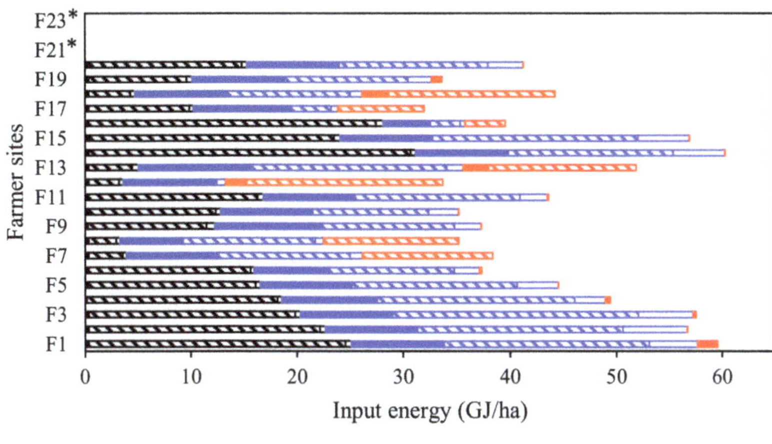

Figure 4. Energy consumption (GJ/ha) from different input sources in rice production. The labels F1–F23 represent different farmers in the study area. * These farmers did not cultivate rice crop in the studied year.

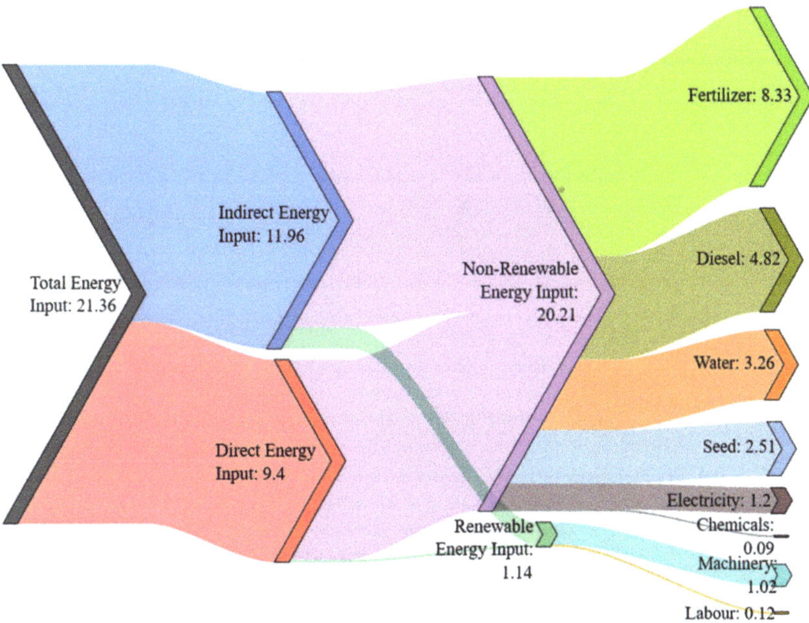

Figure 5. Classification of the average energy consumption (GJ/ha) in wheat production based on direct/indirect and renewable/non-renewable energy use.

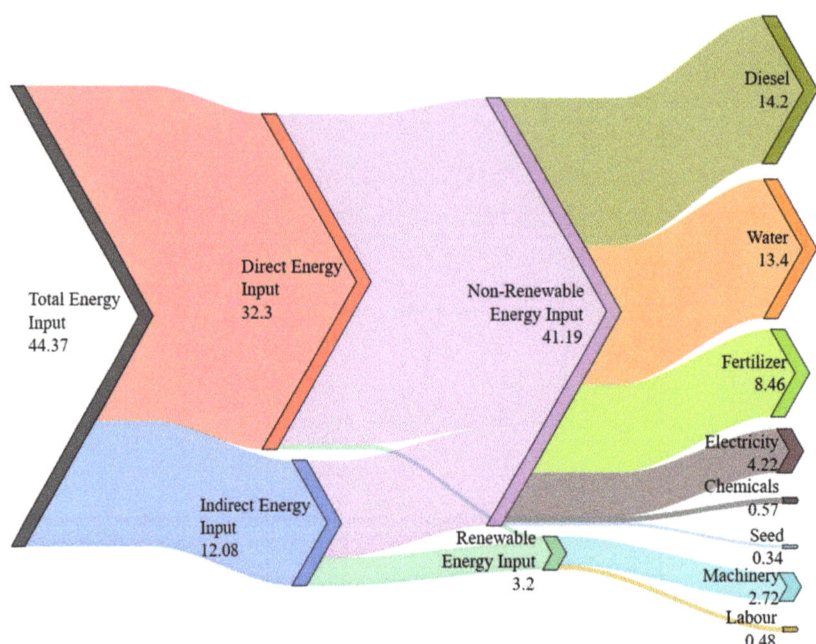

Figure 6. Classification of the average energy consumption (GJ/ha) in rice production based on direct/indirect and renewable/non-renewable energy use.

Similarly, the energy consumption pattern in rice production is shown in Figure 6. The average total energy input was calculated as 44.37 GJ/ha, which is much higher than energy consumed in wheat production. Contrary to wheat production, direct energy consumption dominated over indirect energy use in rice production and constitutes 72.8% of total energy input. This is due to the higher water requirement of rice crops that increased the direct energy input in the form of water, diesel, and electricity. Besides, the average non-renewable energy was calculated as 41.19 GJ/ha and constitutes 92.8% of the total energy input. It is evident from the figure that diesel was the highest energy input source in rice production and constitute 32% of the total energy input. Diesel was mainly used by tractors in pumping irrigation water from tube-wells to meet the higher crop water requirement in rice production. The second highest energy source was water, contributing 30.2% to total energy consumption, and seed contributed only 0.77% to the total energy input and was found to be the lowest energy input source. Komleh et. al. [13] also found similar results in their study on the evaluation of energy consumption in rice production. However, Khan et. al. [52] found fertilizer to be the highest energy input source in their study on energy consumption analyses in rice production in Pakistan. The main reason for this contradiction in results is the difference in climatic conditions of the study areas. Their study was conducted in a comparatively humid region with lesser water-pumping requirements, while the present study was conducted in an arid region. Higher water pumping for irrigation increases the use of machinery and thus diesel consumption in the present scenario. Ultimately, diesel and water were the main sources of energy in rice production and total energy input can be controlled by optimizing these input energies.

Operational energy is the energy from input sources consumed in management operations i.e., sowing, tillage, irrigation, harvesting and threshing, and spraying and fertilizing. The operational energy distribution in wheat and rice cultivation is shown in Figure 7. The average operational energy in rice production was calculated as 20.07 GJ/ha which is 2.7 times higher than that of wheat production. This was due to the higher energy input in irrigating rice production. The average irrigation energy input in rice production was found to be 16.14 GJ/ha and was the highest among all the management operations. Similarly, irrigation was also the highest energy-consuming operation in wheat production, however, the magnitude of irrigation energy in wheat production was almost five times lower than that of rice production, due to the higher water requirement of rice crops over wheat. Tillage was the second-highest energy-consuming operation in both crops. The average tillage energy in wheat production was 2.34 GJ/ha (31.8% of the total operational energy) while the average tillage energy in rice production was found to be 2.6 GJ/ha, which is 13% of the total operational energy. Due to the similarity in the cultivation process, there were no significant differences in the magnitude of tillage energies in both crops. On the other hand, sowing was the lowest energy-consuming operation in wheat cultivation, while energy input in spraying and fertilizing was the lowest in rice production. The average sowing energy in wheat production was 0.00129 GJ/ha, constituting 0.02% of total operational energy, while the average spraying and fertilizing energy in rice production was 0.02 GJ/ha. These results are in accordance with the previous literature [13,15,41]. Ashraf et. al. [15] conducted a study on energy evaluation of wheat crops in the same province and found similar trends in energy consumption between management operations. Similarly, Safa et. al. [41] and Komleh et. al. [13] also found similar results in their studies on wheat and rice production, respectively. Finally, irrigation is the crucial operation in terms of energy consumption in wheat and rice production, and more attention is needed to optimize the energy use in this operation.

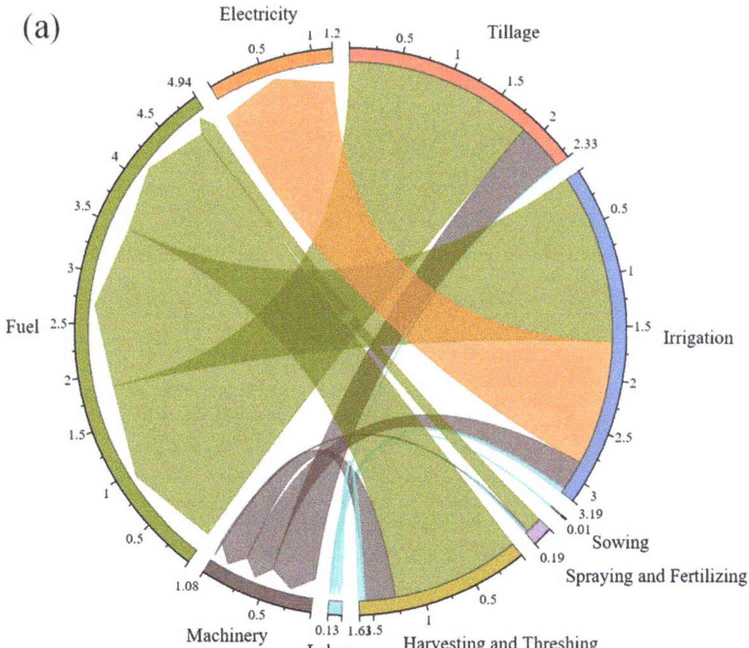

Figure 7. *Cont.*

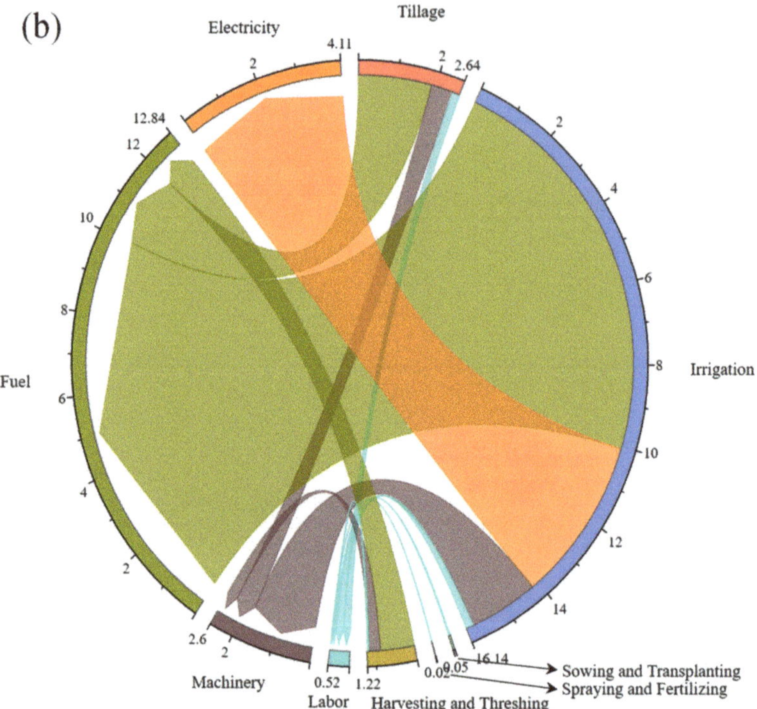

Figure 7. Distribution of energy consumption from different sources in different management operations in (**a**) wheat production and (**b**) rice production.

The overall energy efficiency in the studied crops was determined using three energy indices i.e., energy use efficiency (η_e), energy productivity (η_p), and net energy (ρ). The average values of these indices are exhibited in Table 5. The values of η_e, η_p, and ρ for the wheat crop were calculated as 5.68, 202.3 kg/GJ, and 100.12 kg/ha, respectively. In rice production, these values were found to be 1.71, 100.71 kg/GJ, and 31.13 GJ/ha, respectively. These results show the higher overall energy efficiency of the wheat crops compared to rice. There are two main reasons for the higher energy efficiency of wheat crops i.e., higher input energy in rice production and energy from rice-crop residue is not considered in this study which ultimately decreases the total energy output from rice crops. However, total rice-grain yields per hectare is higher than wheat-grain yields and recorded as 4441.24 kg/ha. Finally, the wheat crop was found to be more energy-efficient and the rice crop was more productive in terms of grain yields. Furthermore, a multiple regression model was developed to predict grain yield for both crops. The comparison between actual and predicted grain yields for wheat and rice crops is shown in Figures 8 and 9, respectively. It is evident that the developed model can predict wheat yield with a coefficient of regression (R^2) value of 0.7733, while the R^2 value for rice yield was 0.7299. These equations might be helpful for the farmers in the study area to predict the response of a specific energy input towards total grain yield before the actual application of that input in the field.

Table 5. Description of average energy consumption, energy outputs, and energy indices in wheat and rice production in the study area.

Parameter	Unit	Crop	
		Wheat	Rice
input energy	GJ/ha	21.36	44.37
output energy	GJ/ha	121.5	75.5
grain yield	kg/ha	4325.38	4441.24
energy use efficiency	-	5.68	1.71
energy productivity	kg/GJ	202.3	100.72
net energy	GJ/ha	100.12	31.13

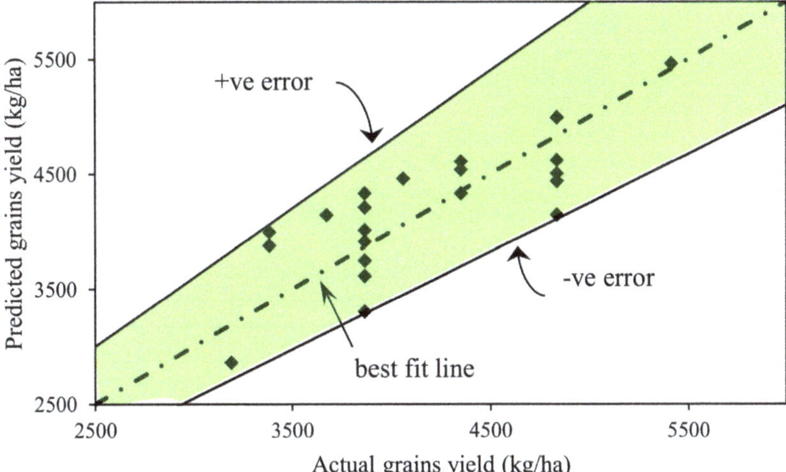

Figure 8. Comparison between actual and predicted grain yields of the studied farms based on the developed multiple regression model for wheat production.

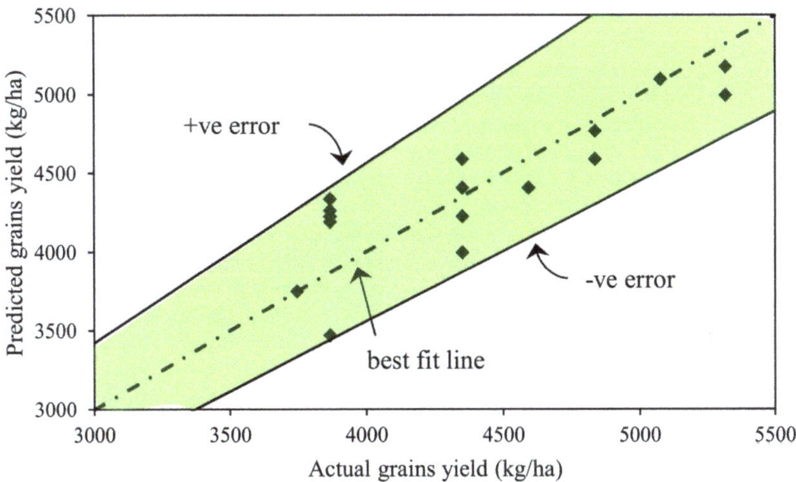

Figure 9. Comparison between actual and predicted grain yields of the studied farms based on the developed multiple regression model for rice production.

Data envelopment analysis (DEA) was also conducted to optimize the energy inputs. Four major energy sources (i.e., diesel, water, fertilizer, and machinery) were considered as controlled inputs while grain yield was taken as output in a CCR-based DEA function. Table 6 exhibits the energy optimization pattern for wheat and rice crops in the study area. The highest reduction potential was observed for machinery energy in both crops i.e., wheat and rice, calculated as −17.48% and −42.97%, respectively. It shows that lower machining operations e.g., minimum/zero tillage can also maintain the same level of output while reducing the total input energy. Moreover, it can enhance the overall energy efficiency/productivity in the wheat–rice crop rotation system. The minimum reduction potential was found in water energy (−2.76%) for wheat crop while diesel energy (−17.25%) in rice production. The lower reduction potential in water energy shows the importance of water for wheat crop in an arid region. Similarly, the reduction potential in water energy for rice crop was also among the lowest and calculated as −17.65%. These results are in accordance with the previous literature [15,38]. Ashraf et al. [15] conducted DEA in wheat production in similar climatic conditions and found that water energy had the least reduction potential. However, they did not consider machinery energy in their study, hence no information was found related to reduction potential in machining operations. Mousavi-Avval et. al. [38] also found similar results in their studies on apple production in Iran. Finally, optimizing the energy use in agricultural production can increase energy use efficiency, energy productivity, and net energy. This would help in reducing energy input while maintaining output, i.e., grain yields, in the current scenario.

Table 6. Energy-use optimization and average technical efficiencies of the studied farmers, calculated by a CCR-based DEA optimization function.

Energy Input	Units	Wheat			Rice		
		Actual	Targeted	Percentage	Actual	Targeted	Percentage
fertilizer	GJ/ha	8.01	7.36	−7.75	8.08	6.46	−18.68
diesel	GJ/ha	4.40	3.78	−11.92	11.43	9.20	−17.25
water	GJ/ha	3.01	2.76	−8.06	10.97	8.68	−17.65
machinery	GJ/ha	0.96	0.76	−17.48	1.86	0.72	−42.97
technical efficiency	%		93.7			82.7	

Carbon dioxide (CO_2) emissions are a byproduct of the production processes of energy sources i.e., machinery, fertilizer, electricity, and chemicals, while on-farm combustion of diesel also emits a large amount of the gas. Figure 10 shows the average total and source wise CO_2 emissions in kg-CO_2/ha units. The average total emissions in wheat and rice production were calculated as 900.9 kg-CO_2/ha and 1762.5 kg-CO_2/ha, respectively. Diesel was the highest emissions source in rice production, constituting 55.34% of total CO_2 emissions. On the other hand, fertilizer was found to be the highest CO_2 emissions source in wheat production, at 49.7% of total emissions. The higher emissions from diesel, in rice production, were due to the higher use of diesel in pumping irrigation water from tractor-operated tube-wells. The lowest emissions calculation belonged to chemical sources for both crops. The average CO_2 emissions from chemicals in rice and wheat production was recorded as 33.2 kg-CO_2/ha and 5.3 kg-CO_2/ha, respectively. These results are in accordance with the results of previous literature [15,34]. Ashraf et. al. [15] and Soltani et. al. [34] found similar results in their studies on wheat production in Pakistan and Iran respectively. Their findings also revealed that CO_2 emissions increase with total input energy. In this regard, a relationship was developed between total input energy and CO_2 as shown in Figure 11. The figure shows that CO_2 emissions increase linearly with total energy input in wheat and rice cultivations. Therefore, emissions from rice production are higher than that from wheat production. To reduce these emissions, there is a need to reduce the total input energy. In this regard, the above-discussed DEA approach can play an important role in reducing the total input energy and, eventually, CO_2 emissions, while maintaining grain yields.

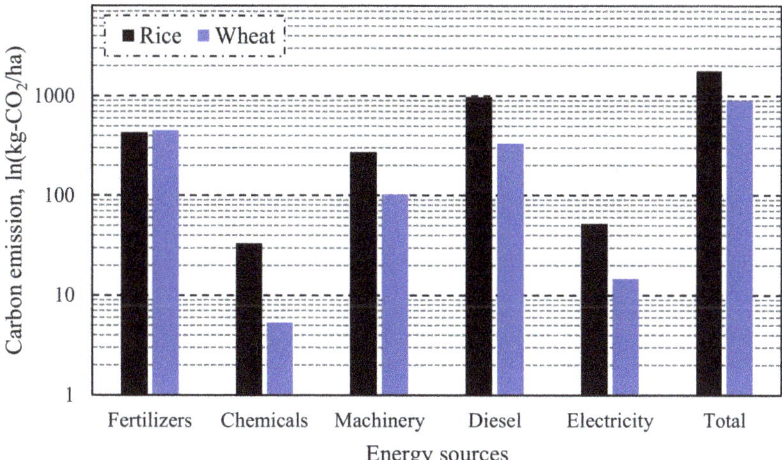

Figure 10. Average carbon dioxide (CO_2) emissions from wheat and rice production in the study area. The vertical axis is the logarithmic scale of carbon emissions in the kg-CO_2/ha unit.

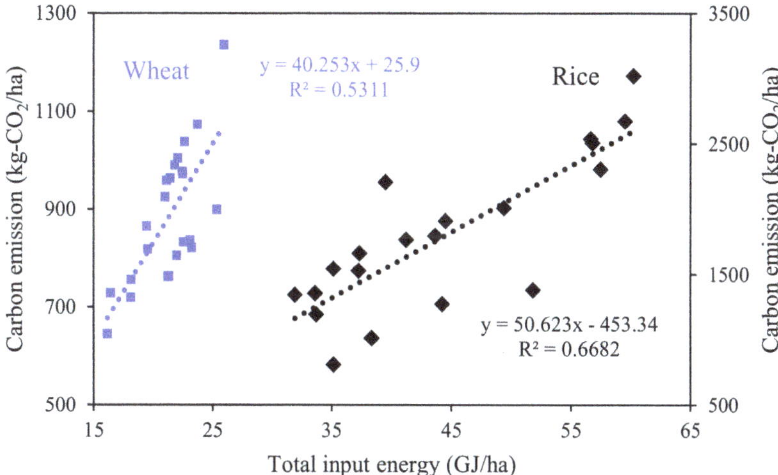

Figure 11. Relationship between total input energy (GJ/ha) and carbon emissions (kg-CO_2/ha) in wheat and rice production. The blue-colored tick markers belong to the primary vertical axis and represent the wheat crop, while black-colored tick markers belong to the secondary vertical axis representing the rice crop.

4. Conclusions

In this study, the input–output energy relationship and CO_2 emissions in a wheat–rice crop rotation system were investigated technically. The data envelopment analysis (DEA) technique was applied to optimize total energy consumption. Based on the key findings of the study, it was concluded that socioeconomic characteristics greatly affect total energy input in agricultural production. For example, an irrigation source could be a key factor influencing total input energy. In this regard, canal-water irrigation was found to be less energy-expensive than tractor-operated and electrical turbine-operated tube-wells. Furthermore, the average total energy consumption in rice production was higher than that of wheat production. It was mainly due to the greater number of irrigations in rice production to meet the higher crop water requirement. It resulted in lowering the energy efficiency/productivity in rice production than that of wheat. However, the data envelopment analysis (DEA) was found to be an efficient approach in reducing the total energy input while maintaining grain yield. This technique is also helpful in reducing total CO_2 emissions by optimized use of energy sources.

Finally, the optimized use of energy sources could increase the overall energy efficiency in the rice–wheat crop-rotation system. In this regard, minimum tillage and high-efficiency irrigation systems could be the best alternatives to conventional management practices. The results of this study showed that irrigation was the highest energy-consuming operation. Therefore, it is important to control energy consumption in this operation to enhance overall energy efficiency.

Author Contributions: Conceptualization, M.N.A. and M.H.M.; methodology, M.N.A.; software, M.N.A.; validation, M.H.M., M.S. and R.R.S.; formal analysis, M.N.A. and M.H.M.; investigation, M.H.M., S.M.I. and M.S.; resources, M.S., R.R.S. and S.M.I.; data curation, M.N.A. and M.H.M.; writing—original draft preparation, M.N.A.; writing—review and editing, M.H.M., M.S. and R.R.S.; visualization, M.S., R.R.S. and S.M.I.; supervision, M.H.M. and M.S.; project administration, M.H.M.; funding acquisition, M.H.M., M.S., S.M.I. and R.R.S. All authors have read and agreed to the published version of the manuscript.

Funding: This work was supported by Researchers Supporting Project number (RSP-2021/100), King Saud University, Riyadh, Saudi Arabia.

Institutional Review Board Statement: Not applicable.

Informed Consent Statement: Not applicable.

Data Availability Statement: The data are contained within the article.

Acknowledgments: The authors acknowledge the help in data collection by Muhammad Imran. This research work has been carried out in the Department of Agricultural Engineering, Bahauddin Zakariya University, Multan-Pakistan. This work was supported by Researchers Supporting Project number (RSP-2021/100), King Saud University, Riyadh, Saudi Arabia. The authors acknowledge the financial support by the Open Access Publication Fund of the Leibniz Association, Germany.

Conflicts of Interest: The authors declare no conflict of interest.

References

1. Cereals Production Quantity. Knoema n.d. Available online: https://knoema.com/atlas/topics/Agriculture/Crops-Production-Quantity-tonnes/Cereals-production (accessed on 24 May 2021).
2. Crops Data. FAOSTAT n.d. Available online: http://www.fao.org/faostat/en/#data/QC/visualize (accessed on 24 May 2021).
3. Giraldo, P.; Benavente, E.; Manzano-Agugliaro, F.; Gimenez, E. Worldwide research trends on wheat and barley: A bibliometric comparative analysis. *Agronomy* **2019**, *9*, 352. [CrossRef]
4. Lizette, K. Food Security & Staple Crops Staple Food Around the World. *IAEA Bull.* **2012**, *53*, 11.
5. Lucas, H. Breakout session P1.1 National Food Security-The Wheat Initiative-an International Research Initiative for Wheat Improvement context–the problems being addressed. In Proceedings of the GCARD-Second Global Conference on Agricultural Research for Development, Punta del Este, Uruguay, 29 October–1 November 2012; pp. 1–3.
6. Mahmood, M.H.; Sultan, M.; Miyazaki, T. Significance of Temperature and Humidity Control for Agricultural Products Storage: Overview of Conventional and Advanced Options. *Int. J. Food Eng.* **2019**, *15*. [CrossRef]
7. Muhammad, H. Mahmood, Muhammad Sultan, Takahiko Miyazaki SK. Study on desiccant air-conditioning system for agricultural product storage in Pakistan. *Intellect. Exch. Innov. Conf. Eng. Sci.* **2015**, *1*, 13–14. [CrossRef]
8. Sultan, M.; El-sharkawy, I.I.; Miyazaki, T.; Baran, B.; Koyama, S. An overview of solid desiccant dehumidification and air conditioning systems. *Renew. Sustain. Energy Rev.* **2015**, *46*, 16–29. [CrossRef]
9. Sultan, M.; El-Sharkawy, I.I.; Miyazaki, T.; Saha, B.B.; Koyama, S.; Maruyama, T.; Maeda, S.; Nakamura, T. Insights of water vapor sorption onto polymer based sorbents. *Adsorption* **2015**, *21*, 205–215. [CrossRef]
10. Sultan, M.; Miyazaki, T.; Mahmood, M.H.; Khan, Z.M. Solar assisted evaporative cooling based passive air-conditioning system for agricultural and livestock applications. *J. Eng. Sci. Technol.* **2018**, *13*, 693–703.

1. Ashraf, H.; Sultan, M.; Shamshiri, R.R.; Abbas, F.; Farooq, M.; Sajjad, U.; Md-Tahir, H.; Mahmood, M.H.; Ahmad, F.; Taseer, Y.R.; et al. Dynamic Evaluation of Desiccant Dehumidification Evaporative Cooling Options for Greenhouse Air-Conditioning Application in Multan (Pakistan). *Energies* **2021**, *14*, 1097. [CrossRef]
2. Conway, R.K.; Stout, B.A. Handbook of Energy for World Agriculture. London and New York: Elsevier, Applied Science, 1990, 504 pp., \$@@-@@135.00. *Am. J. Agric. Econ.* **1991**, *73*, 1302. [CrossRef]
3. Pishgar-Komleh, S.H.; Sefeedpari, P.; Rafiee, S. Energy and economic analysis of rice production under different farm levels in Guilan province of Iran. *Energy* **2011**, *36*, 5824–5831. [CrossRef]
4. Kardoni, F.; Ahmadi, J.A.; Bakhshi, M.R. Energy Efficiency Analysis and Modeling the Relationship between Energy Inputs and Wheat Yield in Iran. *Int. J. Agric. Manag. Dev.* **2015**, *5*, 321–330.
5. Ashraf, M.N.; Mahmood, M.H.; Sultan, M.; Banaeian, N.; Usman, M.; Ibrahim, S.M.; Butt, M.U.B.U.; Waseem, M.; Ali, I.; Shakoor, A.; et al. Investigation of Input and Output Energy for Wheat Production: A Comprehensive Study for Tehsil Mailsi (Pakistan). *Sustainability* **2020**, *12*, 6884. [CrossRef]
6. Maraseni, T.; Chen, G.; Banhazi, T.; Bundschuh, J.; Yusaf, T. An Assessment of Direct on-Farm Energy Use for High Value Grain Crops Grown under Different Farming Practices in Australia. *Energies* **2015**, *8*, 13033–13046. [CrossRef]
7. Ozkan, B.; Akcaoz, H.; Fert, C. Energy input-output analysis in Turkish agriculture. *Renew. Energy* **2004**, *29*, 39–51. [CrossRef]
8. Safa, M.; Samarasinghe, S.; Mohssen, M. Determination of fuel consumption and indirect factors affecting it in wheat production in Canterbury, New Zealand. *Energy* **2010**, *35*, 5400–5405. [CrossRef]
9. Fami, H.S.; Ghasemi, J.; Malekipoor, R.; Rashidi, P.; Nazari, S.; Mirzaee, A. Renewable energy use in smallholder farming systems: A case study in tafresh township of Iran. *Sustainability* **2010**, *2*, 702–716. [CrossRef]
10. Esengun, K.; Gündüz, O.; Erdal, G. Input-output energy analysis in dry apricot production of Turkey. *Energy Convers. Manag.* **2007**, *48*, 592–598. [CrossRef]
11. Singh, G.; Lee, J.; Karakoti, A.; Bahadur, R.; Yi, J.; Zhao, D.; AlBahilyc, K.; Vinu, A. Emerging trends in porous materials for CO_2 capture and conversion. *Chem. Soc. Rev.* **2020**, *49*, 4360–4404. [CrossRef]
12. US Department of Commerce, NOAA GML. Global Monitoring Laboratory—Carbon Cycle Greenhouse Gases n.d. Available online: https://gml.noaa.gov/ccgg/trends/mlo.html (accessed on 5 April 2021).
13. Snoeckx, R.; Bogaerts, A. Plasma technology—A novel solution for CO_2 conversion? *Chem. Soc. Rev.* **2017**, *46*, 5805–5863. [CrossRef]
14. Grant, J.; Low, L.P.; Unsworth, S.; Hornwall, C.; Davies, M. The Low Carbon Economy Index 2018—Time to Get on with It. Available online: https://www.pwc.co.uk/lowcarboneconomy (accessed on 5 April 2021).
15. Eckstein, D.; Künzel, V.; Schäfer, L.; Winges, M. Global Climate Rate Index 2020; Germanwatch e.V., Germany. 2020. Available online: www.germanwatch.org/en/cri (accessed on 5 April 2021).
16. USAID. Greenhouse Gas Emissions Factsheet: Pakistan | Global Climate Change. 2018. Available online: https://www.climatelinks.org/resources/greenhouse-gas-emissions-factsheet-pakistan (accessed on 5 April 2021).
17. Singh, H.; Singh, A.K.; Kushwaha, H.L.; Singh, A. Energy consumption pattern of wheat production in India. *Energy* **2007**, *32*, 1848–1854. [CrossRef]
18. Saad, A.A.; Das, T.K.; Rana, D.S.; Sharma, A.R.; Bhattacharyya, R.; Lal, K. Energy auditing of a maize–wheat–greengram cropping system under conventional and conservation agriculture in irrigated north-western Indo-Gangetic Plains. *Energy* **2016**, *116*, 293–305. [CrossRef]
19. Mani, I.; Kumar, P.; Panwar, J.S.; Kant, K. Variation in energy consumption in production of wheat-maize with varying altitudes in hilly regions of Himachal Pradesh, India. *Energy* **2007**, *32*, 2336–2339. [CrossRef]
20. Chaudhary, V.P.; Gangwar, B.; Pandey, D.K.; Gangwar, K.S. Energy auditing of diversified rice-wheat cropping systems in Indo-gangetic plains. *Energy* **2009**, *34*, 1091–1096. [CrossRef]
21. Singh, H.; Mishra, D.; Nahar, N.M.; Ranjan, M. Energy use pattern in production agriculture of a typical village in arid zone India: Part II. *Energy Convers. Manag.* **2003**, *44*, 1053–1067. [CrossRef]
22. Singh, K.P.; Prakash, V.; Srinivas, K.; Srivastva, A.K. Effect of tillage management on energy-use efficiency and economics of soybean (Glycine max) based cropping systems under the rainfed conditions in North-West Himalayan Region. *Soil Tillage Res.* **2008**, *100*, 78–82. [CrossRef]
23. Tabatabaeefar, A.; Emamzadeh, H.; Varnamkhasti, M.G.; Rahimizadeh, R.; Karimi, M. Comparison of energy of tillage systems in wheat production. *Energy* **2009**, *34*, 41–45. [CrossRef]
24. Soltani, A.; Rajabi, M.H.; Zeinali, E.; Soltani, E. Energy inputs and greenhouse gases emissions in wheat production in Gorgan, Iran. *Energy* **2013**, *50*, 54–61. [CrossRef]
25. Shahan, S.; Jafari, A.; Mobli, H.; Rafiee, S.; Karimi, M. Effect of Farm Size on Energy Ratio for Wheat Production: A Case Study from Ardabil Province of Iran. *Energy* **2008**, *3*, 604–608.
26. Yousefi, M.; Khoramivafa, M.; Mondani, F. Integrated evaluation of energy use, greenhouse gas emissions and global warming potential for sugar beet (Beta vulgaris) agroecosystems in Iran. *Atmos. Environ.* **2014**, *92*, 501–505. [CrossRef]
27. Ra, S.; Hashem, S.; Avval, M.; Mohammadi, A. Modeling and sensitivity analysis of energy inputs for apple production in Iran. *Energy* **2010**, *35*, 3301–3306. [CrossRef]
28. Mousavi-Avval, S.H.; Ra, S.; Mohammadi, A. Optimization of energy consumption and input costs for apple production in Iran using data envelopment analysis. *Energy* **2011**, *36*, 909–916. [CrossRef]

39. Khoshroo, A.; Mulwa, R.; Emrouznejad, A.; Arabi, B. A non-parametric Data Envelopment Analysis approach for improving energy ef fi ciency of grape production. *Energy* **2013**, *63*, 189–194. [CrossRef]
40. Safa, M.; Samarasinghe, S. CO$_2$ emissions from farm inputs "case study of wheat production in Canterbury, New Zealand" *Environ. Pollut.* **2012**, *171*, 126–132. [CrossRef]
41. Safa, M.; Samarasingh, S.; Mohssen, M. A field study of energy consumption in wheat production in Canterbury, New Zealand *Energy Convers. Manag.* **2011**, *52*, 2526–2532. [CrossRef]
42. Safa, M.; Samarasinghe, S. Determination and modelling of energy consumption in wheat production using neural networks: "A case study in Canterbury province, New Zealand". *Energy* **2011**, *36*, 5140–5147. [CrossRef]
43. Kuswardhani, N.; Soni, P.; Shivakoti, G.P. Comparative energy input-output and financial analyses of greenhouse and open field vegetables production in West Java, Indonesia. *Energy* **2013**, *53*, 83–92. [CrossRef]
44. Soni, P.; Taewichit, C.; Salokhe, V.M. Energy consumption and CO$_2$ emissions in rainfed agricultural production systems of Northeast Thailand. *Agric. Syst.* **2013**, *116*, 25–36. [CrossRef]
45. Yilmaz, I.; Akcaoz, H.; Ozkan, B. An analysis of energy use and input costs for cotton production in Turkey. *Renew. Energy* **2005**, *30*, 145–155. [CrossRef]
46. Erdal, G.; Esengün, K.; Erdal, H.; Gündüz, O. Energy use and economical analysis of sugar beet production in Tokat province of Turkey. *Energy* **2007**, *32*, 35–41. [CrossRef]
47. Ozkan, B.; Fert, C.; Karadeniz, C.F. Energy and cost analysis for greenhouse and open-field grape production. *Energy* **2007**, *32*, 1500–1504. [CrossRef]
48. Canakci, M.; Topakci, M.; Akinci, I.; Ozmerzi, A. Energy use pattern of some field crops and vegetable production: Case study for Antalya Region, Turkey. *Energy Convers. Manag.* **2005**, *46*, 655–666. [CrossRef]
49. Snyder, C.S.; Bruulsema, T.W.; Jensen, T.L.; Fixen, P.E. Review of greenhouse gas emissions from crop production systems and fertilizer management effects. *Agric. Ecosyst. Environ.* **2009**, *133*, 247–266. [CrossRef]
50. Khan, S.; Hanjra, M.A. Footprints of water and energy inputs in food production—Global perspectives. *Food Policy* **2009**, *34*, 130–140. [CrossRef]
51. Yuan, S.; Peng, S.; Wang, D.; Man, J. Evaluation of the energy budget and energy use efficiency in wheat production under various crop management practices in China. *Energy* **2018**, *160*, 184–191. [CrossRef]
52. Khan, M.A.; Awan, I.U.; Zafar, J. Energy requirement and economic analysis of rice production in western part of Pakistan. *Soil Environ.* **2009**, *28*, 60–67.
53. Khan, S.B.; Khan, F.; Sadaf, S.; Kashif, R.H. *Crops Area and Production (By Districts) (1981–82 To 2008–09)*; Pakistan Bureau of Statistics: Islamabad, Pakistan, 2009; p. 214.
54. Banaeian, N.; Omid, M.; Ahmadi, H. Energy and economic analysis of greenhouse strawberry production in Tehran province of Iran. *Energy Convers. Manag.* **2011**, *52*, 1020–1025. [CrossRef]
55. Kitani, O.; Jungbluth, T.; Peart, R.M.; Ramdani, A. *CIGR Handbook of Agricultural Engineering (Energy and Biomass Engineering)*; ASAE Publication: St. Joseph, MI, USA, 1999; Volume 5.
56. ASAE. *Agricultural Machinery Management Data*; American Society of Agricultural and Biological Engineers: St. Joseph, MI, USA, 2011.
57. Wells, D. *Total Energy Indicators of Agricultural Sustainability: Dairy Farming Case Study*; Technical Paper; Ministry of Agriculture and Forestry: Wellington, New Zealand, 2001; ISBN 0-478-07968-0.
58. Singh, P.; Singh, G.; Sodhi, G.P.S. Applying DEA optimization approach for energy auditing in wheat cultivation under rice-wheat and cotton-wheat cropping systems in north-western India. *Energy* **2019**, *181*, 18–28. [CrossRef]
59. Charnes, A.; Cooper, W.W.; Rhodes, E. Measuring the efficiency of decision making units. *Eur. J. Oper. Res.* **1978**, *2*, 429–444. [CrossRef]
60. Cooper, W.W.; Seiford, L.M.; Tone, K. *Data Envelopment Analysis: A Comprehensive Text with Models, Applications, References and DEA-Solver Software*, 2nd ed.; Springer: New York, NY, USA, 2007. [CrossRef]
61. Saunders, C.; Barber, A.; Taylor, G. *Food Miles - Comparative Energy/emissions Performance of New Zealand's Agriculture Industry*; Agribusiness & Economics Research Unit, Lincoln University: Lincoln, New Zealand, 2006.
62. Gündogmus, E.; Bayramoglu, Z. Energy input use on organic farming: A comparative analysis on organic versus conventional farms in Turkey. *J. Agron.* **2006**, *5*, 16–22. [CrossRef]

Article

Solar-Hybrid Cold Energy Storage System Coupled with Cooling Pads Backup: A Step towards Decentralized Storage of Perishables

Anjum Munir [1,*], Tallha Ashraf [1], Waseem Amjad [1,*], Abdul Ghafoor [2], Sidrah Rehman [1], Aman Ullah Malik [3], Oliver Hensel [4], Muhammad Sultan [5] and Tatiana Morosuk [6,*]

1 Department of Energy Systems Engineering, University of Agriculture, Faisalabad 38000, Pakistan; talhaashraf313@gmail.com (T.A.); sidrareahman3@yahoo.com (S.R.)
2 Department of Farm Machinery and Power, University of Agriculture, Faisalabad 38000, Pakistan; abdul.ghafoor@uaf.edu.pk
3 Institute of Horticultural Sciences, University of Agriculture, Faisalabad 38000, Pakistan; malikaman1@uaf.edu.pk
4 Department of Agricultural & Biosystems Engineering, University of Kassel, 37213 Witzenhausen, Germany; ohensel@uni-kassel.de
5 Department of Agricultural Engineering, Bahauddin Zakariya University, Multan 60800, Pakistan; muhammadsultan@bzu.edu.pk
6 Institute for Energy Engineering, Technische Universität Berlin, Marchstr. 18, 10587 Berlin, Germany
* Correspondence: anjum.munir@uaf.edu.pk (A.M.); waseem_amjad@uaf.edu.pk (W.A.); tetyana.morozyuk@tu-berlin.de (T.M.)

Citation: Munir, A.; Ashraf, T.; Amjad, W.; Ghafoor, A.; Rehman, S.; Malik, A.U.; Hensel, O.; Sultan, M.; Morosuk, T. Solar-Hybrid Cold Energy Storage System Coupled with Cooling Pads Backup: A Step towards Decentralized Storage of Perishables. *Energies* **2021**, *14*, 7633. https://doi.org/10.3390/en14227633

Academic Editor: Donato Morea

Received: 1 October 2021
Accepted: 11 November 2021
Published: 15 November 2021

Publisher's Note: MDPI stays neutral with regard to jurisdictional claims in published maps and institutional affiliations.

Abstract: Post-harvest loss is a serious issue to address challenge of food security. A solar-grid hybrid cold storage system was developed and designed for on-farm preservation of perishables. Computational Fluid Dynamic analysis was performed to assess airflow and temperature distribution inside the cold chamber. The system comprises a 21.84 m^3 cubical cold storage unit with storage capacity of 2 tonnes. A hybrid solar system comprising 4.5 kWp PV system, 5 kW hybrid inverter, and 600 Ah battery bank was used to power the entire system. A vapor-compression refrigeration system (2 tonnes) was employed coupled with three cooling pads (filled with brine solution) as thermal backup to store cooling (−4 °C to 4 °C). Potatoes were stored at 8 °C for a period of three months (May 2019 to July 2019) and the system was tested on grid utility, solar, and hybrid modes. Solar irradiation was recorded in range of 5.0–6.0 kWh/(m^2 × d) and average power peak was found to be 4.0 kW. Variable frequency drive was installed with compressor to eliminate the torque load and it resulted about 9.3 A AC current used by the system with 4.6 average Coefficient of Performance of refrigeration unit. The average energy consumed by system was found to be 15 kWh with a share of 4.3 kWh from grid and 10.5 kWh from solar, translating to 30% of power consumption from grid and 70% from solar PV modules. Overall, cold storage unit efficiently controlled total weight loss (7.64%) and preserved quality attributes (3.6 ^{0}Brix Total soluble solids, 0.83% Titratable acidity, 6.32 PH) of the product during storage time.

Keywords: solar cooling; post-harvest food losses; decentralized food storage; cooling pads

1. Introduction

Food security is a challenging task in developing countries, despite high production of food crops. It could be related to high post-harvest losses and uncertainty in agriculture and food policies of the governments. The seasonal character of agricultural produce is an important factor to be considered to save spoilage. Some fruits and vegetables are available for a short period of time in the year and need special care to stop their spoilage, especially when stored. Post-harvest losses exist due to lack of storage and transport infrastructures at a farm level and frequent power interruptions during food processing and handling [1,2]. Lack of the post-harvest processing facilities at farm level leads to spoilage of a significant quantity of perishable products. The most popular and conventional modes for the storage of perishables are drying and cooling [3–5]. The process of drying removes the moisture from the perishables; thus, the microbial activity decreases and spoilage of food is prevented [6]. However, the drying process changes the color, taste, and the texture of the product [7–10]. In case of cooling or freezing as a method of preservation, a product is stored at a low temperature to decrease the microbial activity. This process is used for the preservation of perishables effectively without product deterioration and shrinkage issues.

Conventional cold storage systems consume a large amount of energy for a refrigeration system. Due to their high energy consumption, it is not being readily adopted by the farming community. Moreover, commercial refrigerated plants (cold storages) are available only at large scale; they have a high capital cost and are not available in all areas, which translates into additional transportation cost [11]. Therefore, the "average" farmers cannot afford to store their products in those plants and must sell their product at low profit. In case of fruits and vegetables, the process of deterioration starts just after their harvest due to high product moisture content, heat of the field, improper handling, etc. [12,13] To avoid such losses, timely storage is an economically best post-harvest technique. For this, decentralized (on-farm) cold storage facilities can significantly reduce losses and integration of such facilities with solar energy would reduce operational cost after payback period is over (usually 4–6 years). An experimental study on the thermodynamic design of alternate systems based on cold storage is reported in [14].

Considering these challenges in the agriculture sector, especially in developing countries, solar-based cold storage facilities for decentralized applications (on-farm) is a viable solution to reduce post-harvest losses in perishables. Use of solar energy for cooling is attractive due to comparative co-occurrence between solar irradiance and cooling requirement. Refrigeration and air conditioning systems operated by solar power have high potential in energy saving, which could reach 40–50% [15]. The post-harvest losses of perishables are high in developing countries. Moreover, in tropical countries such as Pakistan, solar energy can be successfully used for various decentralized preservation and processing facilities where most of the villages have not access to grid supply. Considering Pakistan, it receives abundant solar radiations throughout the year [16]. There is an average of 5–7 kWh/(m^2 × d) of solar irradiation available in most of the country and receives 19 MJ of energy on average throughout the year [17]. Therefore, solar energy would be a great alternative to the conventional grid system to run the cold storage facilities.

Although extensive work has been reported on refrigeration (types of systems, energy demand, etc.), little research has been done on the potential of solar energy for cooling used materials, and storage options. Otanicar et al. [18] conducted a study of solar cooling in terms of economic and environmental impacts. It was concluded that one of the main hurdles in the extensive commercialization of solar cooling technologies is a high initial cost in case of standalone system due to the large size of battery backup, which decreases the competitiveness of these technologies. However, forecasts predicted that solar electric cooling would involve the least capital investment in 2030 because of the high COP values of vapor-compression refrigeration and high cost saving goals for PV technology. As the COP of refrigeration system is already high, coupling the refrigeration systems with efficient PV technology would, definitely, reduce operation cost. Thus, commercialization of solar cooling would be easy once PV technology becomes cheaper. An off-grid PV cooling system having two-stage energy storage (TSES) consisting of a battery bank and cold-water storage system has been proposed [19]. It was concluded that the key influence for battery capacity and capacity of cold storage system was a chiller schedule. A TSES system could increase the efficiency of the whole system by 6.73% and 10.27% based on single battery storage and single cold-water storage tank, respectively, under the most effective state in convex refrigeration capacity. A solar-powered cold storage system (6–8 tonne capacity, with battery backup and a vapor-compression refrigeration (2.5 TR) was reported in [20]

The system was able to maintain a temperature of 5–25 °C and a relative humidity of 65–95% inside the storage chamber. Although backup was provided, a battery bank cannot maintain required cooling conditions in case of prolonged cloudy weathers [20]. The use of a solar-based solar refrigerator (capacity of 150 L) along with a thermal bank has been reported in [21]. For the cooling bank, a solution of distilled water and propylene glycol was used, which provided a freezing temperature below zero to -10 °C for approximately 16 h. The performance of a solar biomass cold storage hybrid system is reported in [22]. This system is suitable for rural areas where biomass is present in a large quantity. No study has been reported on the decentralized use of solar energy in hybrid mode with cooling pads as a cooling backup, requiring less operational and maintenance cost in comparison to a system with a water-based cooling backup. Water-based cooling systems work on the maximum air conditioning load, leading to a very low efficiency once the machine is in idle mode. On the other hand, an air conditioning unit coupled with solar energy can be operated under low refrigeration capacity period, which helps to improve system energy efficiency [23].

Keeping in view the previous work reported, the potential of solar energy to be used for cooling and to provide a decentralized storage facility with thermal backup, a solar cold storage unit was designed and developed with a salient feature of its thermal backup (cooling pads, e.g., brine pads) for a time period of 36–48 h, depending on the ambient conditions. A sample product (potatoes) was stored for a period of three months (5 May 2019 to 30 July 2019). The system is operated on solar energy, but has an additional feature to be used in hybrid mode (operated on battery and grid). Moreover, this could be connected as a grid-tied or hybrid system in those areas where grid connectivity is available to improve efficiency and to decrease the operational cost of the system. To operate cold storage entirely on solar energy, during sun hours, some of the cooling from thermal pads is used to maintain the set cold storage temperature while the rest is consumed to maintain the brine solution at sub-zero temperature. Therefore, a temperature gradient always present between brine pads and cold storage chamber and cooling is transferred from brine pads to the product stored during night due to that temperature gradient. The research hypothesis is to develop a cold storage unit which can run continuously on solar energy for decentralized preservation of perishables by employing a solar grid hybrid system which automatically utilizes the full potential of available solar energy (as a first priority and switches to battery and utility to batteries or utility for extra energy demand if required) as well as using cooling as a backup source to store extra cooling produced by a cooling machine and maintain the quality of a stored product during storage time.

2. Materials and Methods

2.1. Capacity and Size of Cold Storage Chamber

The size of cold storage depends on the type and quantity of food to be stored, method of handling, and the dimensions of each basket/box used. Food is stored in baskets and these boxes are stacked on a pallet. The number of baskets, mass of food to be stored in a basket, and stacking of pallets depend on the type of food, storage time, and handling. Total capacity of the cold storage unit is calculated as:

$$V_C = a_f h_C \tag{1}$$

where a_f is the area of the floor and h_c is the height of cold storage chamber;

$$h_c = h_p + c_a \tag{2}$$

where h_p is the height of each pellet and its value is 1.53 m and c_a is the (vertical) clearance for aeration and its value was taken as 1.01 m. The floor area of cold storage chamber (a_f) is calculated by using Equation (3):

$$a_f = n_p a_p + a_{sp} + a_{sw} + a_{sd} + a_{se} \tag{3}$$

where n_p is the total no of pallets in cold storage chamber, a_p is the area of one pallet (horizontal), and the product of n_p and a_p is also called net area required for the agri-product (which is equal to 3.11 m^2 in this study), a_{sp} is the space area between pellets (clearance space between each pallet is required for easy handling and proper air circulation; mostly inter pallet spacing is set to 75 mm) and its value was 0.133 m^2, a_{sw} is the space area for walking/forklift etc. and its value was taken as 41.405 m^2, a_{sd} is the area of the space for door and its value was taken as 0.27 m^2, and a_{se} is the area for the space for the evaporator and brine pads, which was 4 m^2 in this study. The number of pellets n_p are calculated as:

$$n_p = \frac{m_{ag}}{m_b n_{b/p}} \tag{4}$$

where m_{ag} is the mass of agri-product to be placed in cold storage unit, m_b is the mass of agri-product in one basket, and $n_{b/p}$ is the number of baskets per pallet (five baskets form one pellet). For two tonnes (2000 kg) of agricultural products, the number of baskets were calculated to be 80 and the number pellets were calculated to be 16 using Equation (4).

By using Equations (1)–(3), the capacity of the cold storage unit was calculated to be 21.84 m^3. A cubical cold storage unit (21.84 m^3) was constructed with a length, width, and height of 3.657 m, 2.438 m, and 2.438 respectively; this is sufficient to store two tonnes of the perishable products (potatoes). Polyurethane (PU) sheets (100 mm thickness) were used to construct cold storage room along with Prepainted Galvinized Iron (PPGI) metal faces of the walls to get better insulation. These metal faces were white in color to deflect the solar radiations and other heat sources to avoid temperature rise of the cold storage room. A PU door (Height: 1.83 m; Width: 1 m) was used to load and unload the agri-product as well as for the inspection of the product placed in cold storage.

2.2. Refrigeration Capacity Calculations

The size of the refrigerator depends on the volume of cold storage, mass, and type of food to be stored as well as time to reach the required temperature. While calculating the refrigeration capacity, all the sources of heat produced or heat losses (cooling losses from cold storage chamber, sensible heat addition of the loaded agricultural product, heat of respiration of the agri-product, heat produced due to human activity as well as heat due to light sand fans) were considered.

The heat transfer through the walls, roof, and floor was calculated by using Equation (5).

Figure 1 shows a 3D CAD and schematic of the developed solar cold storage unit. The entire solar cold storage system consists of three main subsystems named as cold storage system with cooling backup (brine pads), refrigeration system, and solar-hybrid system.

$$Q_T = \frac{UA\ (T_o - T_i) \times 24}{1000} \tag{5}$$

where Q_T is the transmission heat load in kWh/d, U is the thermal transmittance value of insulation in W/(m^2 × K), A is the surface area of the walls, roof, and floor in m^2, T stands for the air temperature inside the cold storage in °C or K, and T_0 is the ambient air temperature in °C or K.

The thermal transmission value of polyurethane insulation (100 mm thickness) was taken as 0.28 W/(m^2 × K), which was used in walls and roof with a total area of 38.65 m^2, while the thermal transmission value of the floor insulation material was taken as 0.86 W/(m^2 × K) which was used in the floor. On average, a 28 °C temperature change occurred lower down the inside temperature of the storage chamber at a set value. The total daily transmission load for the complete storage chamber was calculated as 12.426 kWh/d (7.27 for walls and roof and 5.156 for the floor area) using Equation (5). The sensible heat load of the agri-product was calculated using Equation (6):

$$Q_P = \frac{m \times cp \times (T_{enter} - T_{store})}{3600} \tag{6}$$

where Q_P stands for the agri-product product load in kWh/d, c_p is the specific heat capacity of the product in kJ/(kg × K), m is the mass of new products each day (kg), T_{enter} is the initial temperature of the entering agri-products in °C or K, T_{store} is the storage temperature of agri-products in °C or K, and 3600 is the energy conversion factor from kJ to kWh.

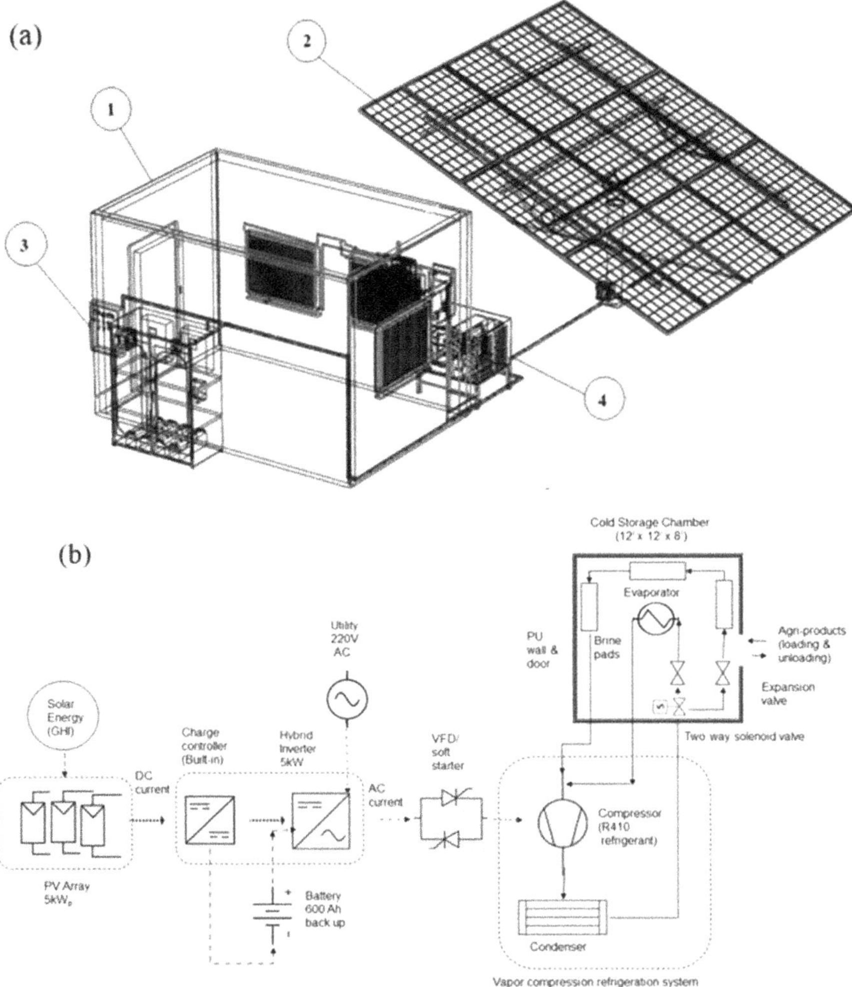

Figure 1. (**a**) 3D CAD of Solar Cold Storage System (1—storage chamber, 2—solar PV system, 3—monitoring and control system, 4—vapor-compression refrigeration system) and (**b**) schematic of solar cold storage system.

For the two tonnes (2000 kg) of agricultural product (potatoes) having specific heat of 3.43 kJ/(kg × K), the sensible refrigeration capacity was calculated to be 53.35 kWh/d; nevertheless, this load is only required during the first day only while loading the cold storage unit. For the rest of the experiments, this component becomes equal to zero, but higher values will be considered during the designing phase. Respiration load of product was calculated as:

$$Q_R = \frac{m_{ag}\, h_r}{24 \times 60 \times 60} \tag{7}$$

where Q_R is the respiration load in kWh/d, m is the mass of product to be stored in storage in kg, and h_r stands for the respiration heat of the product in kJ/kg.

The respiration heat of the potatoes is 310 kJ/kg and the respiration load for 2000 kg potatoes was calculated to be 7.17 kWh/d. Internal heat loads of the persons working inside, the lighting, and the fans were calculated by considering the daily average heat produced and were found to be 6.0, 0.3, and 1.4 kWh/d, respectively. The total heat load from all the sources (transmission heat load, sensible heat load, agri-product respiration heat load, workers heat load, lights and fan heat load) was calculated to be 81.126 kWh/d for storing 2000 kg of agri-products (postposes sample product) in cold storage unit. The theoretical refrigeration capacity (net refrigeration effect) was calculated to be 6.76 kW or 1.93 Tonnes of Refrigeration (TR) using 12 h/d time of operation of the refrigeration unit for safe calculations. Thus, a refrigerator of 2 tonnes of refrigeration was used based on these calculations.

A vapor-compression refrigeration system, having a refrigeration capacity of 2 tonnes of refrigeration (TR) using R404a as a refrigerant, was installed. The refrigeration system consists of a compressor, condenser, expansion valve, and evaporator. Three cooling pads (brine pads) were used to provide cooling backup for operation during nighttime (no sunshine hours). Each brine pad has a 30.5 m copper refrigerant coil, which provides cooling from refrigerant to brine solution. The brine solution had a 1:2 concentration of $CaCl_2$ with water and each cooling pad contains 75 L of brine solution. These pads store cooling equivalent to 0.65 tonnes of refrigeration while the refrigeration system is in operation and transfer cooling to stored product during operation at a rate of 2.275 kJ/s. Photographic views of the cold storage unit and brine pads are shown in Figure 2.

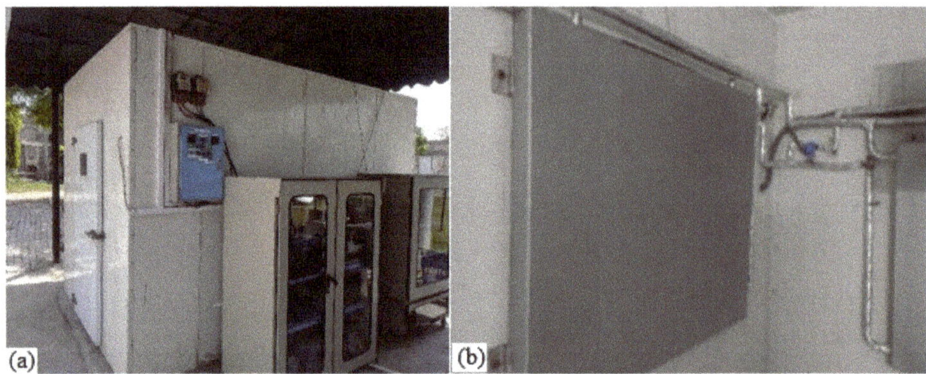

Figure 2. Photographic view of the solar cold storage unit. (**a**) Cold storage and data monitoring system and (**b**) inner view of the cold storage and the brine pads.

2.3. Size of the Solar-Hybrid PV System

The main advantage of using a hybrid solar system is the possibility of decentralized application to overcome the problem of power interruption and automatic switching to utilize solar energy to keep the system running with minimal battery backup. Unlike the grid-tied and standalone systems, a hybrid system does not require continuous supply of utility or a large-sized battery bank. Thus, a hybrid solar system was used in this study for the operation of the solar cold storage unit.

Peak power for the hybrid solar system (P_p) in kWp for cold storage system is calculated as [24]:

$$P_p = \frac{L_e I_p}{H_{avg} \eta_I \eta_B T_{CF}} \tag{8}$$

where L_e stands for the electric load to operate the compressor, I_p stands for the solar irradiance (peak) and for all the calculations, its value was taken 1 kW/m, H_{avg} represents the average Global Horizontal Irradiance (kWh/(m$^2 \times$ d)), and the average value of GHI at the experimental site (Solar Park, University of Agriculture, Faisalabad, Pakistan; 31.4456° N, 73.1356° E) was 5–6 kWh/(m$^2 \times$ d).

For the calculation, the average value of GHI was taken as 5.5 kWh/(m$^2 \times$ d), η_I is the efficiency of inverter (0.95–0.98) and has an average value of 0.965, η_b stands for the battery efficiency and has an average value of 90%, T_{CF} stands for the temperature correction factor (0.92) and is calculated using the values of loss factor (0.4% per °C), and the change in PV temperature at Nominal Operating Cell Temperature (NOCT) was taken as 45 °C. Standard testing conditions for crystalline Silicon were used. This means that power will reduce to about 0.4% for each increment in temperature from 25 °C at standard testing conditions. By substituting all these values in the equations, the value of the PV peak was found to be 4.40 kW$_p$. For 4400 W power using 320 W$_p$ PV panels/modules, the number of panels were calculated to be 13.75. Practically, 14 numbers of solar PV modules were required to operate the system and, in this way, the total installed power of the system became 4480 W$_p$.

Normally, the size of the inverter should be 1.10–1.25 bigger than the product of peak power of the Solar PV array and the surge factor. Surge factor ranges from 1–3 depending upon the appliances used. In case of a soft starter of VFD, its value is taken as 1. Thus, the theoretical value of the inverter was calculated as 4928 and a 5 kW inverter was used for this study, which was easily available on the market.

As the hybrid system is connected to the grid, the battery capacity C_{bat} in Ah required in this system is less than that of a standalone system, which is given by:

$$C_{Bat} = \frac{P\, t_{back}}{D_d\, \eta_{bat}\, V_{bat}} \tag{9}$$

where D_d is the depth of discharge in fraction, η_{bat} is the battery efficiency, which ranges from 0.90 to 0.98, V_{bat} is the nominal voltage of the battery, and t_{back} is the minimum backup time in hours of the appliances.

The solar-hybrid system consisted of solar PV modules, solar-hybrid inverter, battery backup, grid, and connection. A PV system of 4.5 kW$_p$ power was installed (14 PV polycrystalline modules with a peak power output of 320 watts each). These modules were connected to a hybrid inverter (5 kW) in two strings of seven modules per string, so the total solar output becomes 4480 W. The hybrid inverter was also used to switch between grid and solar system depending on the availability. A battery backup of 600 Ah capacity (four batteries of 150 Ah each with 12 V attached in series to form a 48 system compatible with the inverter) was also incorporated in the system to provide backup in case of electrical shortfall.

2.4. Power Estimation

The total power available from solar energy (P_s) is basically the total energy or power (at a certain instant) available from the sunlight [25]. This value starts from zero early in the morning and reaches its peak at solar noon and then decreases to zero at sun set. The instantaneous value was calculated as:

$$P_s = I_s A_e \eta_m \tag{10}$$

where I_s is the instantaneous value solar intensity or also called as Global Horizontal Irradiance (GHI), A is the effective area of PV panels, and η_m is the module efficiency.

The total power consumed by the cold storage (P_c) is obtained as:

$$P_c = 0.9\, A_{ac} V_{ac} \tag{11}$$

where A_{ac} is the current provided by the inverter, V_{ac} is the voltage of the cold storage, and 0.9 is the power factor (PF).

The power utilized from the solar system (P_{um}) was calculated as:

$$P_s = I_{dc}V_{dc} \tag{12}$$

The power utilized from the grid (P_g) was calculated using Equation (13). The equations were used for load calculations during the experiments.

$$P_g = P_t - P_s \tag{13}$$

2.5. Experimental Setup

A data logger (Agilent 34970A) was used to record real time data through attached gadgets and sensors, as shown in Figure 3. For a complete performance analysis, experiments were conducted for a period of three months (5 May to 30 July 2019), under no load as well as under load conditions (2000 kg) using potato as the sample stored product. The cold storage system was operated on grid mode, solar mode, as well as on a hybrid mode (grid plus solar) to check the system performance and to compare energy consumption under different modes of operation. For this, SUB (solar-utility-battery) or SBU (solar-battery-utility) modes of hybrid inverter were set. A data logger (Agilent, 20 channel cartridge) was incorporated for real-time monitoring and data acquisition throughout the storage period. Several parameters were recorded, namely, solar irradiance, cold storage temperature, ambient temperature, compressor inlet temperature, compressor outlet temperature, expansion valve temperature, brine pads temperature, DC current provided by the solar PV modules, DC voltage of the solar PV modules, AC voltage of the solar-hybrid inverter, and the AC current provided by the inverter. For these parameters, thermocouples K type, error < 0.1 °C), hygrometer Pyranometer (SP Lite; response type < 1 s), and DC and AC voltmeters and ammeters were used and connected to data logger. For the assessment of product quality, one physical (weight loss) and three chemical (total soluble solids, titratable acidity, and PH) quality parameters were measured at the end of the stored period.

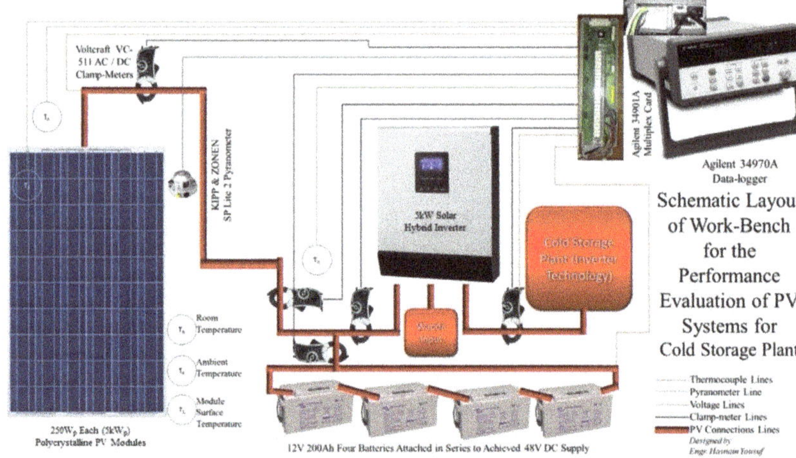

Figure 3. Schematic of the experimental setup.

3. Results and Discussion

3.1. CFD Simulation

Product weight loss, i.e., moisture loss during the storage period, is a result of mass transfer, which depends mainly on the heat transfer; thus, distribution of cold air (working fluid) inside the storage chamber is important [26]. For this, Computational Fluid Dynamics (CFD) software was used in ANSYS Workbench-18. It provides a comprehensive suite for modeling fluid flow and other related physical phenomena. As the profile of airflow and temperature distribution was of concern in the design assessment, only the influencing part of the system (cold storage chamber) was modeled and simulated. The purpose of the 3D model was also to assess flow variation and distribution in the depth of cooling chamber as well along with axial direction, which was not assessable in the 2D model. A 3D model was developed in ANSYS Design Modeler (number of elements: 65143, grid type: 3D, tetrahedral, structured). Wooden boxes and product within these boxes were modeled such that the chamber partially and completely filled with the product. The boxes were perforated from all the sides so that cooled air can thoroughly circulate through the product. The place of evaporator in actual design was taken as the inlet for cooled air to assign inlet boundary conditions (air velocity and temperature). As air has to pass through boxes, a turbulent flow regime was expected, and for this, a k-ε standard turbulence model was applied along with a no-slip condition for the Wall friction model.

Figure 4 shows the profiles of the temperature and air distribution inside the cooling chamber. Cooled air enters into the cold chamber through an evaporator (inlet for the chamber) and circulates throughout the cold chamber uniformly, as can be observed from the velocity profiles (Figure 4c,d). It shows that the applied boundary conditions for inlet air were appropriate to overcome flow resistance due to product and buckets, for the uniform air distribution inside the chamber. The temperature distribution is linked with the air distribution. As soon as cooled air enters the chamber, it starts to exchange heat with the stored product to reduce its temperature. The difference in temperature of cooled air and the stored product can be observed in Figure 4a,b, which normally occurs during the early stages of the storage period. Products lying in front of the evaporator start to decease their temperature, which continues until all the products are at the same temperature. Air is continuously circulated inside the cold chamber and cools down through the evaporator to maintain the set stored temperature for the respective product.

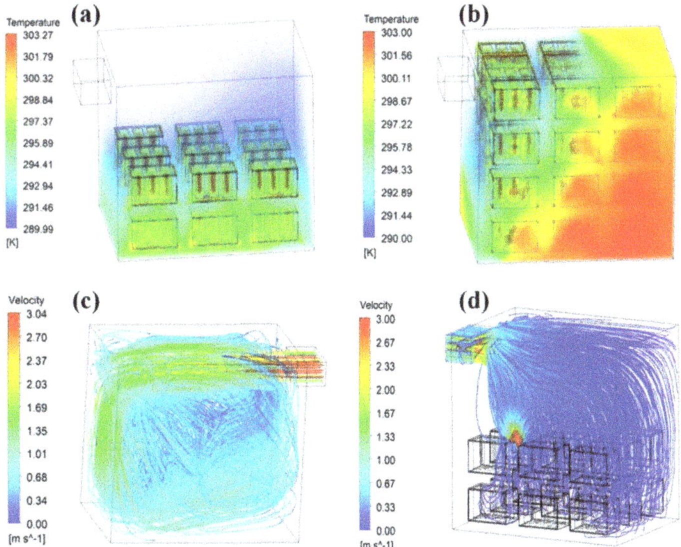

Figure 4. (**a**) Simulated profiles of temperature inside the cold storage chamber under partially loaded conditions and (**b**) fully loaded conditions. (**c**) Velocity distribution profiles inside the cold storage room under empty conditions and (**d**) partially loaded conditions.

3.2. Solar Irradiance Profile

The developed cold storage system aimed to operate on solar energy; solar irradiance is an important parameter which needs to be investigated. The data for solar irradiance were monitored throughout the storage period. Figure 5 shows solar irradiance and total power available from sun for a sample representative storage time. The irradiance was measured using a Pyranometer which was mounted on the solar PV tracking stand parallel to the same plane of the PV panels.

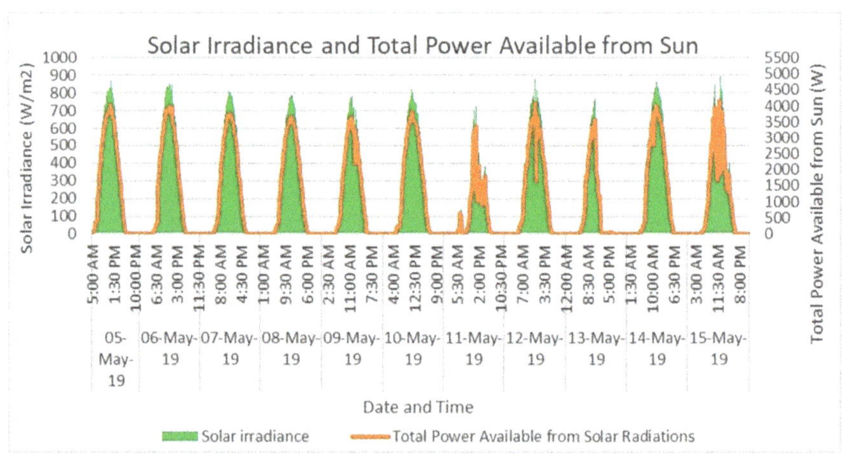

Figure 5. Solar irradiance and total power available from the sun.

It can be observed that values of solar irradiance remained nearly the same throughout the observed data. The data were recorded for three months (5 May 2019 to 30 July 2019) with the help of a data logger and the average value of solar irradiation was recorded between 5–6 kWh/(m^2 × d) during the month of May and June, while the daily average was found to be 210–250 W/m^2. The solar irradiance was observed with a value of 700 W/m^2, while the highest peak solar irradiance was observed to be 897 W/m^2 during the experimental period. It was also observed that the about 4 kW power was available during the peak solar irradiance from the 4.5 kW$_p$ system. The solar irradiation available during the summer season irradiance and available power profiles show that sufficient solar energy is available at site to run the cold storage unit for the preservation of agricultural products.

3.3. AC Current Profile

The solar-based cold storage unit is designed in such a way that it can be operated on utility, solar, as well as on both options simultaneously to maintain the required storage conditions under adverse weather conditions. For this purpose, a hybrid inverter has been employed to switch at any mode on requirement, but the principal objective of the study was to run the cold storage unit on solar energy with a cooling/brine pad backup for night cooling. The research work started in different phases to optimize the system in steps. First trials were made using a conventional vapor-compression refrigeration system running on a utility grid, as shown in Figure 6.

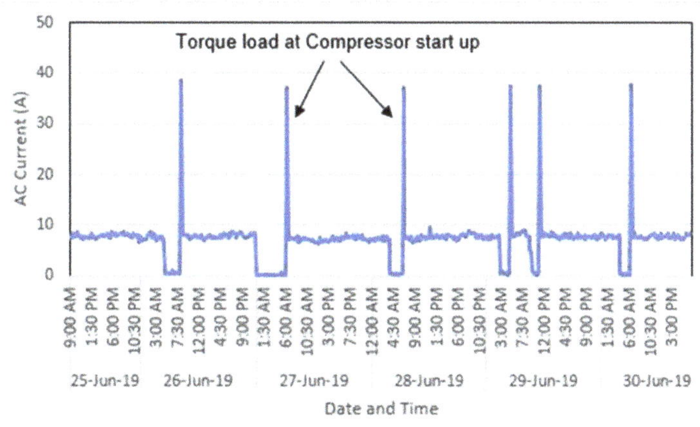

Figure 6. Variation of cold storage AC current (without VFD).

Figure 7 depicts the variation in cold storage AC current consumption during sample representative storage time. It can be observed that AC current dropped to zero when the compressor tripped down, mainly at nighttime when the cold storage temperature dropped to a set value. Thereafter, when the compressor restarted, a sudden surge in current can be seen in the figure. This surge current was due to torque load of the compressor [25]. This surge current is considered the main hurdle to run a cold storage system on solar energy. The surge current due to torque load could be reduced by employing a Variable Frequency Drive (VFD) or soft starter. The incorporation of VFD in the system enables the system to be operated entirely on solar PV system. Then, a VFD was incorporated in the system to reduce the surge current and variation in current (Amps), as shown in Figure 6.

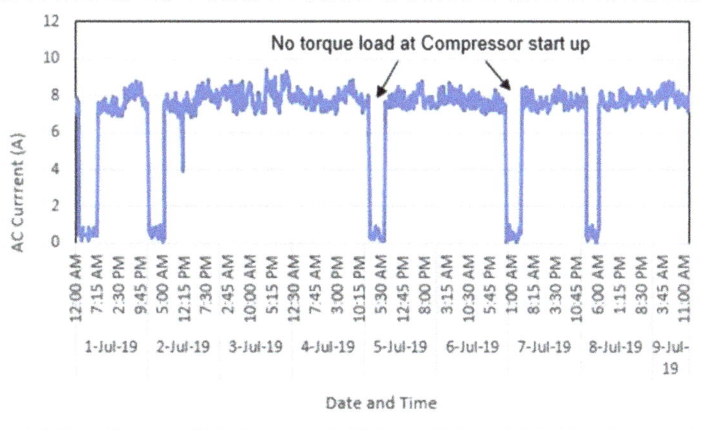

Figure 7. Variation of cold storage AC current (with VFD).

It can be perceived that there was no surge current when the compressor restarted after tripping. This is due to the application of VFD, which gradually increases the frequency of the compressor to the actual operating frequency, while keeping the AC current constant. The variation of the total available power and the power utilized from solar PV modules and grid with and without the application of VFD are demonstrated in Figures 8–10, respectively.

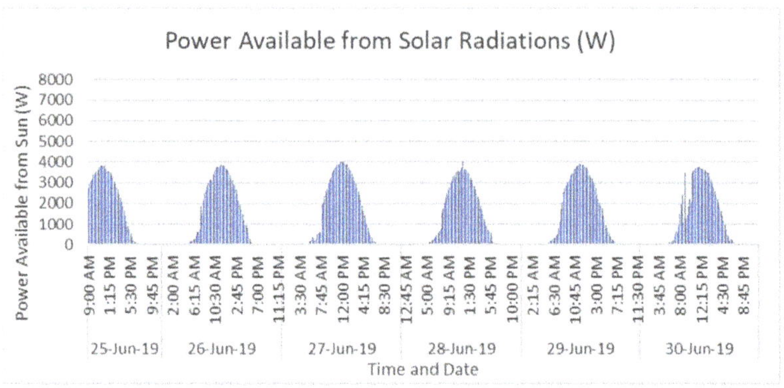

Figure 8. Variation of the power available from solar radiations.

Figure 9. Power utilized from solar and grid (without VFD).

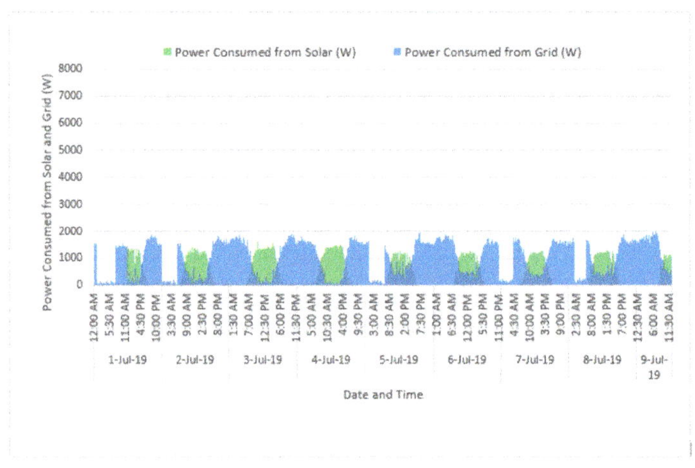

Figure 10. Power utilized from solar and grid (with VFD).

It can be observed from Figures 8 and 9 that a lot of power available from solar radiations was not utilized by the cold storage system. The power available from solar modules went up to 4000 W (Figure 8), but the power utilized by cold storage was less than 2000 W (Figure 9). As the hybrid inverter only utilizes the DC current available from solar modules according to the power requirement of the attached load, the remaining portion of power is wasted and not utilized properly. That wasted power could be harnessed by utilizing the grid-tied inverter and the energy could be sold to the grid by net-metering or by employing a battery bank to store the extra energy when required. From the Figures 9 and 10, a surge in power can also be observed. This is due to the surge current produced by the torque load of the compressor. From Figure 9, it can be observed that the power to compensate the surge current due to torque load of the compressor was obtained from the grid, as the solar system was not sufficient to generate that amount of power. Therefore, the system was unable to be operated solely on solar PV power. The problem was resolved by installing VFD, which controlled the torque load of the compressor (Figure 10). This device gradually increases the frequency of the compressor while keeping the current constant at the normal operating range. Due to this reason, the surge current avoided and the compressor started at normal operating current as shown in Figure 10. It can be observed that the surge in power was minimized with VFD. The peak power requirement at torque load without VFD was 7800 W (Figure 9), which reduced to 1800 W (Figure 10) after the installation of VFD, translating into 77% power saving which was consumed during the surge load of the compressor. After installation of VFD, the system became feasible to be operated solely on the solar PV power, which was the core objective of the current study. The decentralized application of the cold storage system is only possible when it could be operated on the solar PV system, as there is uncertainty in the grid at farm level.

3.4. Operation of a Cold Storage Unit Using Cooling Pads as Backup

The key objective of the current study was to design and develop a cold storage unit operated on solar energy, and to do so, the addition of cooling pads was one of the tasks to enable the system to maintain storage chamber temperature at nighttime storage. A 5 kW$_\mathrm{p}$ was enough to run the compressor integrated with VFD to eliminate the torque load during daytime sunny conditions. A 600 Ah battery backup (using 12 V and provides 3.6 kWh energy at 50% depth of discharge) is sufficient to ensure the continuous operation of a cold storage unit for approximately two hours during partially cloudy condition or during nighttime. One of the principal advantages of developing the solar cold storage unit is that the higher and lower energy demands in summer and winter are automatically met with the seasonal sunlight fluctuation; higher in summer (solar irradiance: 700–1000 and solar irradiation: 5.5–7.0 kWh/(m^2 × d) and lower in winter (solar irradiance: 500–700 W/m^2 and solar irradiation: 4.0–5.5). This best energy match evolved the idea of solar cold storage system. In addition to cooling pads, an evaporator fan is also employed for uniform cooling during day and nighttime and can be run on demand. During functioning, the system automatically runs when the irradiance is high enough to run the compressor. In this way, the cold storage system automatically starts cooling inside the cold storage chamber. As the intensity of solar radiation increases, it lowers the temperature of brine pads automatically throughout the day and the cooling is distribution to the products inside the brine pads. Additionally, the battery is charged with additional energy to get ready for electricity backup in case of solar fluctuation. Figure 11 explains the cooling profiles of the cold storage plant (shown by bars) operated on the cooling coil (curve shown by the line diagram) during a trial of two days (readings were recorded with the help of a data logger).

During the nighttime, the lower temperature of brine pad automatically transfers the cooling to the products placed for the preservation, and hence, the temperature of brine pads starts increasing during the evening and nighttime. The perfect insulation of polyurethane used for the structure (walls, roof, door, and the floor) of the cold storage unit and lower temperature during the nighttime ensure that the temperature does not increase too much, and cold temperatures are successfully stored in bthe rine pads during sunny hours to maintain the cold storage temperature. The detail of two days trials (from 9:15 to 19:00 h) has been shown in Figure 11.

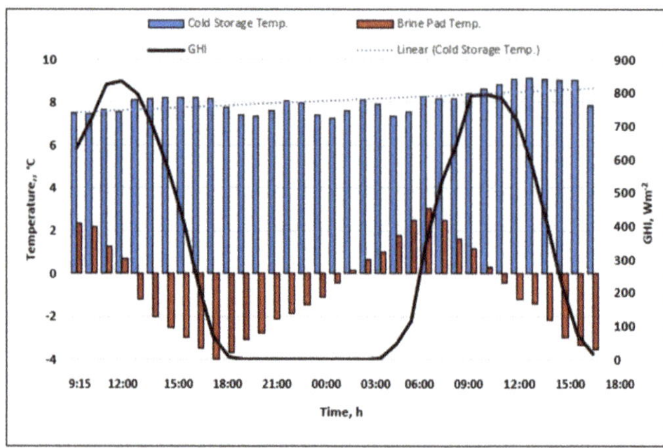

Figure 11. Temperature profiles of cold storage chamber and cooling pads (14–15 May 2019).

It has been observed that the cooling pads temperature decreases during daytime along with the cooling provided to the cold storage unit. During nighttime, this cooling is utilized to cool the cold storage unit at storage temperature using temperature differential. This trial was successfully run for potatoes where storage temperature requirement was 8 °C. For other products which have storage temperature less than this figure (8 °C), a battery backup can be utilized during the nighttime if required without compromise on storage temperature and the quality of the products. For the trials with potatoes, the relationships between AC current and AC voltage and DC current and DC voltage versus time are drawn simultaneously for day and night times in Figure 12. It is evident that the cold storage unit used solar energy to maintain the cooling inside the storage chamber and also charged the cooling pads for nighttime operation, while no electricity was consumed from any source during the nighttime, as the cooling pads were sufficient to maintain the storage temperature.

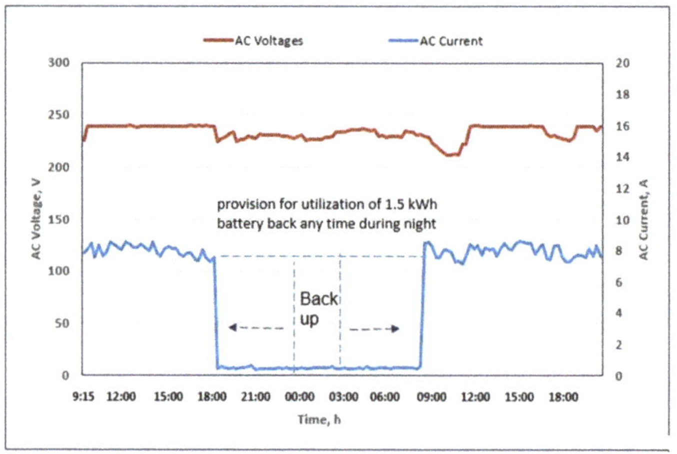

Figure 12. Relationships between AC current and AC voltage and DC current and DC voltage versus time (14–15 May 2019).

3.5. Relation between the Cold Storage Temperature and Relative Humidity

The inside temperature of a cold storage chamber is the most crucial parameter for the safe storage of perishable food items. Moreover, relative humidity is also a very important parameter and it should be maintained above 80% for the preservation of perishables. The set temperature of the cold storage depends on the product stored in the cold storage. In the current experiment, potatoes were used as stored product and the optimum temperature for storing potatoes is 7–8 °C. Therefore, the cold storage temperature was set to 7 °C throughout the experiment. Figure 13 shows temperature and humidity profiles within the cold storage chamber during a sample representative storage time. It can be observed clearly that the storage temperature was maintained at a set value, and the relative humidity was also above 80%.

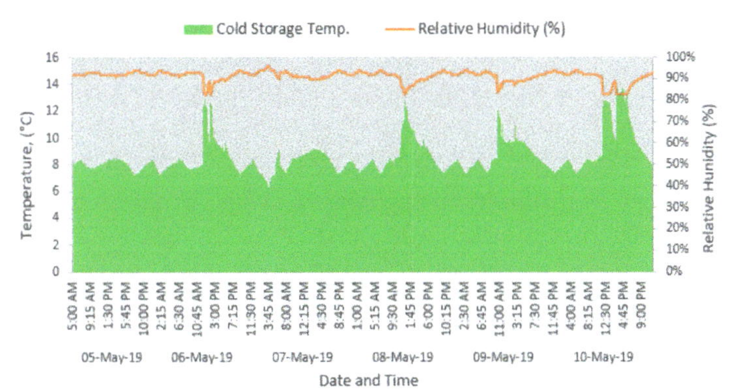

Figure 13. Variation in the cold storage temperature and relative humidity.

3.6. Cold Storage Temperature and Brine Pad Temperature

For cooling backup, three cooling pads were installed inside the cold storage chamber. The temperature of the cooling pads is usually less than 0 °C, although it varied to keep the cold storage temperature constant. Figure 14 shows the temperature variation in the cooling pads and storage chamber to assess the contribution of the cooling pads while maintaining storage temperature. This figure shows the data for a trial of one week where battery and utility were also used when required. It can be observed that the cold storage temperature remained 8 °C for most of the storage time, while brine pads temperature varied in range of −4 °C to 4 °C. The rise in the cooling pads temperature was because of the fact that the cold storage temperature increased due to some activity (e.g., opening of the door) and the cooling pads in turn counteract to reduce the change in temperature by providing cooling to the storage space to maintain the set temperature.

3.7. Coefficient of Performance (COP)

A vapor-compression refrigeration unit was installed in the system. The COP of the refrigeration system is the most important parameter for the performance evaluation of refrigeration system [27]. The COP is basically the ratio of output of refrigeration system to the input provided. The COP of Vapor-compression Refrigeration System is calculated by the ratio of change in enthalpies. The formula for calculation of COP is given by Equation (14):

$$COP_R = \frac{h_1 - h_{f3}}{h_2 - h_1} \tag{14}$$

where h_1 is the vapor enthalpy of refrigerant before entering the compressor (average value = 361.4 kJ/kg), h_2 is the vapor enthalpy of refrigerant after the compression (average value = 386 kJ/kg), and h_{f3} is the liquid enthalpy of the refrigerant after leaving the condenser (average value = 247.9 kJ/kg). Those enthalpies were tabulated by comparing the corresponding temperatures from enthalpy sheet of R404a refrigerant gas. The average COP of the system throughout the week was approximately 4.61.

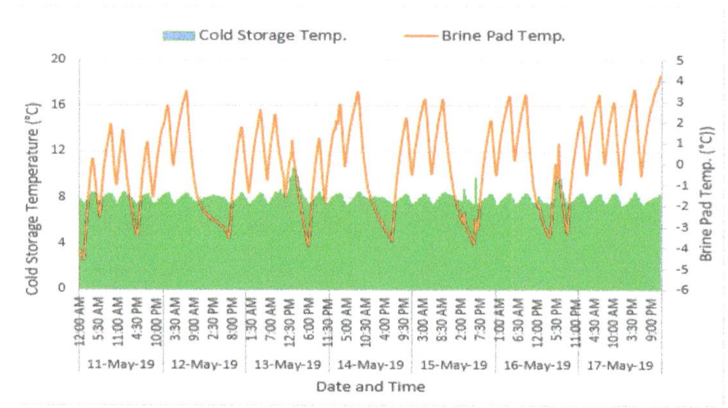

Figure 14. Variation in the cold storage temperature and brine pads temperature.

3.8. Energy Consumed

Energy consumption is a major parameter for the operation of a cold storage system. Considerable saving in the energy is possible through optimization of cold storage operation in terms of heat loads and the operation of the refrigeration system [28]. Energy meters were used to measure the daily energy consumption. The daily consumed units or kWh of energy were recorded, and the total energy consumption of 10 days were observed. Figure 15 shows energy consumption of the cold storage. It can be observed that the minimum units utilized by the cold storage were 10 kWh and the maximum utilized units were 19 kWh. The average units consumed were 15 kWh. Meanwhile, the average kWh consumed from grid and solar were 4.3 kWh and 10.5 kWh, respectively, which translates to 30% of power consumption from the grid and 70% from the solar PV modules.

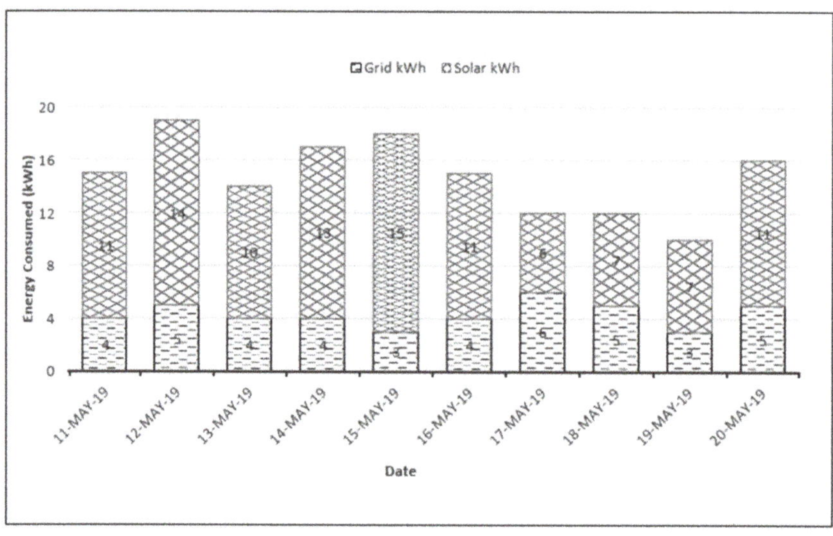

Figure 15. Energy consumed by the solar cold storage.

3.9. Effect of the Cold Storage on Potato Weight Loss and Quality

The effectiveness of cold storage was also demonstrated through measuring its efficiency in terms of quality preservation of stored potato. For this purpose, potatoes stored in cold storage for three months (5 May 2019 to 30 July 2019) at 7 ± 1 °C were evaluated for weight loss and other quality parameters, at fortnightly intervals. Quality assessments were carried out at Postharvest Research and Training Centre, Institute of Horticultural Sciences, UAF. Weight loss was measured by taking the total weight of eight buckets (initially 25 kg each) and calculated as the percentage of weight loss [29]. Total soluble solids were recorded by using a digital refractometer (ATAGO Co. Ltd., Minato-ku, Tokyo, Japan) and expressed as ^{0}Brix. The titratable acidity of potato was estimated from extracted juice, as detailed by Malik et al. [30]. The pH of potato pulp juice was measured by a benchtop digital pH meter (HI98130-Hanna Instruments Inc. Romania). Results are shown in Figure 16.

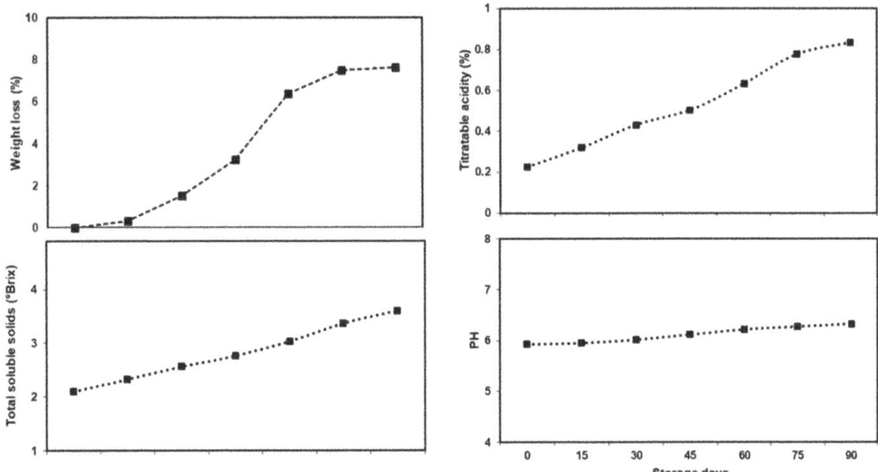

Figure 16. Effect of cold storage on weight loss and various quality parameters of stored potatoes.

As shown in Figure 16, potato weight loss was 0.32% after the first 15 days, and 1.53% in the first month. The total weight loss over three months was 7.64%. In a previous study, Khurana et al. [31] reported that out of 11 potato cultivars stored, three cultivars had less than 10% weight loss, while eight cultivars had more than 10%, with an overall range of 8% to 23%. Thus, the total weight loss (7.64%) in the current study is lower/in range with comparison to the product stored in a grid supply centralized cold storage facility showing the effectiveness of cooling of the developed unit during the product storage time. Total soluble solids (TSS) showed the conversion of starch to sugar content in potato. At day 0, the TSS was 2.1 ^{0}Brix, which increased gradually in storage, reaching 3.6 ^{0}Brix at the end of storage. These results are in line with the potato storage study by Khurana et al. [32]. Titratable acidity (TA) mostly affects the pH of the product. At the start of the experiment, the titratable acidity of the potatoes was 0.22%, which increased as storage proceeded, finally reaching 0.83%. The pH is used to measure the acidity or basicity of the solution or food product. At day 0, the pH of the potatoes was 5.92, which slightly increased, reaching 6.32 at the end of the experiment. Overall, in this experiment, cold storage efficiently controlled weight loss and preserved potato quality, demonstrating its cooling efficiency.

4. Conclusions

The current study was conducted to develop a solar-grid hybrid cold storage system for on-farm storage of perishables to reduce post-harvest losses at production sites. The main body of cold storage room was made of polyurethane material (100 mm thick) to minimize the heat load. The system is operated on solar energy, employing a solar hybrid inverter with a battery and utility backup to run the system in case of adverse weather conditions. As a backup source, cooling pads were used to provide cooling during nighttime and system took approximately 9–12 h to reach set temperature of 7 °C from 25 °C when operated on cooling pads. These pads maintained the temperature of the cold storage chamber for more than 12 h with a 5 °C rise in temperature in 24 h. The system is capable to meet higher and lower energy demand in summer and winter automatically with the seasonal sunlight fluctuation. The use of a variable frequency drive (VFD) with a compressor saved 77% energy comparative to energy consumed without installing VFD. Due to hybrid nature of the system and its priority to use the full potential of the available solar energy, 70% of the consumed energy was shared from solar PV modules. Effective utilization of solar energy with enough cooling backup and minimum dependency on grid utility made the system energy efficient and enabled it to store product without deterioration (total weight loss was 7.64%, TSS increased from 2.1 ^{0}Brix to 3.6 ^{0}Brix, and TA increased from 0.22% to 0.83%, while the pH was 5.92 to 6.32 at the end of the experiment). Based on the outcomes, the following conclusions were drawn:

- The research concluded that 2 tonnes of potatoes could be stored in a 20 m^3 capacity of cold storage constructed of polyurethane sheets (100 mm thick) and coupled with 2 tonnes of vapor using a compression refrigeration system (VRF technology) employing a 5 kW photovoltaic hybrid system.
- It is also concluded that cooling pads maintained the required storage temperature by charging and discharging during day- and nighttime, respectively, eliminating the nighttime operation of refrigeration machine due to less temperature difference. Therefore, integration of such cooling pads is useful technology for a cold storage system run by solar energy.
- This solar-hybrid technology can play a vital role in addressing the decentralized storage of various agricultural products to reduce losses with minimum energy requirements for the addition of value and generation of income.

In order to reduce food losses and to develop cold storage facilities, the following recommendations are made:

- The current system was developed for handling of a specific quantity of product (2 tonnes), showing a solar energy contribution of more than 70%. Thus, in case of large commercial units, the scaled-up replications of such units would play a vital role to reduce the high operational cost. Therefore, it is recommended to scale up the system for commercial applications.
- The design and placement of the opening door is very important in terms of losses. We recommended that a separate chamber is added at the entrance to minimize these losses.

Author Contributions: Conceptualization, A.M., T.A. and W.A.; methodology, A.G. and A.U.M.; software, S.R. and W.A.; validation, A.M. and O.H.; formal analysis, T.A., W.A., A.G. and T.M.; investigation, W.A., A.G., S.R. and T.M.; resources, A.M. and O.H.; data curation, A.M., T.A. and W.A; writing—original draft preparation, A.M. and T.A.; writing—review and editing, W.A., M.S. and T.M.; visualization, A.U.M., M.S. and T.M.; supervision, A.M. and O.H.; project administration, A.M.; funding acquisition, A.M. All authors have read and agreed to the published version of the manuscript.

Funding: This work was supported by the Higher Education Commission (HEC) of Pakistan under its program "Technology Development Fund" (Project number TDF-082).

Institutional Review Board Statement: Not applicable.

Informed Consent Statement: Not applicable.

Data Availability Statement: Not applicable.

Acknowledgments: This work was supported by the Higher Education Commission (HEC) of Pakistan under its program "Technology Development Fund" (Project number TDF-082).

Conflicts of Interest: The authors declare no conflict of interest.

References

1. Godfray, H.C.J.; Beddington, J.R.; Crute, I.R.; Haddad, L.; Lawrence, D.; Muir, J.F.; Pretty, J.; Robinson, S.; Thomas, S.; Toulim, C.; et al. Food security: The challenge of feeding 9 billion people. *Science* **2010**, *327*, 812–818. [CrossRef] [PubMed]
2. Maphosa, B. Investigation on Post-Harvest Processing of Fruits Using a Solar-Bio-Energy Hybrid Dryer 2020. North-West University South Africa Report. Available online: https://repository.nwu.ac.za/bitstream/handle/10394/35592/Maphosa_B.pdf?sequence=1 (accessed on 30 September 2021).
3. Raza, H.M.U.; Ashraf, H.; Shahzad, K.; Sultan, M.; Miyazaki, T.; Usman, M.; Shamshiri, R.; Zhou, Y.; Ahmad, R. Investigating applicability of evaporative cooling systems for thermal comfort of poultry birds in Pakistan. *Appl. Sci.* **2020**, *10*, 4445. [CrossRef]
4. Lal Basediya, A.; Samuel, D.V.K.; Beer, V. Evaporative cooling system for storage of fruits and vegetables-A review. *J. Food Sci. Technol.* **2013**, *50*, 429–442. [CrossRef] [PubMed]
5. Olosunde, W.A.; Igbeka, J.C.; Olurin, T.O. Performance evaluation of absorbent materials in evaporative cooling system for the storage of fruits and vegetables. *Int. J. Food Eng.* **2009**, *5*, 1–15. [CrossRef]
6. Hasan, M.; Shang, Y.; Akhter, G.; Jin, W. Application of VES and ERT for delineation of fresh-saline interface in alluvial aquifers of Lower Bari Doab, Pakistan. *J. Appl. Geophys.* **2019**, *164*, 2200–2213. [CrossRef]
7. Hasan, M.U.; Malik, A.U.; Ali, S.; Imtiaz, A.; Munir, A.; Amjad, W.; Anwar, N. Modern drying techniques in fruits and vegetables to overcome postharvest losses: A review. *J. Food Process. Preserv.* **2019**, *43*, e14280. [CrossRef]
8. Hanif, S.; Sultan, M.; Miyazaki, T.; Koyama, S. Investigation of energy-efficient solid desiccant system for the drying of wheat grains. *Int. J. Agric. Biol. Eng.* **2019**, *12*, 221–228. [CrossRef]
9. Amjad, W.; Crichton, S.O.J.; Munir, A.; Hensel, O.; Sturm, B. Hyperspectral imaging for the determination of potato slice moisture content and chromaticity during the convective hot air drying process. *Biosyst. Eng.* **2018**, *166*, 170–183. [CrossRef]
10. Ramos, I.N.; Brandão, T.R.S.; Silva, C.L.M. Structural changes during air drying of fruits and vegetables. *Food Sci. Technol. Int.* **2003**, *9*, 201–206. [CrossRef]
11. Ajiboye, A.O.; Afolayan, O. The impact of transportation on agricultural production in a developing country: A case of kolanut production in Nigeria. *Int. J. Agric. Econ. Rural Dev.* **2009**, *2*, 49–57.
12. Sipahioglu, O.; Barringer, S.A. Dielectric properties of vegetables and fruits as a function of temperature, ash, and moisture content. *J. Food Sci.* **2003**, *68*, 234–239. [CrossRef]
13. Ali, S.D.; Ramaswamy, H.S.; Awuah, G.B. Thermo-physical properties of selected vegetables as influenced by temperature and moisture content. *J. Food Process. Eng.* **2002**, *25*, 417–433. [CrossRef]
14. Zhang, H.; Zhou, R.; Lorente, S.; Ginestet, S. Thermodynamic design of cold storage-based alternate temperature systems. *Appl. Therm. Eng.* **2018**, *144*, 736–746. [CrossRef]
15. Diaconu, B.M.; Varga, S.; Oliveira, A.C. Numerical simulation of a solar-assisted ejector air conditioning system with cold storage. *Energy* **2011**, *36*, 1280–1291. [CrossRef]
16. Ghafoor, A.; Rehman Tur Munir, A.; Ahmad, M.; Iqbal, M. Current status and overview of renewable energy potential in Pakistan for continuous energy sustainability. *Renew. Sustain. Energy Rev.* **2016**, *60*, 1332–1342. [CrossRef]
17. Sheikh, M.A. Energy and renewable energy scenario of Pakistan. *Renew. Sustain. Energy Rev.* **2010**, *14*, 354–363. [CrossRef]
18. Otanicar, T.; Taylor, R.A.; Phelan, P.E. Prospects for solar cooling–An economic and environmental assessment. *Sol. Energy* **2012**, *86*, 1287–1299. [CrossRef]
19. Wang, D.; Hu, L.; Liu, Y.; Liu, J. Performance of off-grid photovoltaic cooling system with two-stage energy storage combining battery and cold water tank. *Energy Procedia* **2017**, *132*, 574–579. [CrossRef]
20. Singh, P.L.; Jena, P.C.; Giri, S.K.; Gholap, B.S.; Kushwah, O.S. Solar-Powered Cold Storage System for Horticultural Crops. In *Energy Environ*; Springer: Berlin/Heidelberg, Germany, 2018; pp. 125–131.
21. Pinto, S.; Madhusudhan, A. Solar Powered Refrigeration System with Cold Bank. *Indian J. Sci. Technol.* **2016**, *9*, 1–5. [CrossRef]
22. Ul, N.; Rather, R.; Moses, S.; Sahoo, U.; Tripathi, A. Performance Evaluation of Hybrid Cold Storage using Solar & Exhaust heat of Biomass Gasifier for Rural Development. *IJRITCC* **2017**, *5*, 563.
23. Zheng, L.; Zhang, W.; Liang, F. A review about phase change material cold storage system applied to solar powered air-conditioning system. *Adv. Mech. Eng.* **2017**, *9*, 1–20. [CrossRef]
24. Ghafoor, A.; Munir, A. Design and economics analysis of an off-grid PV system for household electrification. *Renew. Sustain. Energy Rev.* **2015**, *42*, 496–502. [CrossRef]
25. Medford, A.J.; Lilliedal, M.R.; Jørgensen, M.; Aarø, D.; Pakalski, H.; Fyenbo, J.; Krebs, F. Grid-connected polymer solar panels: Initial considerations of cost, lifetime, and practicality. *Opt. Express* **2010**, *18*, A272–A285. [CrossRef]
26. Chourasia, M.K.; Goswami, T.K. Simulation of Transport Phenomena during Natural Convection Cooling of Bagged Potatoes in Cold Storage, Part II: Mass Transfer. *Biosyst. Eng.* **2006**, *94*, 207–219. [CrossRef]
27. Liu, Y.; Vittal, V.; Undrill, J.; Eto, J.H. Transient model of air-conditioner compressor single phase induction motor. *IEEE Trans. Power Syst.* **2013**, *28*, 4528–4536. [CrossRef]

28. Evans, J.A.; Hammond, E.C.; Gigiel, A.J.; Fostera, A.M.; Reinholdt, L.; Fikiin, K.; Zilio, C. Assessment of methods to reduce the energy consumption of food cold stores. *Appl. Therm. Eng.* **2014**, *62*, 697–705. [CrossRef]
29. Razzaq, K.; Khan, A.S.; Malik, A.U.; Shahid, M.; Ullah, S. Role of putrescine in regulating fruit softening and antioxidative enzyme systems in 'Samar Bahisht Chaunsa'mango. *Postharvest. Biol. Technol.* **2014**, *96*, 23–32. [CrossRef]
30. Malik, A.U.; Singh, Z. Improved fruit retention, yield and fruit quality in mango with exogenous application of polyamines. *Sci. Hortic.* **2006**, *110*, 167–174. [CrossRef]
31. Khurana, S.M.P.; Minhas, J.S.; Pandey, S.K. *The Potato: Production and Utilization in Sub-Tropics*; Mehta Pub: New Delhi, India, 2003.
32. Amjad, A. *Screening Potato Cultivars for Sugar Accumulation in Response to Cold Storage and Their Mitigation Through Reconditioning*; University of Agriculture: Faisalabad, Pakistan, 2017.

 *energies*

Article

A Time-Dependent Model for Predicting Thermal Environment of Mono-Slope Solar Greenhouses in Cold Regions

Shuyao Dong [1], Md Shamim Ahamed [2], Chengwei Ma [3] and Huiqing Guo [1,*]

[1] Department of Mechanical Engineering, University of Saskatchewan, Saskatoon, SK S7N 5E2, Canada; shd491@mail.usask.ca
[2] Department of Biological and Agricultural Engineering, University of California, Davis, CA 95616-5270, USA; mahamed@ucdavis.edu
[3] Key Laboratory of Agricultural Engineering in Structure and Environment of Ministry of Agriculture, China Agricultural University, Beijing 100083, China; macwbs@cau.edu.cn
* Correspondence: huiqing.guo@usask.ca; Tel.: +1-306-966-5350

Abstract: Most greenhouses in the Canadian Prairies shut down during the coldest months (November to February) because of the hefty heating cost. Chinese mono-slope solar greenhouses do not primarily rely on supplemental heating; instead, they mostly rely on solar energy to maintain the required indoor temperature in winter. This study focuses on improving an existing thermal model, entitled RGWSRHJ, for Chinese-style solar greenhouses (CSGs) to increase the robustness of the model for simulating the thermal environment of the CSGs located outside of China. The modified model, entitled SOGREEN, was validated using the field data collected from a CSG in Manitoba, Canada. The results indicate that the average prediction error for indoor and relative humidity is 1.9 °C and 7.0%, and the rRMSE value is 3.3% and 11.5%, respectively. The average error for predicting the north wall and ground surface temperature is 4.2 °C and 2.3 °C, respectively. The study also conducted a case study to analyze the thermal performance of a conceptual CSG in Saskatoon, Canada. The energy analysis indicates the heating requirement of the greenhouse highly depends on the availability of solar radiation. Besides winter, the heating requirement is relatively low in March to maintain 18 °C indoor temperature when the average outdoor temperature was below −4 °C, and negligible during May–August. The results indicate that vegetable production in CSGs could save about 55% on annual heating than traditional greenhouses. Hence, CSGs could be an energy-efficient solution for ensuring food security for northern communities in Canada and other cold regions.

Keywords: Chinese solar greenhouse; thermal model; north wall; cold regions; heating requirement

Citation: Dong, S.; Ahamed, M.S.; Ma, C.; Guo, H. A Time-Dependent Model for Predicting Thermal Environment of Mono-Slope Solar Greenhouses in Cold Regions. *Energies* **2021**, *14*, 5956. https://doi.org/10.3390/en14185956

Academic Editor: Donato Morea

Received: 28 August 2021
Accepted: 17 September 2021
Published: 19 September 2021

Publisher's Note: MDPI stays neutral with regard to jurisdictional claims in published maps and institutional affiliations.

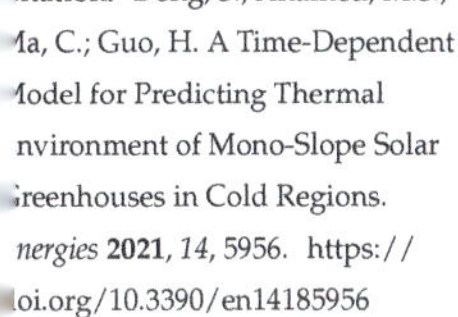

1. Introduction

The extreme cold outdoor temperature is a major barrier to greenhouse production in cold regions. Supplemental heating and dehumidification have been commonly practiced in Canadian greenhouses. The heating cost could be around 70% to 85% of the total operating cost of crop production in greenhouses in high latitudes, excluding the cost associated with labor [1]. The heating cost in the Canadian greenhouse accounts for 10–35% of the total production cost [2,3]; therefore, most greenhouses in the Canadian Prairies shut down during the coldest months. The demand for local produce has increased over the last decade as more and more people acknowledge the health benefits of fresh vegetables and the need to reduce carbon emissions caused by long-distance transportation of fresh vegetables. Research indicates that the vegetables lose 15% to 77% of their vitamin C within a week of harvest [4]. Chinese solar greenhouses with a south mono-slope (CSGs) do not primarily rely on supplemental heating; instead, they are designed to maximize solar energy gain and minimize heat loss to maintain suitable indoor temperatures. Fortunately, the Canadian Prairies have extensive sunshine hours in a year, which provide favorable conditions for adapting CSGs and reducing heating cost as predicted by Ahamed et al. [5,6]

and Beshada et al. [7] through thermal modeling. Beshada et al. [7] conducted a field experiment in Elie, Manitoba, Canada, showing that CSGs is more energy-efficient for winter vegetable production in cold regions than traditional gutter-connected greenhouses. Ahamed et al. [8] simulated the heating requirement in a CSG located in Saskatoon, Canada and found that it is about 50% of a gutter-connected greenhouse.

Many mechanistic models have been developed for simulating the thermal environment (temperature, RH, light intensity) of CSGs using MATLAB, CFD, FORTRAN, and VC++ methods [6,9–14]. Most of these models are used to predict the diurnal variation of temperature [12,13,15] and humidity [14,16] inside the greenhouses. Zhang et al. [17] and Huang et al. [18] developed mathematical models to simulate the light environment (solar radiation) inside of CSGs. A few studies developed models for specific purposes such as investigating the potential of renewable energy (solar and wind) to power the solar greenhouse [19], heat-pipe heating system [20], optimizing design parameters [21,22] etc. However, these models are not ready to use by industry, researchers, and growers. CFD (computation fluid dynamic)-based models [10,23] are reliable but require high computation capacity as well as a need to define the complicated boundary conditions [24]. A few studies [25–27] used commercial software, such as TRNSYS and EnergyPlus, to simulate the thermal environment of various types of solar greenhouses. However, most of these studies were conducted without considering the plant mass due to the complexity of modeling the dynamic nature of plant evapotranspiration, which is a primary component in energy balance. These commercial software packages are designed for building energy simulation, and thus, they need complicated modification to model the actual operation of greenhouses [25]. Some studies developed black-box models for CSGs using different approaches, such as the convex bidirectional extreme learning machine algorithm [16] and least squares support vector machine (LSSVM) with a particle swarm technique (PSO) for parameter optimization [28]. Zhang et al. [29] developed a model with a detailed 3D tomato canopy structure and simulated the micro light environment using a functional structural plant model (FSPM). However, these black-box models have the limitation in terms of computation time and flexibility, and sometimes need to input some complicated unknown parameters. For example, the 3D model developed by Zhang et al. [29] needs a total running time of approximately 20 h to simulate results for 8 h using an Intel Core I7 CPU and 16 GB RAM. Additionally, many thermal models [5,30–32] have been developed to simulate the microclimates of conventional greenhouses, which are not suitable for mono-slope greenhouses due to the differences in structures and environment control practices. Hence, a readily available mechanistic model (user-friendly software) with reasonable accuracy, low computational time, and various outputs (indoor temperature and RH, temperature of soil and north wall surface, and heating and cooling loads) would be a valuable tool for the design, optimization, and estimation of energy loads for CSGs.

Ma et al. [33] developed a software entitled RGWSRHJ for the dynamic simulation of the thermal environment of CSGs using finite difference numerical methods. This software could be a valuable tool for designers to compare and optimize CSG designs in China, however, it has many limitations for use outside of China. The major problem is that it is designed for Chinese CSGs, which usually lacks automated temperature control systems, so the indoor air temperature is constantly changing. The goal of the software is to predict the indoor thermal environment as a result of the passive impact of outdoor climatic conditions. In modern greenhouses, the practice is that the indoor temperature is controlled at its set point by heating and ventilation systems, and the goal of the thermal simulation is to analyze energy consumption under various indoor temperature and relative humidity (RH) set-points and ambient conditions, which the RGWSRHJ cannot do. Hence, the objective of this study was to modify the RGWSRHJ model into a new model named SOGREEN to predict the thermal environments (temperature and relative humidity) as well as the heating requirement of CSGs and validate the model using the data collected from a CSG in Canada. The SOGREEN model had many improvements to make it applicable to cold regions such as Canadian Prairies. Besides allowing supplemental heating prediction for maintaining the required set-point temperature, the other improvements include allowing polystyrene pellet insulation and double-layer inflated front-roof cover, resetting weather data type, modifying hourly simulation, adding sun blocker and internal thermal screen function, and converting the software in English. Finally, the SOGREEN model was used to analyze the thermal environment and heating requirement in a case study for a CSG in Saskatoon, Canada.

The article is organized into five sections. The literature review and knowledge gaps are presented in the introduction (Section 1). The model theory with the detailed mathematical formulation is described in Section 2. Section 3 describes the model formulation and steps for defining the building structures and control schedules, and the model validation is presented in Section 4. The results from the case study for the year-round operation of CSGs in the cold region (Saskatoon) are analyzed in Section 5. The final section contains the conclusions and future research direction for further development.

2. Model Theory

The SOGREEN model consists of five modules: the wall, ground, front roof covering material, ventilation, and evapotranspiration. The model was developed based on the heat balance of indoor air with consideration of all possible heat sources and sinks in CSGs. The differential Equation (1) reflects the sensible heat balance of the indoor air as affected by the greenhouse modules [21]:

$$\rho_a C_p V \frac{dt_i}{d\tau} = Q_w + Q_s - Q_g - Q_v - Q_e \tag{1}$$

where ρ_a is the air density (kg/m^3); C_p is the air specific heat (J/kg·°C); V is the greenhouse volume (m^3); t_i is the indoor air temperature (°C); τ is the time (s); Q_w is the heat gain (negative value is loss) from the walls and north roof (W); Q_s is the heat released from the ground (W); Q_g is the heat loss from the front roof (W); Q_v is the heat loss through air exchange (W); and Q_e is the heat loss by evapotranspiration (W).

A two-dimensional unsteady state (dynamic) model that describes heat transfer through the composite envelopes (the walls, north roof, and ground) is given in Equation (2). As compared with the commonly used one-dimensional models, it reflects the heat transfer process more accurately:

$$\rho_a C_p \frac{\partial t}{\partial \tau} = \frac{\partial}{\partial x}\left[\lambda \frac{\partial t}{\partial x}\right] + \frac{\partial}{\partial y}\left[\lambda \frac{\partial t}{\partial y}\right] + S \tag{2}$$

where λ is the heat conductivity coefficient (W/m·°C) and S is the heat from solar radiation per volume of the envelope material (W/m^3).

Solar radiation is the primary heat source in the solar greenhouse. When sunshine strikes the walls, back roof, and ground surface in the daytime, their surface temperatures rise. After discretizing the control equation with the appropriate transformation, Equation (2) can be expressed as the following differential linear equations:

$$\begin{cases} P_{0,j}t_{0,j} - A_{0,j}t_{1,j} = K_{0,j}t_{0,j,0} + SS_{0,j} \\ -A_{i-1,j}t_{i-1,j} + P_{i,j}t_{i,j} - A_{i,j}t_{i+1,j} = K_{i,j}t_{i,j,0} + SS_{i,j} \\ -A_{n,j}t_{n,j} + P_{n+1,j}t_{n+1,j} = K_{n+1,j}t_{n+1,j,0} + SS_{n+1,j} \end{cases} \tag{3}$$

where $A_{i,j} = \dfrac{\lambda_{i,j}(\Delta y)_j}{(\delta x)_i}$; $A_{i-1,j} = \dfrac{\lambda_{i,j}(\Delta y)_j}{(\delta x)_{i-1}}$;

$B_{i,j} = \dfrac{\lambda_{i,j}(\Delta x)_i}{(\delta y)_j}$; $B_{i,j-1} = \dfrac{\lambda_{i,j}(\Delta x)_i}{(\delta y)_{j-1}}$; $K_{i,j} = \dfrac{\rho_{i,j}C_{i,j}(\Delta x)_i(\Delta y)_j}{\Delta \tau}$;

$P_{i,j} = B_{i,j-1} + A_{i-1,j} + K_{i,j} + A_{i,j} + B_{i,j}$;

$SS_{i,j} = S_{i,j}(\Delta x)_i(\Delta y)_j + B_{i,j}t_{i,j+1} + B_{i,j-1}t_{i,j-1}$;

$\Delta \tau$ is the simulation step length (s); $(\Delta y)_j$ is the height of the node (i,j) (m); $(\Delta x)_i$ is the width of the node (i,j) (m); $(\delta y)_j$ is the center distance between the j and $(j + 1)$ segment (m); $(\delta x)_i$ is the center distance between the node (i,j) and $(i + 1,j)$ (m); $\lambda_{i,j}$ is the heat conductivity coefficient of the node (i,j), (W/m·°C); $\rho_{i,j}$ is the density of the node (i,j) (kg/m^3); $C_{i,j}$ is the specific heat of the node (i,j), (J/(kg·°C)); and $SS_{i,j}$ is the heat source of the node (i,j) (W/m^3).

The above linear equations set (3) can be written in the following matrix form:

$$\begin{bmatrix} P_{0,j} & -A_{0,j} & 0 & 0 & 0 & 0 \\ -A_{0,j} & P_{1,j} & -A_{1,j} & 0 & 0 & 0 \\ 0 & -A_{1,j} & P_{2,j} & -A_{2,j} & 0 & 0 \\ \cdots & \cdots & \cdots & \cdots & \cdots & \cdots \\ 0 & 0 & 0 & -A_{n-1,j} & P_{n,j} & -A_{n,j} \\ 0 & 0 & 0 & 0 & -A_{n,j} & P_{n+1,j} \end{bmatrix} \begin{bmatrix} t_{0,j} \\ t_{1,j} \\ t_{2,j} \\ \cdots \\ t_{n,j} \\ t_{n+1,j} \end{bmatrix} = \begin{bmatrix} K_{0,j}t_{0,j,0} + SS_{0,j} \\ K_{1,j}t_{1,j,0} + SS_{1,j} \\ K_{2,j}t_{2,j,0} + SS_{2,j} \\ \cdots \\ K_{n,j}t_{n,j,0} + SS_{n,j} \\ K_{n+1,j}t_{n+1,j,0} + SS_{n+1,j} \end{bmatrix} \quad (4)$$

This tridiagonal matrix can be efficiently solved using the simplified form of Gaussian elimination known as the Thomas algorithm [34]. The solution of the above equations gives the transient temperature field of the wall, back roof, and ground. Based on the estimated temperature of the heat-storage component surfaces, the heat transfer from the wall, north roof, and ground to indoor air can be calculated by the flowing equations:

$$Q_w = \sum \alpha_i (\Delta y)_j L_{gh} (t_{wsj} - t_i) \quad (5)$$

$$Q_s = \sum \alpha_i (\Delta y)_j L_{gh} (t_{ssj} - t_i) \quad (6)$$

where α_i is the convective heat transfer coefficient between indoor air and surfaces (W/m^2·°C); $(\Delta y)_j$ is the height of node (i,j) (m); L_{gh} is the total length of greenhouse (m); t_{wsj} is the surface temperature of the wall and back roof (°C); and t_{ssj} is the surface temperature of ground (°C).

The heat loss through the transparent front roof, air exchange from infiltration and ventilation, and evapotranspiration can be estimated as follows [33]:

$$Q_g = KA_f (t_i - t_o) \quad (7)$$

$$Q_v = \rho_a C_p \dot{V} (t_i - t_o) \quad (8)$$

$$Q_e = rA_s K_v (p_{ws} - p_w) \quad (9)$$

where K is the heat transfer coefficient of the front roof (plastic film covering) (W/m^2·°C); A_f is the total area of the front roof (m^2); $\dot{V}$ is the air exchange rate through ventilation and infiltration (m^3/s); p_{ws} is the saturated vapor pressure at room temperature (Pa); p_w is the water vapor pressure at room temperature (Pa); K_v is the coefficient of evapotranspiration per square meter of the ground area (kg/m^2·s·Pa); r is the latent heat of water vaporization (2442 kJ/kg); and A_s is the indoor ground area (m^2).

The saturated vapor pressure and the water vapor pressure are estimated according to the ASHARE fundamental [35]. The air exchange rate is input through scheduling based on the time, crop species, and growth stages. Evapotranspiration is a dynamic process that depends on several factors including plant species, growth stages, indoor temperature, RH and light intensity. The evapotranspiration coefficient of the crops needs to be selected based on crop density.

3. Model Development and Operation

As shown in Figure 1, the simulation process starts with the input of basic simulation conditions, including the location and weather conditions, physical and thermal properties of greenhouse structural materials, and operation schedule. Based on the input parameters, the model starts calculating the indoor thermal field for the next simulation step and recording each step's thermal parameters. When the simulation process comes to the final moment, the model compares the specified tracking point's temperature to the corre sponding initial temperature and replaces the initial temperature with the predicted final temperature. After several simulation cycles, the fluctuation in indoor thermal parameters comes within the acceptable range (10^{-5}, and the model will output the simulation results.

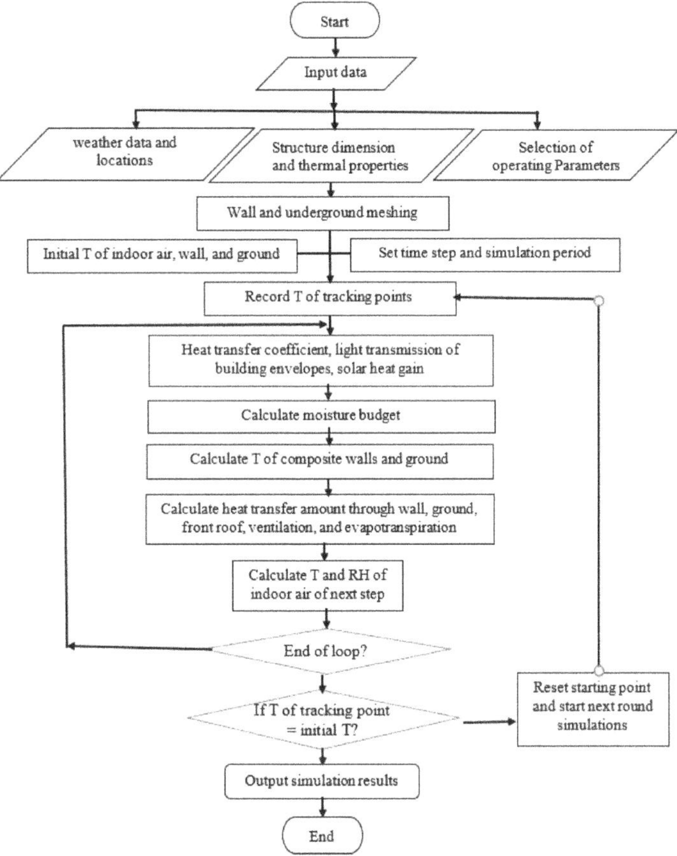

Figure 1. Simulation flow chart of the SOGREEN model.

Figure 2 shows the initial interface of the simulation model. It contains weather condition settings, greenhouse constructional designs, materials selection for the envelope, indoor environment control parameters, etc. A save/load design function is added to simplify the operational process of the model. Different schedule modules are added for setting the operation schedules of the thermal blanket, supplemental heating, and ventilation systems. Moreover, a sun blocker function is added in the indoor control interface to reduce solar radiation entering the greenhouse in summer. An interior thermal screen function is added to reduce heat loss when the exterior thermal blanket fails to work because of mechanical failure or frozen blankets. The sun blocker and interior thermal screen are essential for cold regions to reduce greenhouse heating or cooling during the summer and winter seasons.

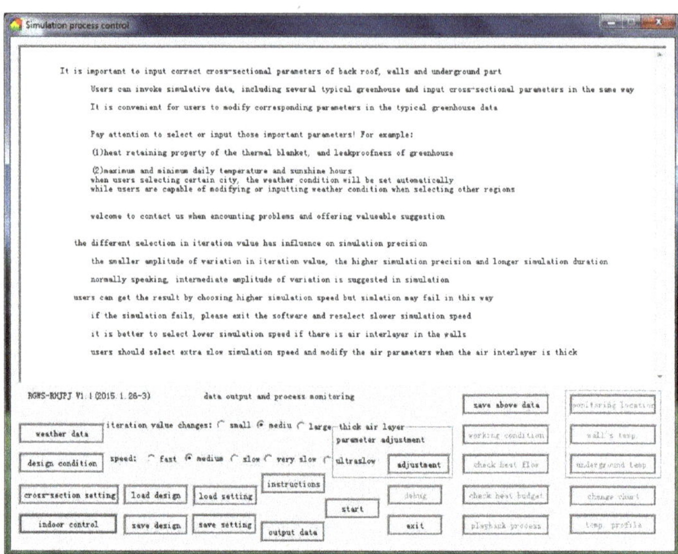

Figure 2. Starting interface of the SOGREEN model.

The first step in simulation is to define the outdoor weather conditions. The user may select one of the three types of weather data, including user-defined weather data, measured weather data, and built-in weather data. The built-in weather data only includes several Chinese cities. The standard weather data, i.e., the typical meteorological year data (TMYD), were added to the user-defined module in this study. The simulation period, location (latitude and longitude), and elevations are also input in the weather condition interface. The SOGREEN model is extended to higher latitudes and longitudes (0°–90° N, −179°–179° E) and increases the wind speed limit up to 17 m/s to suit Canadian weather conditions.

The next step is to set the solar greenhouse design conditions, including the structural information, plant information, and building materials properties. For the south roof besides plastic films, the modified model allows the option for glass, polycarbonate board, and other newly developed energy-efficient glazings, such as polystyrene (PS) pellet insulation and double-layer inflated film. The heat transfer coefficient of the PS pellet insulation is considered at 4.0 W/m^2·K for the daytime and 0.3 W/m^2·K at night. The overall heat transfer coefficient of a 6-cm thick double-layer inflated plastic film cover is 4.45 W/m^2·K, which slightly higher than the value (4.0 W/m^2·K) from the ASABE standard [36]. The aging factor may affect the rate of light transmission through the glazing materials; thus, the SOGREEN model allows users to select the degree of aging based on their judgment. The thermal blanket is an important measure to reduce the nighttime heat loss from the greenhouse, so a wide range of thermal properties for the thermal blanket is provided in the model from 3.0 to 0.5 RSI. The ground floor condition and the wetness level can affect the evaporation rate and the indoor thermal environment; therefore, users also need to select floor types, such as soil, mulch, concrete, and wetness level. Based on the plant condition, users need to choose plant density options: very sparse, sparse, ordinary, dense, and very dense. The plant transpiration is estimated based on the plant mass per unit area, which depends on the plant species and density information. Finally, the plant height and distance from the north wall are used to calculate the north wall's shaded area.

The next step is to define the physical and thermal properties of constructional materials for the greenhouse. The dimensions and building materials are set for the north roof, north wall, and floor. As shown in Figures 3 and 4, users input the segment lengths in the north wall and the building material's thermal properties, including density, thermal conductivity, and specific heat. The model also has a built-in material library that facilitates users to find commonly used materials. Figure 4 shows the interface of the greenhouse's detailed physical and thermal properties.

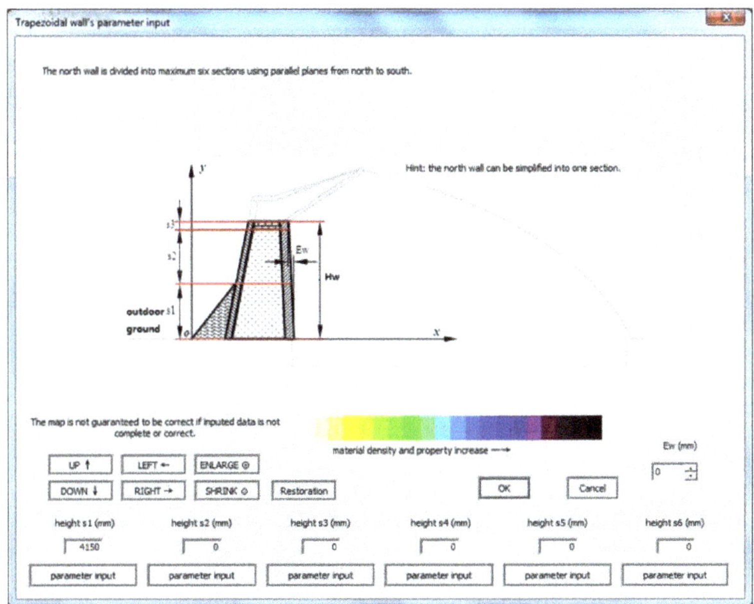

Figure 3. The interface for defining the physical properties of the north wall.

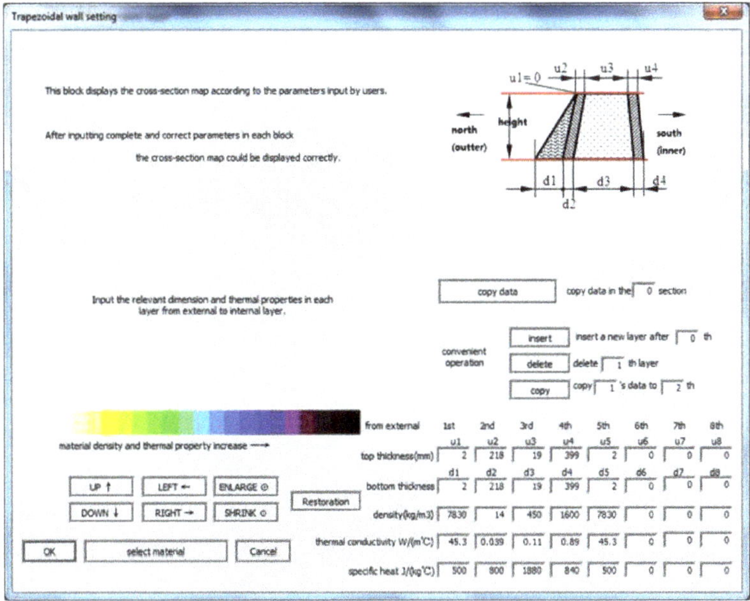

Figure 4. The interface for defining the thermal properties of the building materials.

Finally, the operational setting for the greenhouse needs to be defined. It includes the operation hour for the thermal blanket, supplemental lighting schedule, temperature and RH set-points, and ventilation schedule. The model can automatically cover and uncover the thermal blanket based on solar radiation availability. For example, the thermal blanket can be set to unveil in the morning when the outdoor solar radiation reaches 80 W/m^2 and cover when it drops below 80 W/m^2. The user can also define the working schedule for each simulation day.

After completing all settings, the user can start the simulation process, which generally takes 15 to 90 min depending on the selected simulation periods and computer speed. The model can then provide the outputs, including supplemental heating and cooling needs and the indoor thermal parameters, such as air temperature and RH and the wall and soil temperature.

4. Model Validation

A commercial solar greenhouse located in Elie, Manitoba (49°55′ N, 97°28′ W) was used for validation of the SOGREEN model. The greenhouse has the classic structure of mono-slope CSGs (Figure 5). Three days of measurement data (28 to 30 March 2017) by Ahamed et al. [6] were obtained to validate the model. The greenhouse is 28 m in length and 6.7 m in width. The north wall and ridge heights are 2.1 m and 3.5 m, respectively and 34° is the angle between the north roof and the horizontal plane. The south roof is covered with a 6-mil single-layer polyethylene film, while the cotton thermal blanket (RSI-1.2) covers the south roof from the outside at night. The north wall is made of wooden stud wall; from outside to inside are corrugated galvanized sheet steel (2 mm), fiberglass insulation (152 mm), plywood (13 mm), sand (152 mm), and corrugated galvanized steel (2 mm). A portion of the north wall interior surface was painted black to absorb solar energy. The north roof is made of corrugated galvanized sheet steel (2 mm), fiberglass insulation (152 mm), plywood (13 mm), and plastic film from inside [6]. Table 1 lists the thermal properties of the materials for the north wall, sidewall, and north roof obtained from the ASHRAE standards [35].

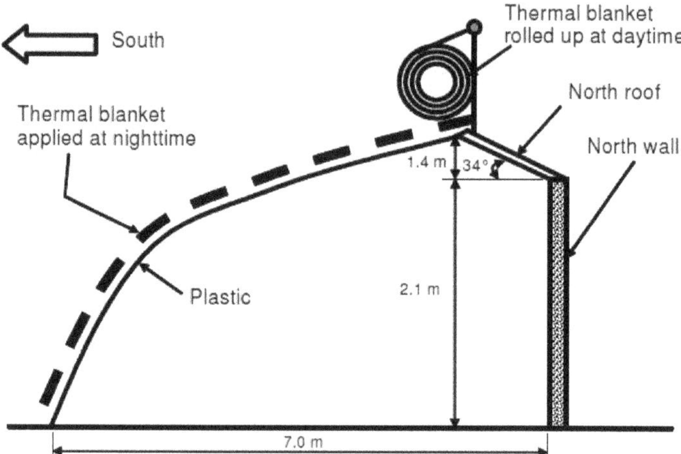

Figure 5. The side view of the study greenhouse used for validation of the model [7].

Table 1. Thermal properties of wall and north roof materials from exterior to interior layers.

Layer of Materials	L1	L2	L3	L4	L5
North wall material	Steel	Fiberglass	Plywood	Sand	Steel
Density (kg/m^3)	7830	14	460	1600	7830
Conductivity (W/m·K)	45.3	0.039	0.093	0.89	45.3
Specific heat (J/kg·K)	500	800	1880	840	500
Sidewall and roof material	Steel	Fiberglass	Plywood	Plastic	-
Density (kg/m^3)	7830	14	460	900	-
Conductivity (W/m·K)	45.3	0.039	0.093	5.5	-
Specific heat (J/kg·K)	500	800	1880	1900	-

The height of the newly transplanted tomato plants was about 14 cm during the experiment. The distance between tomato plants was about 31 cm, and the space between the north wall and the bed was 96 cm. The outdoor temperature ranged from -2.0 °C at night and 10.0 °C at noon (Figure 6). An electrical heater (3.6 kW) controlled by a thermostat was used to heat the greenhouse when the indoor temperature dropped below 12 °C, and the heater turned off when the temperature reached 18 °C. The cotton thermal blanket was rolled down to cover the south roof before the sunset at 5:30 p.m. and rolled up to uncover the roof after the sunrise at around 7:00 a.m. The vent was opened manually to reduce the indoor air temperature at noon when the indoor temperature was very high. For validation, the air exchange by infiltration (0.156 m^3/s) was considered most of the time (2:30 p.m. to 11:00 a.m.) because no intentional ventilation was allowed to the greenhouse. The vent was typically opened between 11:00 a.m. to 2:00 p.m. and the air exchange by ventilation was estimated at 0.52 m^3/s. A detailed description of the experimental setup can be found in Ahamed et al. [6].

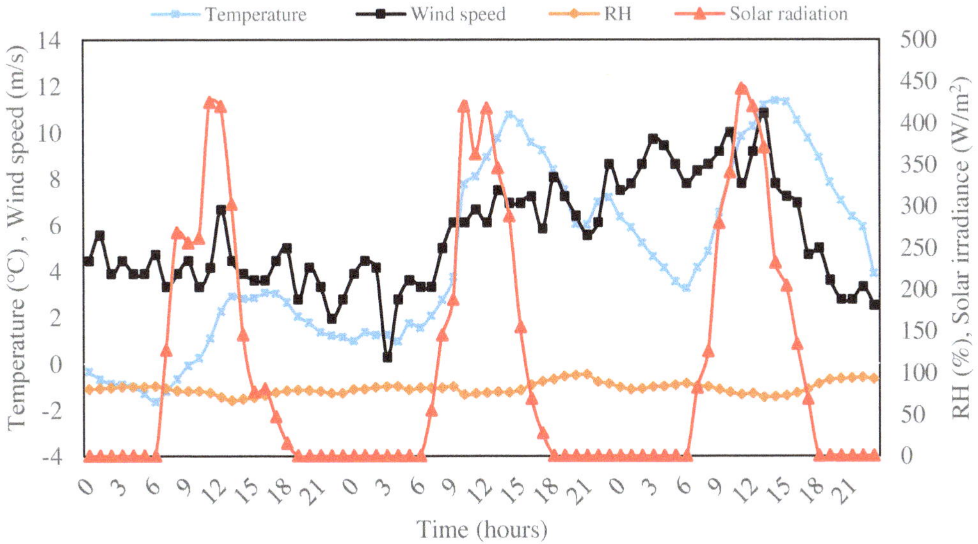

Figure 6. The measured outdoor weather conditions during the experimental period.

Figure 7 shows the comparisons of the predicted and measured indoor air temperature in the greenhouse. The predicted temperature closely followed the measured values with 1.8 °C of average discrepancy and a maximum difference of about 8.8 °C. Large discrepancies often occurred at noon; the vent's opening for cooling caused the sharp measured temperature drops at noon. Figure 7 also shows that the predicted temperature increased sharply at 8:00 a.m. on the first day, while the measured temperature increased relatively slowly, which could be due to the unwanted ventilation (not considered in simulation) caused by the greenhouse entrance door's frequent openings.

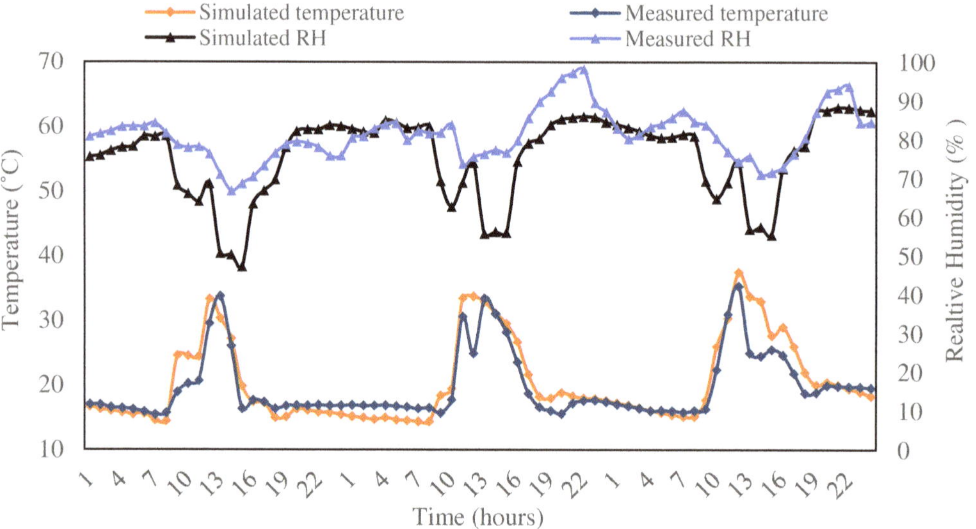

Figure 7. The comparison of the measured and predicted indoor air temperature and RH in the greenhouse.

Similarly, the overall trend of the predicted indoor RH is close to the measured one except during the noon ventilation periods; the average discrepancy is about 7.0%. The high discrepancy could be caused due to two factors. Firstly, actual plant transpiration was very small as the plants were transplanted a few days before the experiment. The evaporation from the wet soil mainly dominated the moisture production at noon. Secondly, the air exchange rate during the ventilation period was based on assumptions. This study used an air exchange rate based on constant input parameters, which might differ from its actual value.

Figure 8 shows the comparison of the measured and predicted temperatures for the north wall surface. In general, the diurnal profiles have a similar trend, but the predicted values are higher than measured ones with an average discrepancy of 4.2 °C and the highest difference of 9.7 °C. The discrepancy could mainly be caused by possibly large discrepancies of the actual values of the thermal properties of the wall materials (sheet steel sand (wetness), wood stud, etc.) with the book values selected for the model. Furthermore, the model is limited in defining the sheet metal's corrugated shape (larger surface area), so it uses a flat surface instead. With some modifications of the thermal properties, the model predictions could be improved, although this study did not do so in order to stick to the original book values.

Figure 9 shows that the predicted soil surface temperature typically followed a similar pattern with higher daytime predictions than measured values, while at night, the two are very close except when the electric resistance cable heater was in operation for the third night, which the model did not consider. The average discrepancy between predictions and measured values is 2.3 °C, and the maximum is 7.9 °C.

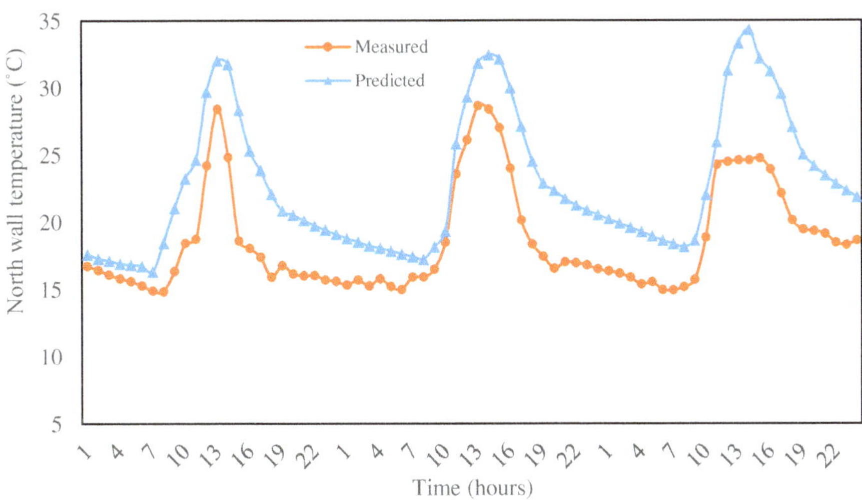

Figure 8. The comparison of measured and predicted temperature of the north wall surface.

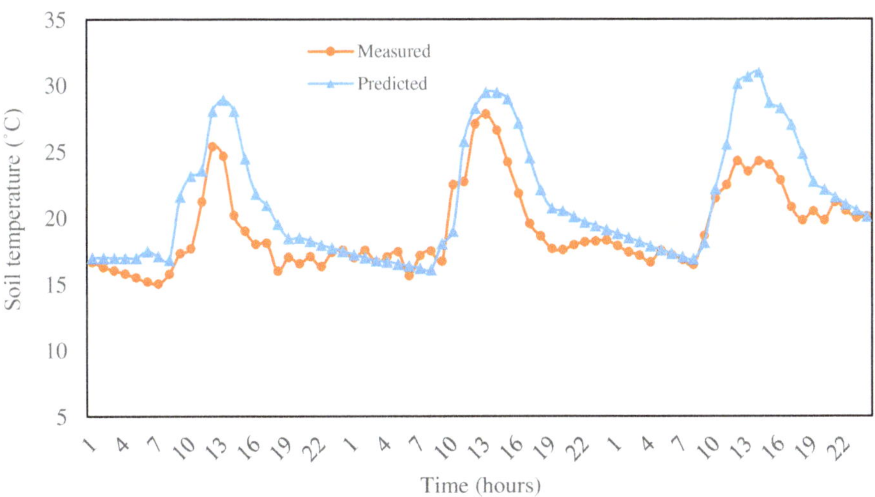

Figure 9. The comparison of the measured and predicted soil temperature near the surface.

Table 2 lists the statistical parameters, including the coefficient of determination (R^2), standard deviations (SD), mean absolute error (MAE), root mean square error (RMSE), and relative root means square error (rRMSE) for the model validation [6,37]. The predicted air, wall, and floor temperatures look satisfactory, with R^2 values 0.79–0.86, but are relatively low for RH. The SD and MAE value is also high for RH and less than 2.30 °C for temperature except the MAE for the north wall. Compared with other similar mechanistic dynamic models, SOGREEN has good accuracy; the RMSE of the room air temperature for a conventional greenhouse is 5.67 °C [32], 5.3 °C for a semi-solar greenhouse [38], 2.82 °C for a passive solar greenhouse [39], 2.6 °C for CSGs [14], and 1.46 °C (RMSE) and 0.89 °C (MAE) for black-box models [16,28]. For RH, the RMSE of previous mechanistic models is ranged between 4.3–14.6% [16,32], and is 2.5% for the black-box model [16]. Compared with previous studies, the SOGREEN model also has good accuracy for predicting the soil surface temperature but is less accurate for the north wall. For soil temperature, an

RMSE of 3.45 °C was reported for a semi-solar greenhouse in Mohammadi et al. [38] and 2.19 °C in Taki et al. [40], but these studies were conducted in a small fully-controlled greenhouse, whereas SOGREEN was validated against data from a commercial greenhouse with multiple uncontrolled factors (frequent door opening, manual control of thermal blanket, and water tanks placed inside the greenhouse). In previous studies, the RMSE of the north wall was reported as 2.0 °C in Mobtaker et al. [39] and 1.8 °C in Ahamed et al. [6], but it is relatively higher (4.68 °C) for SOGREEN. It needs to be noted that the north wall is the most complicated component to model in CSGs, as many factors affect its temperature variation. Previous studies [30,41] also indicated that a rRMSE value of around 10% is reasonably acceptable for hourly simulation. In general, the results suggest that the model performs very well to predict indoor air temperature but needs to address the high errors for simulating the surface temperatures of the north wall and floor in future studies. Additionally, the limitations of the experimental greenhouses noted above need to be overcome in future studies to obtain accurate test data for further model validation and improvement. The actual thermal properties of the materials should be measured and used in the model instead of the book values, especially ground soil wetness, wall sand density, and moisture content, which greatly impact heat transfer. It also needs to be noted that the model prediction would be much closer to the measured values if the CSGs had an automated heating and ventilation system to control temperature and RH. The manual ventilation and uneven heating could cause unpredictable variations of the indoor thermal environment. Additionally, evapotranspiration is a dynamic process that needs to be addressed in future studies for more accurate results, as the evapotranspiration coefficient is defined based on user input for plant density.

Table 2. Statistical indices for evaluating the model performance for various thermal parameters of the greenhouse.

Thermal Parameters	R^2	SD (°C)	MAE (°C)	RMSE (°C)	rRMSE (%)
Indoor temperature	0.85	1.92	1.85	2.66	3.28
Indoor RH	0.52	6.19	7.02	9.34	11.49
Ground/soil temperature	0.79	2.01	2.30	3.04	15.84
North wall temperature	0.86	2.10	4.19	4.68	25.15

5. Case Study Greenhouse

The SOGREEN model was used to simulate the annual thermal performance of a conceptually designed CSGs located in Saskatoon, Canada (52.13° N, 106.62° W) to predict the heating requirement. The greenhouse is 100 m × 12 m and 5.62 m high in the ridge (Figure 10). Similar structural materials and dimensions as the greenhouse used in validation are considered except for wall materials (the insulation fiberglass: 218 mm; interior plywood: 19 mm; and heat storage material sand: 399 mm). A large growing space and thicker walls are considered to be able to produce on a commercial scale, increase heat storage capacity, and reduce heat loss. The angle between the south roof and the horizontal plane is set at 40° to minimize the shading from the south roof. The perimeter insulation consisting of a fiberglass board and plywood is considered beneath the walls and the south end of the roof to reduce the heat loss through the ground. The south roof is considered to be covered with a double-layer inflated film. The indoor ground is covered with landscaping fabric, making a better working place for greenhouse growers. Figure 10 shows the cross-section of the conceptual CSGs.

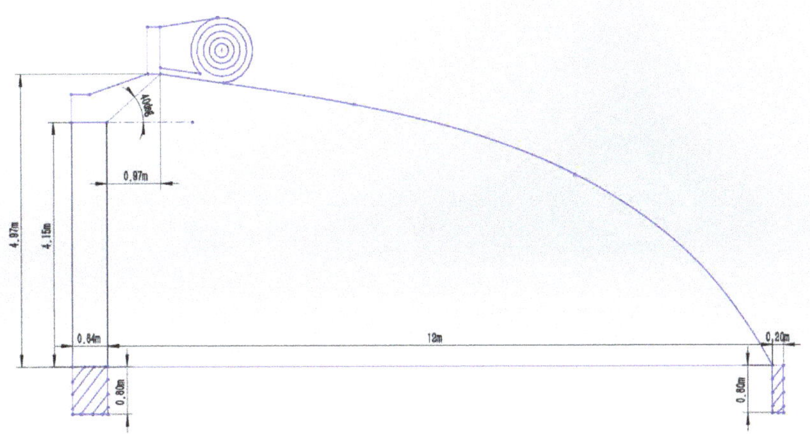

Figure 10. Cross-section of a conceptual CSG in Saskatoon.

A different schedule for the thermal blanket and ventilation rate is considered based on sunrise and sunset times, weather conditions, and likely vent opening for temperature control for each month. No interior thermal screen is considered in the simulation. The greenhouse is equipped with a heating and ventilation system to control the air temperature and RH. The daytime set-point temperature should be around 22 °C for optimal production of tomatoes; however, the set-point temperature is set at 18 °C for daytime and nighttime for two reasons. Firstly, the daytime temperature in CSGs is most likely higher than 22 °C (except for extremely low outdoor temperatures) when solar radiation is available. Secondly, a lower set-point temperature will reduce the heating cost. During the cold months, a low air exchange rate based on infiltration is set at night and in the morning to reduce CO_2 loss through ventilation, and moderate ventilation rates are arranged at noon for mild months. High ventilation rates with a long ventilation duration are applied during the warmer months. A sunshade screen is considered to block half of the total solar irradiance from June to August when it is needed. Table 3 shows the work schedules in the study greenhouse.

Table 3. The schedule for the thermal blanket and ventilation in the study greenhouse.

	January	February	March	April	May	June
Thermal Blanket	16:30–10:00	17:00–9:30	8:30–18:00	7:00–19:00	6:30–19:30	6:00–21:00
Ventilation Rate (m^3/s)	12:00–13:00: (0.67)	11:00–12:00: (1.0); 13:30:15:00: (1.67)	10:00–11:00: (1.0); 13:00–14:00: (2.67)	8:30–9:00: (1.0); 10:30–12:30: (5.67); 12:30–18:00: (5.0)	7:30–10:00: (3.33); 10:00–12:30: (8.33); 12:30–18:30: (6.67)	6:30–10:00: (3.33); 10:00–12:00: (8.33); 12:00–17:00: (3.33); 17:00–21:00: (8.33)

	July	August	September	October	November	December
Thermal Blanket	5:30–24:00	6:30–23:00	7:40–18:00	8:30–17:00	10:00–16:00	10:30–16:00
Ventilation Rate (m^3/s)	6:30–9:00: (3.33); 9:00–22:00: (8.33)	7:30–9:00: (3.33); 9:00–11:30: (8.33); 11:30–18:00: (10.0); 18:00–20:00: (5.0)	9:30–12:00: (1.0); 12:00–17:00: (5.0)	10:00–11:00: (1.0); 12:00–15:00: (4.0)	11:30–12:00: (1.0)	12:00–12:30: (1.0)

5.1. Annual Thermal Simulation

The hourly thermal environment and heating demand in the study greenhouse were simulated for a year, and Figure 11 shows the monthly average heating requirement, indoor temperature, and RH in 2017; the monthly average ambient temperature was below 0 °C from November to March (lowest −15.6 °C in January). The average temperature peak was 18.8 °C in July. The simulated results indicate that the heating demand fluctuates with the weather conditions, with the highest daily heating load of 7546 MJ in December and 4093 MJ in March. The heating demand is low in April, September, and October with an above-zero average temperature, while no supplemental heating is needed from May to August, although the daily low temperature often fell below 10 °C, showing the benefit of the heat storage capacity of the solar greenhouse.

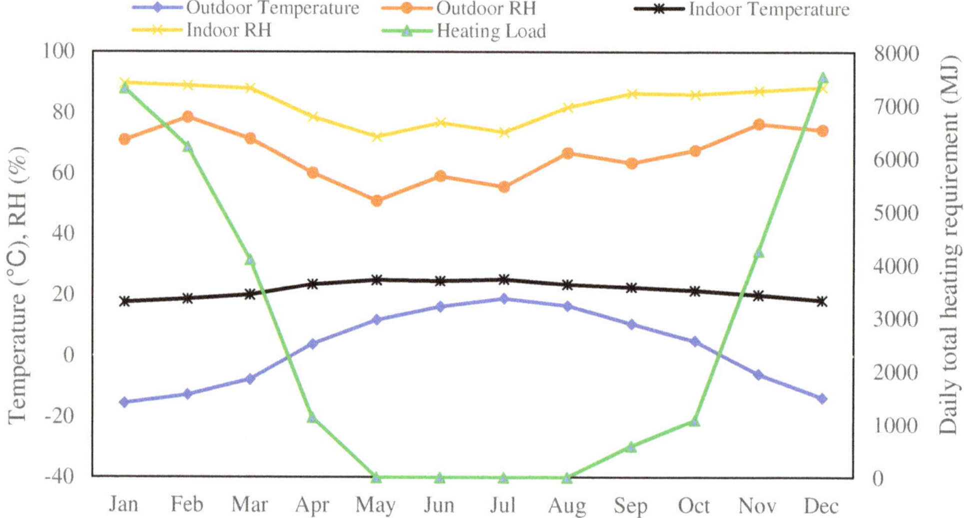

Figure 11. Monthly average thermal environment and heating requirement in the CSG located in Saskatoon.

Three months (December, March, and July) were selected to analyze the daily thermal performance of CSGs in the cold, mild, and warm seasons in Saskatoon. As shown in Figure 12, the daily average ambient temperature fluctuated between −30 °C and 0 °C in December, while the indoor temperature and RH are set at 18 °C and 90%, respectively. On 5 December, the simulated heating load was about 13,000 MJ when the daily total solar irradiance was only 135 W/m^2, while it was below 6500 MJ on 9 and 10 December with solar irradiance of 1145 W/m^2. As the sunshine duration is very short in the coldest months (January and December) at the high northern latitudes, the north wall is insufficient to provide heating as a thermal battery, resulting in high heating demand.

Most traditional greenhouses in the Canadian Prairies start operation in March, so comparing CSGs and conventional greenhouses starting from March is more appropriate. As shown in Figure 13, the daily average ambient temperature fluctuated between −21 °C and 2 °C in March. The average daily total solar irradiance almost quadrupled that of December so that the heating load could be reduced significantly.

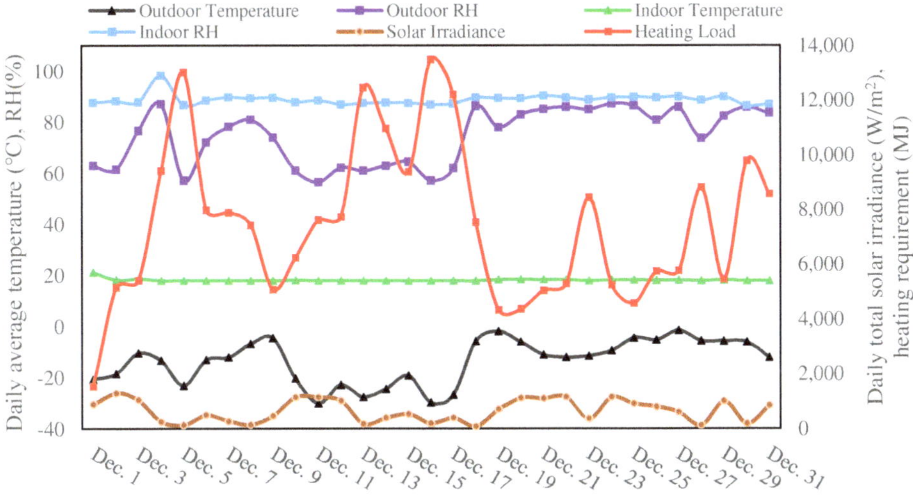

Figure 12. Daily average indoor conditions, solar irradiance, and heating loads in December.

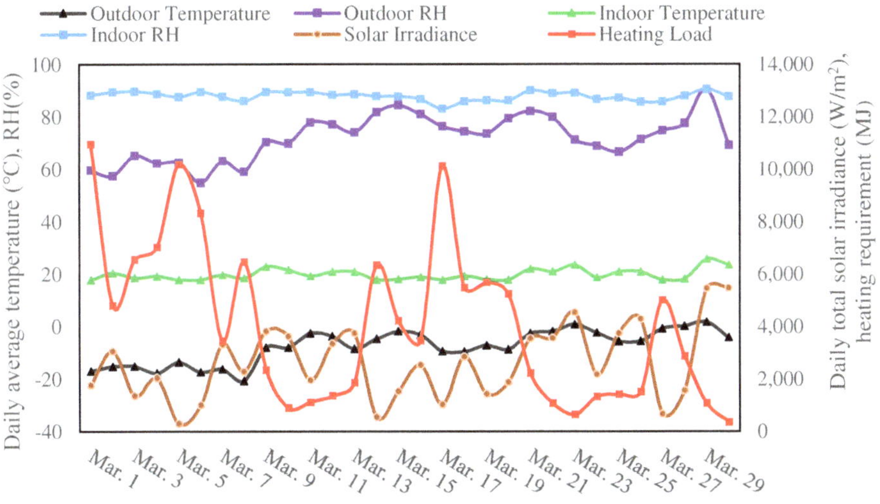

Figure 13. Daily average indoor conditions, solar irradiance, and heating loads in March.

In July, as shown in Figure 14, the average daily indoor temperature ranged between 22 to 28 °C, and RH fluctuated in an extensive range due to the air exchange rate changing from infiltration at night to high ventilation rate during the daytime. The daily total solar irradiance was high, and a sun blocker is considered to protect the tomato plants. Supplemental heating is not required except for the early morning on 25 July when the ambient temperature is below 10 °C, and the thermal blanket is rolled up.

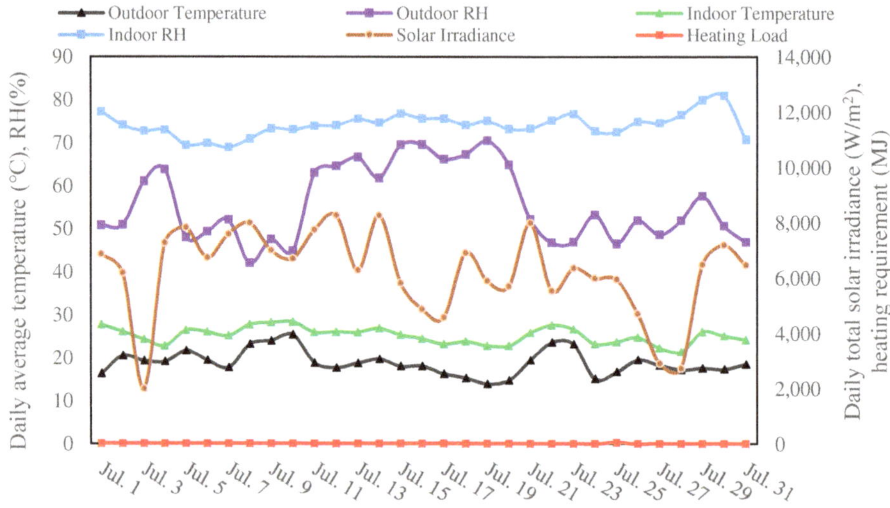

Figure 14. Daily average indoor conditions, solar irradiance, and heating loads in the study greenhouse in July.

5.2. Heating Cost Analysis

Based on the annual simulation results, the annual heating cost of the solar green house was calculated. This study selected electricity, natural gas, and coal as the heating fuels, and the corresponding heating costs were compared. The cost was estimated as described below:

✓ Electricity: The heating cost using electricity was estimated based on the electrical rate from the City of Saskatoon, which is 14.52 ¢/kWh for the first 14,500 kWh and 7.67 ¢/kWh after that.

✓ Natural gas: The cost for using natural gas was estimated based on the standard heating value (0.0373 GJ) per unit volume (1 m^3) of natural gas [42]. The natural gas burner efficiency is considered to be 92%. The commercial rate of natural gas from SaskEnergy is $38.5 for the basic monthly charge, and the delivery charge is $0.0743/m^3.

✓ Coal: The heating value is considered 18 GJ per unit tone of coal for a coal-fired heating system. We assumed the coal price is $45 per ton and a high-efficiency coal boiler with 90% efficiency. This study neglects the cost for residual ash disposal (2.8–6.6 tons).

Table 4 provides a comparison of the heating cost of different energy sources. The results indicate that electricity is the most expensive heating source while natural gas and coal have a comparable annual cost, one-tenth of the cost of using electricity. Natural gas has some unique advantages compared to coal, such as low capital costs and a better thermal control system. In addition, the natural gas combustion process does not produce solid by-products, such as ash, which needs additional cost for disposal.

Table 4. Annual heating cost using different energy sources.

	Electricity	Natural Gas	Coal
Total energy consumption (kWh)	261,292	261,292	261,292
Total energy consumption (GJ)	940.6	940.6	940.6
Amount of fuel	261,292 kWh	27,409 m^3	58.0 tons
Annual cost ($)	26,379	2499	2610

Finally, the heating cost of the conceptually designed CSG was compared with a traditional multi-span commercial greenhouse located near Saskatoon, which is heated by natural gas boilers. This greenhouse was closed during the coldest months (December and January), and the heating cost was compared for the remaining ten months (February to November), which is also a limitation for this study for annual evaluation. The heating cost in Table 4 for the conceptual CSG and the prorated natural gas bill for heating Grandora Gardens are presented in Table 5. The solar greenhouse has a much lower energy consumption and achieves 55% of cost-saving. The preliminary results indicate that CSGs could be an energy-saving solution for winter vegetable production in remote northern communities of Canadian Prairies. Therefore, the policymaker or stakeholders need to fund further research and development of this technology for the extreme cold regions and provide financial support for dissemination in the northern communities.

Table 5. Heating cost comparison between Grandora Gardens and the conceptual solar greenhouse.

	Grandora Gardens	Study Solar Greenhouse
Total growing area (m^2)	3252	1200
Natural gas cost ($)	15,250	2498
Natural gas cost per unit area ($/m^2$)	4.7	2.1

6. Conclusions

In this study, an existing CSG simulation model (RGWSRHJ) was modified into a new model entitled SOGREEN, which is aimed to predict the heating requirement of CSGs. The new functions of SOGREEN model allow the prediction of the hourly thermal environment of CSGs over the year. Additionally, the modified model allows the temperature and RH control and many different energy-saving technologies, such as insulation, indoor thermal blanket and screen, and additional functions required in modern greenhouses. The model was validated with field data measured in a solar greenhouse in Manitoba, Canada. The average discrepancies between the measurement and prediction are 1.9 °C, 7.0%, 4.2 °C, and 2.3 °C for the indoor temperature, RH, north wall surface temperature, and soil surface temperature, respectively. The model predictions could be improved by using measured thermal properties of the wall and ground materials instead of book values. Finally, a case study was conducted for a conceptually designed CSG (100 m × 12 m) in Saskatoon, Canada, to produce tomatoes. The annual simulation indicates that the daily average heating in the coldest month (January) could be two times higher (6.3 MJ/m^2·day) compared with March (3.4 MJ/m^2·day). Low supplemental heating is required to grow tomatoes in April, September, and October, when the average daily outdoor temperature is between 3.8 °C to 10.5 °C, and no supplemental heating is needed from May to August. Comparing the study of CSGs with a traditional local greenhouse, the heating cost of CSGs is about 55% less than the even-span gable roof greenhouse. Hence, this study concluded that a CSG has the potential for energy-efficient year-round vegetable production for cold regions, such as the Canadian Prairies, to ensure the food security of these remote communities.

Finally, it could be concluded that the SOGREEN model can predict the environment parameters (temperature and RH) and the heating and cooling loads of CSGs with reasonable accuracy, but the model needs to be improved further to minimize the error in predictions, especially for the north wall. The model was validated using three days of data at the end of winter, so further validation could be accomplished against the data from the other seasons. The limitations of the previous experiment (uncontrolled factors in commercial CSGs) also need to be addressed in future studies. In addition, the case study used the monthly gas bill to estimate the heating cost for the conventional greenhouses; the energy-saving comparison with conventional greenhouses needs to be further studied using more accurate short-term data, such as transient heating data.

Author Contributions: Conceptualization, S.D., M.S.A. and H.G.; methodology, S.D. and H.G.; software, S.D. and C.M.; validation, S.D., M.S.A. and H.G.; formal analysis, S.D. and M.S.A.; investigation, S.D. and M.S.A.; resources, H.G.; data curation, S.D. and M.S.A.; writing—original draft preparation, S.D.; writing—review and editing, M.S.A. and H.G.; visualization, S.D. and M.S.A.; supervision, H.G.; project administration, H.G.; funding acquisition, H.G. All authors have read and agreed to the published version of the manuscript.

Funding: This research received no external funding.

Institutional Review Board Statement: Not applicable.

Informed Consent Statement: Not applicable.

Data Availability Statement: The data are contained within the article.

Acknowledgments: Authors would like to thank Grandora Gardens for their support in the study by sharing greenhouse energy data.

Conflicts of Interest: The authors declare no conflict of interest.

References

1. Rorabaugh, P.; Jensen, M.; Giacomelli, G. Introduction to Controlled Environment Agriculture and Hydroponics. *Control. Environ. Agric. Cent.* **2002**, 1–130.
2. Spencer, R. Starting a Commercial Greenhouse Business in Alberta. Available online: http://www1.agric.gov.ab.ca/$department/deptdocs.nsf/all/opp11207 (accessed on 12 July 2016).
3. Ahamed, M.S.; Guo, H.; Taylor, L.; Tanino, K. Heating demand and economic feasibility analysis for year-round vegetable production in Canadian Prairies greenhouses. *Inf. Process. Agric.* **2019**, *6*, 81–90. [CrossRef]
4. Rickman, J.C.; Barrett, D.M.; Bruhn, C.M. Nutritional comparison of fresh, frozen and canned fruits and vegetables. Part 1 Vitamins C and B and phenolic compounds. *J. Sci. Food Agric.* **2007**, *87*, 930–944. [CrossRef]
5. Ahamed, M.S.; Guo, H.; Tanino, K. A quasi-steady state model for predicting the heating requirements of conventional greenhouses in cold regions. *Inf. Process. Agric.* **2018**, *5*, 33–46. [CrossRef]
6. Ahamed, M.S.; Guo, H.; Tanino, K. Development of a thermal model for simulation of supplemental heating requirements in Chinese-style solar greenhouses. *Comput. Electron. Agric.* **2018**, *50*, 235–244. [CrossRef]
7. Beshada, E.; Zhang, Q.; Boris, R. Winter performance of a solar energy greenhouse in southern Manitoba. *Can. Biosyst. Eng.* **2006**, *48*, 1–8.
8. Ahamed, M.S.; Guo, H.; Tanino, K.K. Modeling of heating requirement in Chinese Solar Greenhouse. In Proceedings of the 2016 American Society of Agricultural and Biological Engineers Annual International Meeting, Orlando, Florida, USA, 17–20 July 2016; ASABE: St. Joseph, MI, USA, 2016.
9. Meng, L.; Yang, Q.; Bot, G.P.; Wang, N. Visual simulation model for thermal environment in Chinese solar greenhouse. *Trans. Chin. Soc. Agric. Eng.* **2009**, *25*, 164–170.
10. Tong, G.; Christopher, D.M.; Li, B. Numerical modelling of temperature variations in a Chinese solar greenhouse. *Comput. Electron. Agric.* **2009**, *68*, 129–139. [CrossRef]
11. Xu, F.; Li, S.; Ma, C.; Zhao, S.; Han, J.; Liu, Y.; Hu, B.; Wang, S. Thermal environment of Chinese solar greenhouses: Analysis and simulation. *Appl. Eng. Agric.* **2013**, *29*, 991–997. [CrossRef]
12. Guo, H.; Li, Z.; Zhang, Z. The dynamic simulation of temperature inside a sunlight greenhouse. *J. Shenyang Agric. Univ.* **1994**, 438–443.
13. Zhou, N.; Yu, Y.; Yi, J.; Liu, R. A study on thermal calculation method for a plastic greenhouse with solar energy storage and heating. *Sol. Energy* **2017**, *142*, 39–48. [CrossRef]
14. Liu, R.; Li, M.; Guzmán, J.L.; Rodríguez, F. A fast and practical one-dimensional transient model for greenhouse temperature and humidity. *Comput. Electron. Agric.* **2021**, *186*, 106186. [CrossRef]
15. Tong, G.; Christopher, D.M.; Tianlai, L.; Tieliang, W. Temperature variations inside Chinese solar greenhouses with external climatic conditions and enclosure materials. *Int. J. Agric. Biol. Eng.* **2008**, *1*, 21–26. [CrossRef]
16. Zou, W.; Yao, F.; Zhang, B.; He, C.; Guan, Z. Verification and predicting temperature and humidity in a solar greenhouse based on convex bidirectional extreme learning machine algorithm. *Neurocomputing* **2017**, *249*, 72–85. [CrossRef]
17. Zhang, X.; Lv, J.; Xie, J.; Yu, J.; Zhang, J.; Tang, C.; Li, J.; He, Z.; Wang, C. Solar radiation allocation and spatial distribution in Chinese solar greenhouses: Model development and application. *Energies* **2020**, *13*, 1108. [CrossRef]
18. Huang, L.; Deng, L.; Li, A.; Gao, R.; Zhang, L.; Lei, W. Analytical model for solar radiation transmitting the curved transparent surface of solar greenhouse. *J. Build. Eng.* **2020**, *32*, 101785. [CrossRef]
19. Riahi, J.; Vergura, S.; Mezghani, D.; Mami, A. Smart and Renewable Energy System to Power a Temperature-Controlled Greenhouse. *Energies* **2021**, *14*, 5499. [CrossRef]
20. Du, J.; Bansal, P.; Huang, B. Simulation model of a greenhouse with a heat-pipe heating system. *Appl. Energy* **2012**, *93*, 268–276. [CrossRef]

21. Wu, X.; Liu, X.; Yue, X.; Xu, H.; Li, T.; Li, Y. Effect of the ridge position ratio on the thermal environment of the Chinese solar greenhouse. *R. Soc. Open Sci.* **2021**, *8*, 201707. [CrossRef]
22. Esmaeli, H.; Roshandel, R. Optimal design for solar greenhouses based on climate conditions. *Renew. Energy* **2020**, *145*, 1255–1265. [CrossRef]
23. Zhang, G.; Fu, Z.; Yang, M.; Liu, X.; Dong, Y.; Li, X. Nonlinear simulation for coupling modeling of air humidity and vent opening in Chinese solar greenhouse based on CFD. *Comput. Electron. Agric.* **2019**, *162*, 337–347. [CrossRef]
24. Boulard, T.; Roy, J.C.; Pouillard, J.B.; Fatnassi, H.; Grisey, A. Modelling of micrometeorology, canopy transpiration and photosynthesis in a closed greenhouse using computational fluid dynamics. *Biosyst. Eng.* **2017**, *158*, 110–133. [CrossRef]
25. Ahamed, M.S.; Guo, H.; Tanino, K. Modeling heating demands in a Chinese-style solar greenhouse using the transient building energy simulation model TRNSYS. *J. Build. Eng.* **2020**, *29*, 101114. [CrossRef]
26. Candy, S.; Moore, G.; Freere, P. Design and modeling of a greenhouse for a remote region in Nepal. *Energy Procedia* **2012**, *49*, 152–160. [CrossRef]
27. Guan, Y.; Bai, J.; Gao, X.; Hu, W.; Chen, C.; Hu, W. Thickness Determination of a Three-layer Wall with Phase Change Materials in a Chinese Solar Greenhouse. *Procedia Eng.* **2017**, *205*, 130–136. [CrossRef]
28. Yu, H.; Chen, Y.; Hassan, S.G.; Li, D. Prediction of the temperature in a Chinese solar greenhouse based on LSSVM optimized by improved PSO. *Comput. Electron. Agric.* **2016**, *122*, 94–102. [CrossRef]
29. Zhang, Y.; Henke, M.; Li, Y.; Yue, X.; Xu, D.; Liu, X.; Li, T. High resolution 3D simulation of light climate and thermal performance of a solar greenhouse model under tomato canopy structure. *Renew. Energy* **2020**, *160*, 730–745. [CrossRef]
30. Vanthoor, B.H.E.; van Henten, E.J.; Stanghellini, C.; de Visser, P.H.B. A methodology for model-based greenhouse design: Part 3, sensitivity analysis of a combined greenhouse climate-crop yield model. *Biosyst. Eng.* **2011**, *110*, 396–412. [CrossRef]
31. Sethi, V.P.; Sharma, S.K. Thermal modeling of a greenhouse integrated to an aquifer coupled cavity flow heat exchanger system. *Sol. Energy* **2007**, *81*, 723–741. [CrossRef]
32. Singh, G.; Singh, P.P.; Lubana, P.P.S.; Singh, K.G. formulation and validation of a mathematical model of the microclimate of a greenhouse. *Renew. Energy* **2006**, *31*, 1541–1560. [CrossRef]
33. Ma, C.W.; Han, J.J.; Li, R. Research and development of software for thermal environmental simulation and prediction in solar greenhouse. *North. Hortic.* **2010**, *15*, 69–75.
34. Zhang, Y.; Cohen, J.; Davidson, A.A.; Owens, J.D. A Hybrid Method for Solving Tridiagonal Systems on the GPU. *GPU Comput. Gems Jade Ed.* **2012**, 117–132. [CrossRef]
35. ASHRAE. *ASHRAE Handbook of Fundamentals*; SI Units, Ed.; American Society of Heating Ventilation Refrigeration and Air-Conditioning Engineers: Atlanta, GA, USA, 2017.
36. ASABE. *Heating, Ventilating, and Cooling Greenhouse. ASABE Standards*, 53rd ed.; ASABE: Michigan, MI, USA, 2006.
37. Ahamed, M.S.; Guo, H.; Tanino, K. Cloud cover-based models for estimation of global solar radiation: A review and case study. *Int. J. Green Energy* **2021**, 1–15. [CrossRef]
38. Mohammadi, B.; Ranjbar, S.F.; Ajabshirchi, Y. Application of dynamic model to predict some inside environment variables in a semi-solar greenhouse. *Inf. Process. Agric.* **2018**, *5*, 279–288. [CrossRef]
39. Mobtaker, H.G.; Ajabshirchi, Y.; Ranjbar, S.F.; Matloobi, M. Simulation of thermal performance of solar greenhouse in north-west of Iran: An experimental validation. *Renew. Energy* **2019**, *135*, 88–97. [CrossRef]
40. Taki, M.; Ajabshirchi, Y.; Ranjbar, S.F.; Rohani, A.; Matloobi, M. Modeling and experimental validation of heat transfer and energy consumption in an innovative greenhouse structure. *Inf. Process. Agric.* **2016**, *3*, 157–174. [CrossRef]
41. Baptista, F.J. Modelling the Climate in Unheated Tomato Greenhouses and Predicting Botrytis Cinerea Infection. Ph.D. Thesis, Evora University, Evora, Portugal, 2007.
42. Canada Energy Regulator Energy Conversion Tables. Available online: https://apps.cer-rec.gc.ca/Conversion/conversion-tables.aspx?GoCTemplateCulture=en-CA#2-3 (accessed on 5 May 2018).

Article

Development and Validation of Air-to-Water Heat Pump Model for Greenhouse Heating

Adnan Rasheed [1], Wook Ho Na [1], Jong Won Lee [2], Hyeon Tae Kim [3] and Hyun Woo Lee [1,4,*]

[1]   Smart Agriculture Innovation Center, Kyungpook National University, Daegu 41566, Korea; adnanrasheed@knu.ac.kr (A.R.); wooks121@hanmail.net (W.H.N.)
[2]   Department of Horticulture Environment System, Korea National College of Agriculture and Fisheries, 1515, Kongjwipatjwi-ro, Deokjin-gu, Jeonju-si 54874, Jeollabuk-do, Korea; leewon1@korea.kr
[3]   Department of Bio-Industrial Machinery Engineering, Institute of Agricultural and Life Sciences, Gyeongsang National University, Jinju 660-701, Korea; bioani@gnu.ac.kr
[4]   Department of Agricultural Engineering, Kyungpook National University, Daegu 702-701, Korea
*   Correspondence: whlee@knu.ac.kr; Tel.: +82-53-950-5736

Abstract: This study proposes a building energy simulation (BES) model of an air-to-water heat pump (AWHP) system integrated with a multi-span greenhouse using the TRNSYS-18 program. The proposed BES model was validated using an experimental AWHP and a multi-span greenhouse installed in Kyungpook National University, Daegu, South Korea (latitude 35.53° N, longitude 128.36° E, elevation 48 m). Three AWHPs and a water storage tank were used to fulfill the heat energy requirement of the three-span greenhouse with 391.6 m^2 of floor area. The model was validated by comparing the following experimental and simulated results, namely, the internal greenhouse temperature, the heating load of the greenhouse, heat supply from the water storage tank to the greenhouse, heat pumps' output water temperature, power used by the heat pumps, coefficient of performance (COP) of the heat pump, and water storage tank temperature. The BES model's performance was evaluated by calculating the root mean square error (RMSE) and the Nash–Sutcliffe efficiency (NSE) coefficient of validation results. The overall results correlated well with the experimental and simulated results and encouraged adopting the BES model. The average calculated COP of the AWHP was 2.2 when the outside temperature was as low as −13 °C. The proposed model was designed simply, and detailed information of each step is provided to make it easy to use for engineers, researchers, and consultants.

Keywords: greenhouse energy modeling; renewable energy; energy-saving screen; greenhouse microclimate control

Citation: Rasheed, A.; Na, W.H.; Lee, W.; Kim, H.T.; Lee, H.W. Development and Validation of Air-to-Water Heat Pump Model for Greenhouse Heating. *Energies* **2021**, *14*, 4714. https://doi.org/10.3390/en14154714

Academic Editor: Muhammad Sultan

Received: 19 July 2021
Accepted: 30 July 2021
Published: 3 August 2021

Publisher's Note: MDPI stays neutral with regard to jurisdictional claims in published maps and institutional affiliations.

1. Introduction

Recently, greenhouse farming has increased rapidly in many countries, including South Korea. The primary objective of greenhouse farming is to obtain a year-round and high crop production in areas with severe climatic conditions unfavorable for crop production, where farming is possible only by maintaining the optimum greenhouse microclimate throughout the production period. Different heating and cooling systems are used to provide a favorable environment to crops inside the greenhouse. Therefore, fossil fuels are being used for heating and cooling, which not only increases production costs [1] but also causes CO_2 emissions and environmental pollution [2]. In South Korea, greenhouse heating costs have increased to 45% of the total production cost [3] because of continuous oil price increases. To reduce CO_2 emissions and cope with the continuously increasing oil price, the South Korean government has promoted the use of new and renewable energy (NRE) sources for different purposes, including in the agriculture sector. The NRE law was executed in 2004, enforcing the installation of the NRE systems [4].

Heat pumps are widely used in commercial as well as residential buildings [5], air-source heat pumps (ASHPs) are the most widespread heating source in commercial and residential building due to the large availability of the external heat source and relatively low investment cost [6]. As in residential and commercial buildings, heat pumps are also being utilized in agricultural greenhouses worldwide. Many greenhouse growers have reported that using heat pumps instead of conventional heating or cooling systems reduces 80% of the fuel cost and 5–8% of total production cost [7]. Researchers have evaluated many configurations to use heat pumps for greenhouse heating under different environmental conditions [7–11]. Two types of heat pumps are used to provide heating and cooling for agricultural greenhouses, namely, ground-source water and air-source heat pumps. Ground-source heat pumps are more efficient than air-source heat pumps [12]. There are listed some studies, who investigated different heat pump systems for greenhouse heating. Seung et al. developed a ground-source heat pump to maintain the greenhouse internal setpoint temperature during winter and reported a 3.25 coefficient of performance (COP) of the heat pump in heating mode [13]. Kim et al. used thermal effluent from a power plant and applied it to a heat pump system to fulfil the heat energy requirement of the greenhouse. Furthermore, the performance of the proposed heat pump system was investigated by comparing the results with those of a conventional boiler [14]. Yildirim et al. conducted an experimental study to evaluate the ground-source heat pump (GSHP) system assisted with solar photovoltaic panels. The study considered the monthly and annual cooling and heating demand of the greenhouse, and economic analyses and payback period were also considered [15]. Boughanmi et al. studied the performance of a new conic helicoidal geothermal heat exchanger with GSHP for greenhouse heating. The analysis of the study focusses on the COP of the system [16]. Hassanien et al., in a recent study, investigated the performance of an evacuated tube solar collector as a solar water heater assisted by an electric heat pump for greenhouse heating. The analyses of this study considered the thermal efficiency of the system and payback period [17].

Many other studies have applied and evaluated ground-source heat pumps for agricultural greenhouse heating [12,18–21]. Because of low cost and ease of installation, the use of air-source heat pumps is increasing rapidly worldwide, and they show great potential for agricultural greenhouse heating [9]. Lu et al. [22] used the TRNSYS program to predict the AWHP system's performance for heating greenhouses in Melbourne, Australia. The study compared the cost of the system with that of a liquid petroleum gas (LPG) heating system and reported a six-year payback period and a 16% reduction in LPG consumption. Another study validated the TRNSYS model using experimentally obtained results for a high-temperature AWHP and thermal energy storage for a residential building [23]. The results correlated well, with a root mean square error (RMSE) of 4.14%. In a study, Moon et al. investigated the AWHP with 7.1 kW of heating capacity, along with the storage tank and heating pipes [24]. The study conducted an experimental investigation of just different heating treatments, including, growing part heating, space heating, and growing part and space heating with different temperature level controls to a small single-span greenhouse. In a recent study, Lim et al. [25] analyzed the heating performance of the air-to-water heat pump (AWHP) for greenhouses. The study focused on the COP and economic analysis of the system. The results showed an average COP of 4.5 and 70% heat energy cost reduction compared with the conventional air heater. There are many AWHP systems with different capacities available in the market. There are also many studies both simulated and experimentally conducted for the evaluation of AWHPs for greenhouse heating and cooling. The results showed the different COPs of the particular AWHPs under specific weather conditions. According to best of our knowledge, previously conducted studies were from specific points of view, and the use of a simulation model for AWHP for the specific purpose of greenhouse heating is lacking. The heat energy demand of the greenhouse and the performance of AWHP may differ depending on the different climatic conditions and the greenhouse construction and control [26]. Ensuring the accurate energy performance of the system is generally a difficult task for an HVAC designer. There is a need to develop a model that is capable of evaluation of a specific AWHP systems' efficiency with desired greenhouse construction design, including shape covering materials, screen materials, and environmental control inside the greenhouse under local weather conditions.

The TRNSYS program is an extremely flexible graphically based environment used to simulate the behavior of transient systems. It focuses on assessing the performance of thermal energy system simulations, including building energy simulation (BES), solar thermal processes, solar applications, geothermal energy, heat pump systems, ground-coupled heat transfer, airflow modeling, wind and photovoltaic (PV) systems, power plants, and energy system calibration. The University of Wisconsin's Solar Energy Lab developed TRNSYS, which has been commercially available since 1975 for simulating thermal systems. However, it has since undergone continuous development to become a hybrid simulator [27]. It is a component-based program simulating complex energy flows in buildings. TRNSYS can be easily connected to other programs for pre- and postprocessing the model. Over the last two decades, TRNSYS has been widely used in industry and research as a reliable tool for BES [28]. Moreover, for the modeling and simulation of agricultural greenhouses, the TRNSYS program shows high flexibility, as many greenhouse models have been developed and validated [1,29–34].

This study proposes a BES model of an AWHP system integrated with a multi-span greenhouse using the TRNSYS-18 program. The proposed model results for the internal greenhouse temperature, heat energy demand, storage tank temperature, and temperature supply and return from the heat pump and greenhouse, were validated using experimental data on the heating mode of the AWHP. Furthermore, the feasibility of the AWHP to fulfill the heat energy demand of the greenhouse was investigated. The proposed model considered all time-varying control factors of the greenhouse, including thermal screen control, ventilation control, internal temperature control, and temperature control on/off heat pump system. All physical factors of the greenhouse, including the design, covering, and screen materials and the specification of the heat pumps, storage tank, and water circulation pumps, were also considered. The study considered dynamic simulation of specific AWHP and multi-span greenhouse with fully controlled conditions under local weather conditions of Deagu, South Korea. The proposed BES model of the AWHP integrated with multi-span greenhouse was designed simply, and detailed information on each step is provided to make it easy to use for engineers, researchers, and consultants. Researcher can use this model for the dynamic thermal simulations of their specific greenhouse design and control requirements according to the local weather conditions. Moreover, AWHP analysis can help to find the feasible solution to increase the COP.

2. Materials and Methods

2.1. Experimental Greenhouse

The experimental greenhouse was a three-span rectangular, north–south (N–S) oriented, Venlo-roofed greenhouse, in which, the roof was covered with horticulture glass (HG, 4 mm) and the side walls were covered with polycarbonate (PC, 16 mm). The experimental greenhouse is in Kyungpook National University, Daegu, South Korea (latitude 35.53° N, longitude 128.36° E, elevation 48 m) (Figure 1). The total floor area of the experimental greenhouse was 391.2 m^2 with dimensions of 24 m × 16.3 m × 7.6 m. The experimental greenhouse was further divided into three equal parts to create different climatic conditions for different experiments. Each part's dimensions were 8 m × 16.3 m × 7.6 m, with a floor area of 130.4 m^2 each. HG 4 mm material was used in the division. Figure 2 shows the locations of the sensors and dimensions of the greenhouse. Weather data were recorded inside and outside the greenhouse from 1 January 2021 to 31 March 2021, during the heating period. The weather data recorded inside the greenhouse were air temperature, relative humidity, and solar radiation to validate the BES model. The weather data recorded outside the greenhouse were air temperature, relative humidity, solar radiation, and wind speed and direction. The ambient pressure data were obtained from the Korean Meteorological Administration (KMA). The outside weather data were obtained to use as input for the

BES model. Table 1 shows all recorded weather variables and their characteristics. Figure 3 shows the mean outside air temperature and solar radiation.

Figure 1. Experimental greenhouse at Daegu (latitude 35.53° N, longitude 128.36° E).

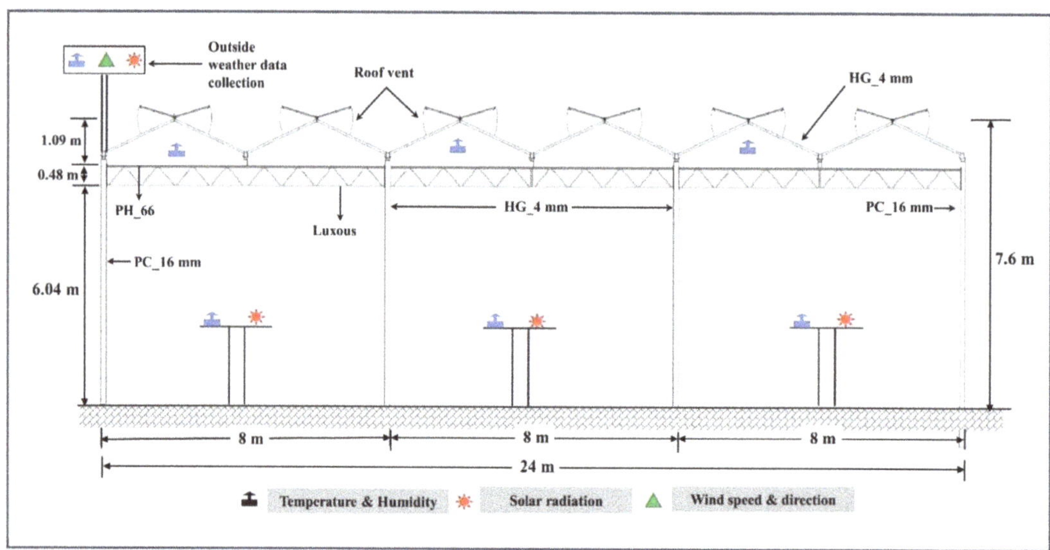

Figure 2. Experimental greenhouse's dimensions and location of the sensors.

Table 1. Characteristics of the recorded weather data.

Weather Parameter	Unit	Time Interval	Sensor	Accuracy of Sensors
Air temperature	°C	10 min	MTV Active, Ridder	±1%
Relative humidity	%	10 min	MTV Active, Ridder	±2%
Solar radiation	$W \cdot m^{-2}$	10 min	SR05-D2A2-TMBL, Hukseflux	IEC 61724-1:2017 standard, Class C, Basic
Wind speed	$m \cdot s^{-1}$	10 min	Clima Sensor US, Thies Clima	±5%
Wind direction	degrees	10 min	Clima Sensor US, Thies Clima	±5% of measured value
Water temperature	°C	10 min	HortiMax Omni, Ridder	±0.5%
Water flow rate	Liter	10 min	FS-WLH 40, FLSTRONIC	±1% of measured value
Ambient pressure	hPa	10 min	PTB-220TS, VAISALA	±5% hPa

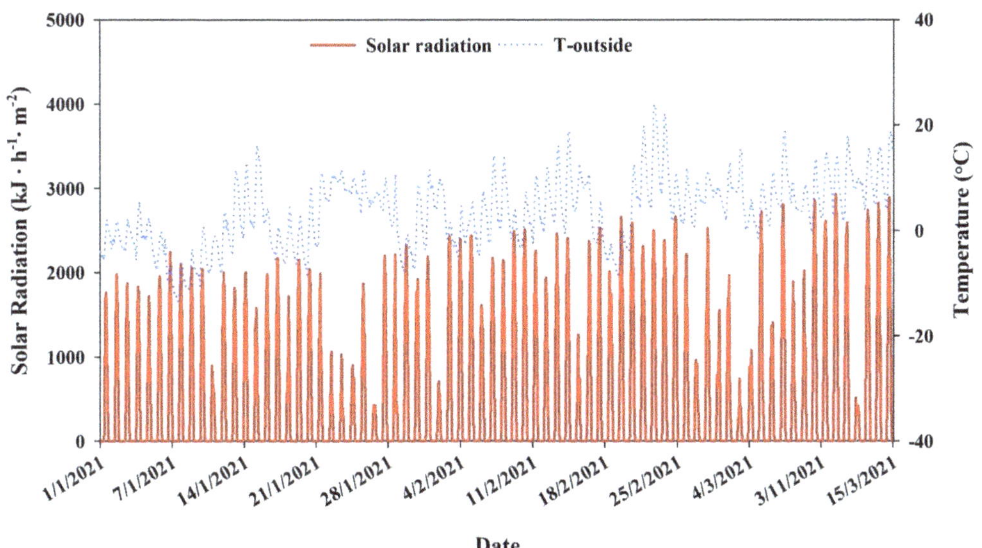

Figure 3. Ambient solar radiation and air temperature.

2.2. AWHP

Three AWHP units with a water storage tank were analyzed for their heating performance to fulfill the heat energy demand of the three-span greenhouse (detailed above). Figure 4a,b shows the experimental AWHP and the water storage tank, respectively, installed in Kyungpook National University, Daegu, South Korea. Figure 5 shows a schematic of the entire process of the AWHP in heating mode and the location of the sensors (Table 1 details the characteristics). The water storage tank stores hot water from the AWHP and supplies it to the greenhouse when heating had to reach the setpoint internal air temperature. Inside the greenhouse, heating pipes and two fan coil systems were installed to exchange heat in the greenhouse (GH3), and two fan coil systems each were used in the other two compartments (GH1 and GH2). Table 2 details the specifications of the AWHP system. We monitored the water flow rate and temperature and various locations mentioned in Figure 5 to calculate the heat energy supply from the AWHP to the water storage tank, and from the water storage tank to the greenhouse using Equation (1)

$$Q = \dot{m} \times cp \times \Delta T \tag{1}$$

where Q is the amount of heat transfer or heating capacity of the AWHP (kJ), ṁ is the mass flow rate (kg·s^{-1}), cp is the specific heat capacity of water (kJ·kg^{-1}·°C^{-1}), and ΔT is the temperature change (°C). Further, the COP of the heat pump was calculated using Equation (2)

$$CP = \frac{Q}{P_{HP}}$$

(2)

where P$_{HP}$ is the power usage of AWHP in kJ.

Figure 4. Field picture of (**a**) AWHP and (**b**) water storage tank.

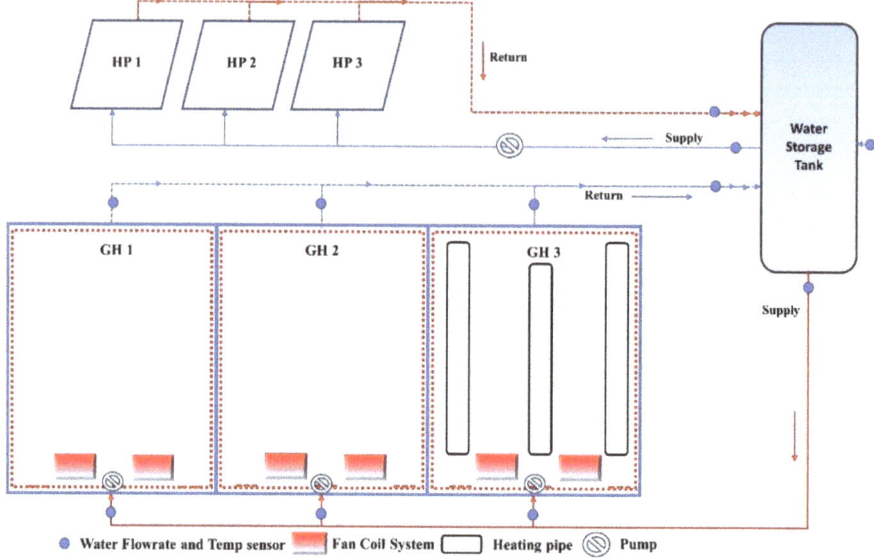

Figure 5. Schematic diagram of AWHP and location of the sensors.

Table 2. Specification of the AWHP.

Component	Properties	Specification
Heat pump	Model	PSET-C60W (MIDEA)
	Heating capacity	70 kW
	Power consumption	21.9
	Voltage	380 V–415 V, 3-phase, 60 Hz
	Refrigerant	R-410a
Heat storage tank	Heat storage fluid	Water
	Capacity	50 m^3
Water circulation pump	Model	Wilo TOP-S 40/7
	Max. fluid temperature	130 °C
	Max. fluid temperature	−20 °C
	Power consumption	390 W
Fan coil unit	Model	IN-FCG0016-L
	Heating capacity	27,000 W
	Airflow rate	83 m^3·m^{-1}

2.3. BES Modeling and Simulation

Designing the proposed BES model was divided into two steps. First, the greenhouse model was developed, and secondly, the heat pump system model was designed. The BES model of the AWHP system integrated, with the multi-span greenhouse with all physical and technical parameters the same as the experimental setup, was designed using the TRNSYS-18 program. Figure 6 shows the simulation studio (the main interface of the TRNSYS program) connecting all model components. Table 3 shows all components (types) and their complete descriptions, as used in the simulation studio of the TRNSYS program, to design the proposed model. First, the 3D model of the multi-span experimental greenhouse was designed in Transys3d, an add-on of Google SketchUp^{TM}, and imported as a .idf file (readable by TRNSYS) into the TRNBuild (a building interface of the TRNSYS program, TYPE 56). The greenhouse covering and screen material properties were introduced into the Lawrence Berkeley National Laboratory Windows 7.7 program, creating the DOE-2 file of the materials readable by TRNBuild. Table 4 shows the covering material and screens material's properties, while in Table 5, steel (greenhouse structure) and ground properties used in the simulation are shown. TRNBuild managed the thermal model of the building to account for natural ventilation into the greenhouse thermal model TRNSFLOW (a ventilation module of the TRNSYS program), coupling the airflow network with the thermal model to simulate the effect of natural ventilation on the greenhouse's thermal environment. Furthermore, in the simulation studio, the weather data and weather data processors were linked to the TRNBuild to simulate the effect of an ambient environment. Table 6 shows the details of the referenced greenhouse, including physical and control strategies used in the BES model simulation.

After modeling the greenhouse, the AWHP system model was prepared. Three Type 941 from the TRNSYS Tess library were used to model the AWHP. This type uses performance data files for heating and cooling provided by the manufacturer and ambient air temperature and relative humidity as input. The heating performance data file used in the simulations, and datasheet of the heat pump provided by the manufacturer can be found in Appendix A. Table 2 shows the rated heating capacity and power consumption values of the heat pump. The heat pumps are controlled with an on/off signal, as they would be in a real system. In Type 649, a water-mixing valve is used to collect hot water from all three heat pumps and deliver it to the water storage tank (Type 4). Type 114 (circulation pump) delivers cold water to the heat pump at a constant speed. From the storage tank to the fan coil unit (Type 928), hot water is provided using Type 3 (variable speed circulation pump) with a Type 22 proportional integral derivative controller to control the mass flow rate with feedback control. The fan coil unit provides hot air to the greenhouse. Moreover, Type 709 (pipe) is used inside the greenhouse to provide heat. Hot water runs into the pipes, and heat loss from the pipes provides heat inside the greenhouse. Figure 7 shows the flow chart diagram of the proposed model, explaining the detail information of the pre-processing, simulation and output.

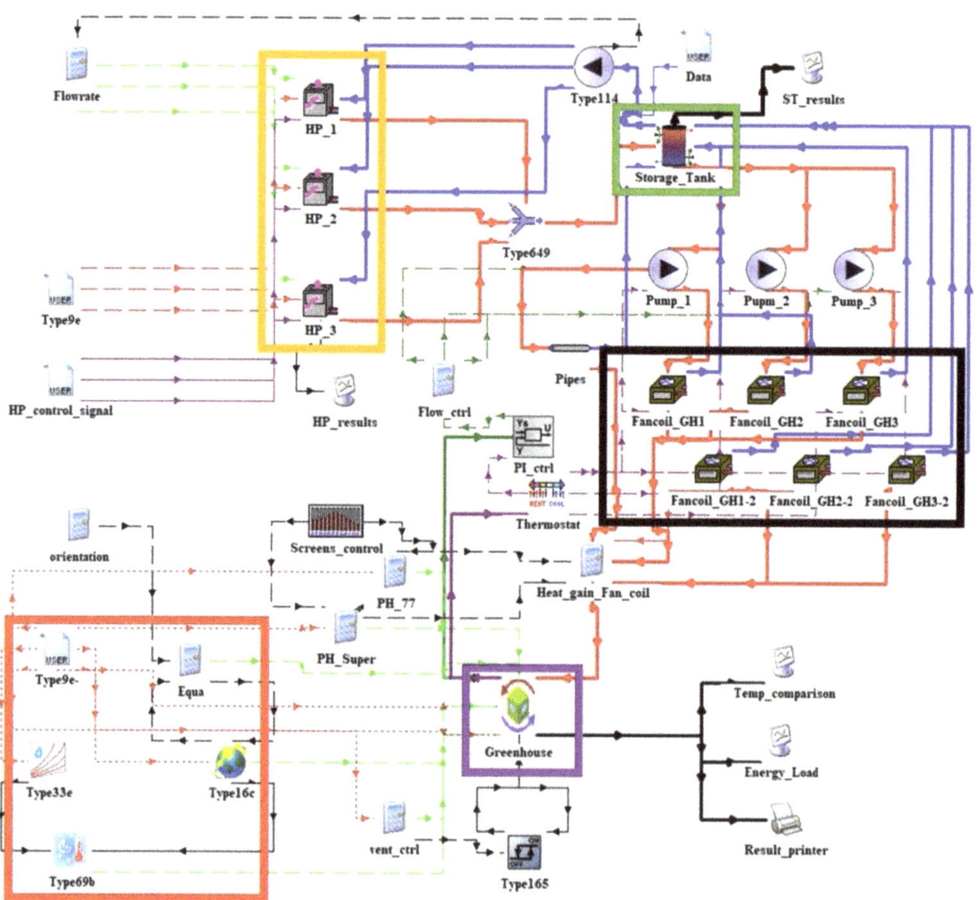

Figure 6. TRNSYS simulation studio showing proposed model's components for air-to-water heat pump system integrated with multi-span greenhouse. Yellow box: heat pump, green box: storage tank, black box: fan coil units, purple box: multi-span experimental greenhouse modeling, TRNBuild, red box: weather data reading and processing.

Table 3. Components of the greenhouse model in TRNSYS 18.

Component	Type	Description
Data reader	9	Reads the user-defined weather data file
Solar radiation processor	16	Uses total direct solar radiation on the horizontal surface as an input and calculates the total, beam, reflected, and diffuse radiation on all greenhouse tilt surfaces
Sky temperature calculator	69	Input: dewpoint temperature, beam, and diffuse radiation on horizontal surface to calculate sky temperature
Psychrometric chart	33	Calculates dewpoint temperature using dry bulk temperature and humidity ratio
Equation editor		Inserts equation in the model

Table 3. Cont.

Component	Type	Description
Greenhouse building model	56-TRNFlow	Uses TRNBuild and processes the thermal behavior of the greenhouse along with the natural ventilation
Air-to-water heat pump (AWHP)	941	This model is based on user-supplied data files containing catalog data for water capacity and power. It takes air-relative humidity and outside temperature as an input
Water storage tank	4	Water storage tank
Pipe	709	Models the fluid flow into the pipe; calculates the heat loss from the pipe
Fan coil unit	928	Takes hot water as an input and provides hot air to the greenhouse for temperature control
Pump	3	Variable-speed water circulation pump
Pump	114	Constant-speed water circulation pump
Valve	649	Water-mixing valve, which combines different liquid streams into a single output mass flow. Combines the output water of three heat pumps and delivers water to the storage tank
Thermostat	108	A five-stage thermostat for the on/off control function. Controls the circulation pump and fan coil unit for the greenhouse's internal temperature setpoint
Controller	165	Controls the natural ventilation of the greenhouse
Monthly forcing function	518	Inputs schedules and screen opening and closing times that change monthly
Printer	25	Prints results on user-provided external files
Plotter	65	This type was used to plot the results.

Table 4. Physical and thermal properties of the greenhouse coverings and thermal screens.

Cover Characteristics	Covering		Screens	
	HG	PC	PH_66	Luxous
Solar transmittance front	0.89	0.78	0.38	0.58
Solar transmittance back	0.89	0.78	0.38	0.57
Solar reflectance front	0.08	0.14	0.50	0.30
Solar reflectance back	0.08	0.14	0.48	0.25
Visible radiation transmittance front	0.91	0.75	0.38	0.58
Visible radiation transmittance back	0.91	0.75	0.38	0.57
Visible radiation reflectance front	0.08	0.15	0.50	0.30
Visible radiation reflectance back	0.08	0.15	0.48	0.25
Thermal radiation transmittance	0.1	0.02	0.35	0.26
Thermal radiation emission front	0.90	0.89	0.48	0.45
Thermal radiation emission back	0.90	0.89	0.55	0.42
Conductivity (W·m^{-1}·K^{-1})	0.1	0.19	0.06	0.05
Air permeability (m^2)	—	—	1.49×10^{-11}	1.33×10^{-11}
Thickness (mm)	4	16	0.24	0.25

Table 5. Opaque materials' properties.

Material	Thickness (m)	Thermal Conductivity (kJ·h^{-1}·m^{-1}·K^{-1})	Thermal Capacity (kJ·kg^{-1}·K^{-1})	Density (kJ·m^{-3})	Convective Heat Transfer Coefficient (kJ·h^{-1}·m^{-2}·K^{-1})	
					Front	Back
Steel	0.04	54	1.8	7800	11	64
Ground	0.100	0.97	0.75	2900	11	0.001

Table 6. Summary of reference three-span greenhouse.

Parameter	Operating Condition
Greenhouse type	Multi-span
Roof type	Venlo
No. of spans	3
Roof glazing	HG, 4 mm
Side glazing	PC, 16 mm
GH portion dividing glazing	HG, 4 mm
Orientation	North–South
Dimension	20.6 m × 16.3 m × 7.6 m
Floor area	391.2 m^2
Volume	2362.8 m^3
Period	1 January 2021 to 28 February 2021
Natural ventilation	roof vents
Natural vents control set point temp	26 °C
Energy screen position	Roof only
Energy screens (1 and 2)	PH-66, Luxous
Thermal screens' control	Ph-66 retract (After sunrise, OR S.R 100 W) Ph-66 Deploy (After sunset, AND S.R 100 W) Luxous_1 retract (After sunrise, OR S.R 150 W) Luxous_1 deploy (After sunset, AND S.R 150 W)
Heating setpoints, GH portion (1, 2, 3)	16, 18, and 17 °C

2.4. Statistical Analysis of BES Model

Statistical analyses were performed to predict the BES model's performance using the Nash–Sutcliffe efficiency (NSE) coefficient and compare the experimentally measured data with the BES model's output. This coefficient quantitatively describes the accuracy of the model results, it indicates how well the plot of observed versus simulated data fits the 1:1 ratio. Its value ranges from $-\infty$ to 1, and values closer to 1 indicate better predictive power of the model. The NSE is mathematically expressed using Equation (3). The performance ratings for NSE values are as follows: NSE > 0.9 = very good, 0.8–0.9 = good 0.65–0.80 = acceptable, and <0.6 = unsatisfactory [35]. Furthermore, Equation (4) for the RMSE was used to quantify the error for units of the variables. The equations are mathematically defined as follows:

$$\text{NSE} = 1 - \left[\frac{\sum_{i=0}^{n} \left(T_i^{exp} - T_i^{sim} \right)^2}{\sum_{i=0}^{n} \left(T_i^{exp} - T_i^{mean} \right)^2} \right] \tag{3}$$

$$\text{RMSE} = \sqrt{ \frac{\sum_{i=0}^{n} \left(T_i^{exp} - T_i^{sim} \right)^2}{n} } \tag{4}$$

where T_i^{exp} are the experimentally values, T_i^{sim} are the simulated values, T_i^{mean} are the mean of the experimental values, and n is the total number of observations.

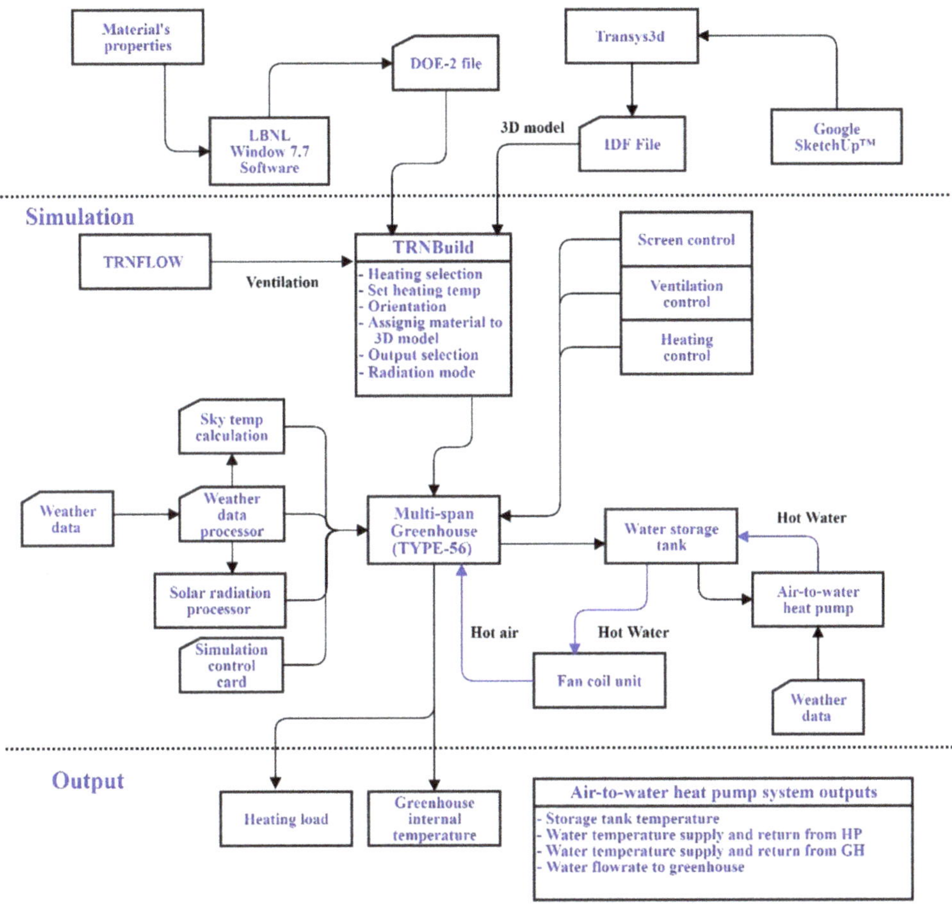

Figure 7. Flow chart of BES modeling using TRNSYS18.

3. Results and Discussion

To validate the proposed AWHP system integrated with the multi-span greenhouse BES model, the computed internal air temperature of the greenhouse, heating load, heat pump output temperature, and storage tank temperature were compared with those obtained experimentally using the same physical and operating conditions. Validation was conducted during the winter from 1 January to 31 March 2021. The normal operation of the heat pump in heating mode was repeated daily; therefore, only some days' analysis results are presented here. Figure 8 shows the greenhouse's internal experimental and simulated temperature along with the ambient temperature from 1–21 January 2021. The validation results are for all three compartments of the greenhouse, where the heating setpoint was different for each compartment. The heating setpoints were 16, 18, and 17 °C for compartments 1, 2, and 3, respectively, whereas the ventilation setpoint was 26 °C for all compartments. The simulated greenhouse internal temperature results correlated well with those of the experimentally measured temperatures. The RMSE values for the validation results were 1.9, 1.8, and 2.0 °C, indicating the maximum temperature difference between the predicted and experimental results. The NSE values were 0.71, 0.70, and 0.65, respectively, which are acceptable. Compartment 3 shows a slightly lower value because, for some days, the experimental temperature was not well controlled to 17 °C. The overall performance analysis results show that the proposed BES model is accurate enough to predict the internal greenhouse temperature.

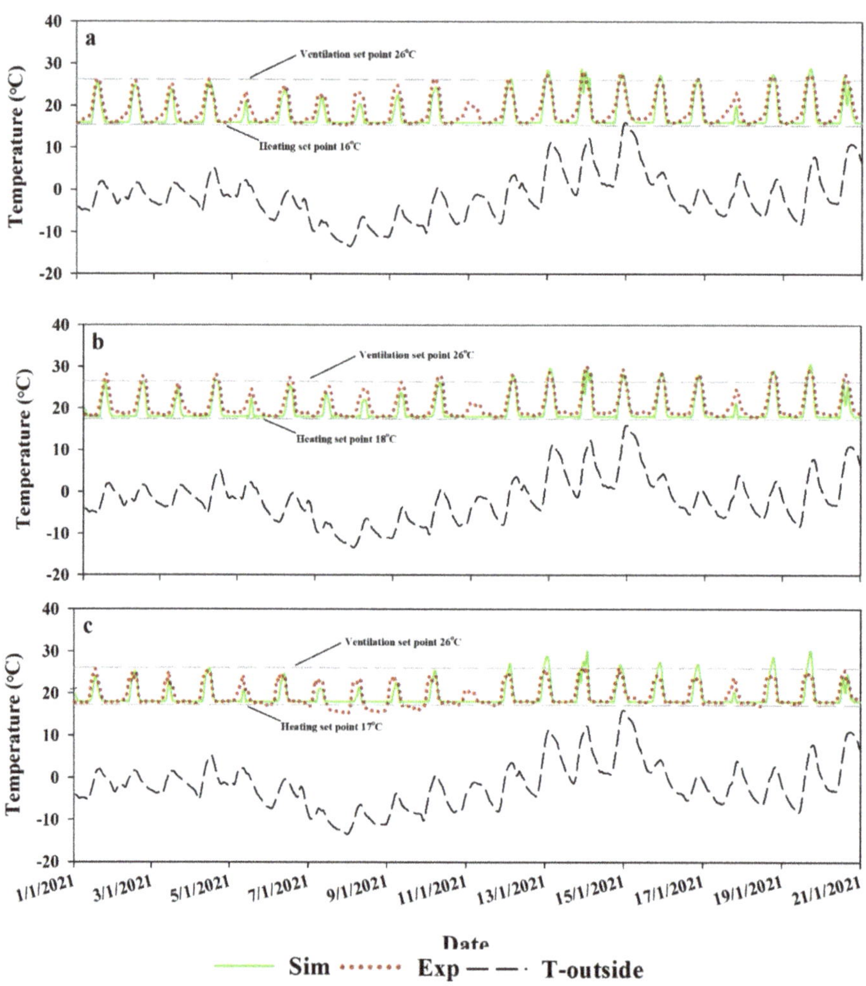

Figure 8. Experimental vs. simulated internal air temperatures (**a**) compartment 1, (**b**) compartment 2, and (**c**) compartment 3.

Figure 9a–c shows the results for experimental and simulated heating loads of all three greenhouse compartments from 1–10 January 2021, while considering the different heating setpoints for each. The maximum heating load of compartments 1, 2, and 3 were 30,000, 25,300, and 26,400 Kcal·h^{-1} on 8 January 2021, when the outside temperature was the lowest for the season (−13 °C). Figure 3 shows the ambient temperature for the entire analysis period, showing 8 January 2021, with the lowest temperature. The middle compartment showed the least heating load, even when the heating setpoint was higher than that of the other two compartments because of the reduction in heat loss. Both sidewalls were adjacent to the other compartments, and less area was exposed to the ambient environment than in the other compartments. Furthermore, compartment 3 had less heating load than that of compartment 1 because its two sidewalls were exposed to the sun and received more solar heat during the day than compartment 1, of which only one sidewall was exposed to the sun.

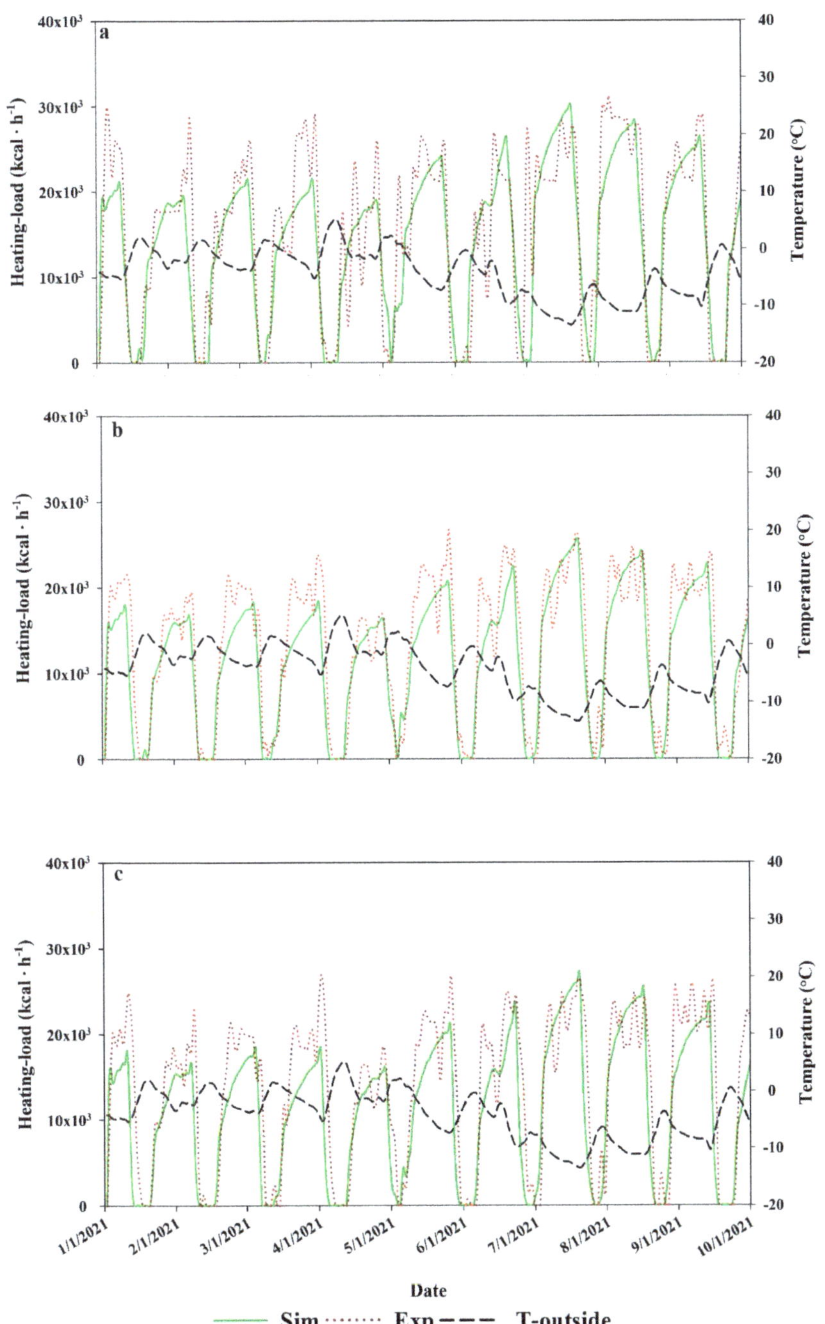

Figure 9. Experimental vs. simulated heating loads of greenhouse (**a**) compartment 1, (**b**) compartment 2, and (**c**) compartment 3.

Furthermore, the model's performance for the results shown in Figure 9 was analyzed using NSE, RMSE, and a scatter plot against a 1:1 line to visually inspect the results. Figure 10a–c shows the scatter plot for the experimental vs. simulated heating loads of the greenhouse against a 1:1 line and the NSE value. The higher performance can result in a scatter plot closer to the 1:1 line. The NSE values of 0.73, 0.81, and 0.67 for greenhouse compartments 1, 2, and 3, respectively, were acceptable. Moreover, to quantify the error for heating load units, the RMSE values of greenhouse compartments 1, 2, and 3 were 5140, 3674, and 5141 Kcal·h^{-1}. The maximum difference between the experimental and simulated results occurred on 4 January 2021, because the experimentally calculated heat load was not in a steady state, resulting in a sudden rise and fall of hot water supply to the greenhouse. Figure 9 shows that the experimentally calculated heat supply was higher on 4 January 2021, even though the outside temperature was higher than on 8 January 2021. However, the simulated results were in a steady state and showed an acceptable heating load trend with the outside temperature.

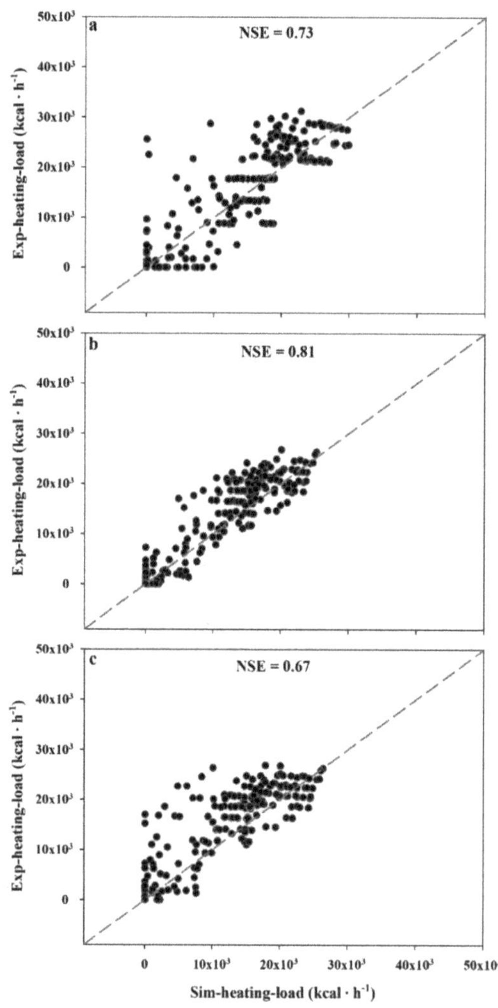

Figure 10. Statistical analysis of measured vs. computed heating loads of greenhouse (**a**) compartment 1, (**b**) compartment 2, and (**c**) compartment 3.

Figure 11 shows the results for the experimentally calculated vs. simulated heating supplied from the water storage tank to the greenhouse along with ambient air temperature from 1 January to 15 March 2021. The experimental values were calculated per unit area of the greenhouse using Equation (1) (Section 2.2) using the measured water flow rate and temperature difference between the supply and return water temperature. The results indicate that the maximum heating supplied was on 8 January 2021, as the ambient temperature was at its lowest value of $-13\ ^\circ$C. The results shown in Table 7 are the heating demand of the experimental greenhouse estimated by TRNSYS without using a particular heating system, and they confirm the maximum supplied load under the same weather conditions. The experimental results show high energy supplied to the greenhouse from 1–12 February 2021, even when the ambient temperature was high, which only occurred for one hour because of the unsteady water flow, whereas simulated results showed linear interpolated results. The overall trend and results correlated well with the validation results. Figure 12 shows that the simulated value is close enough to the experimental values, as the scattered values follow the 1:1 line. Moreover, the statistical analysis of the results shows an acceptable NSE value of 0.70 and RMSE value of 20 Kcal$\cdot$h$^{-1}\cdot$m^{-2}.

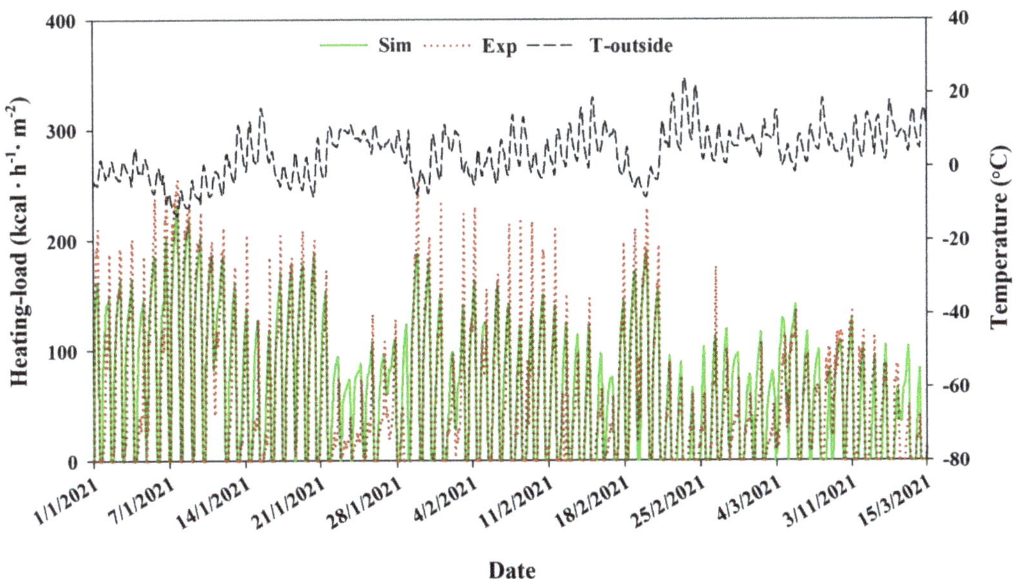

Figure 11. Experimental vs. simulated heating supply from water storage tank to greenhouse.

Table 7. TRNSYS estimated maximum heating load of the greenhouse.

Lowest Outside Air Temp (°C)	Greenhouse Setpoint Temp (°C)	Greenhouse Heating Area (m^2)	Max. Heating Load (kcal$\cdot$h^{-1})	Max. Heating Load per Unit Area (kcal$\cdot$h$^{-1}\cdot$m^{-2})
-13	18	391.2	97,800	250

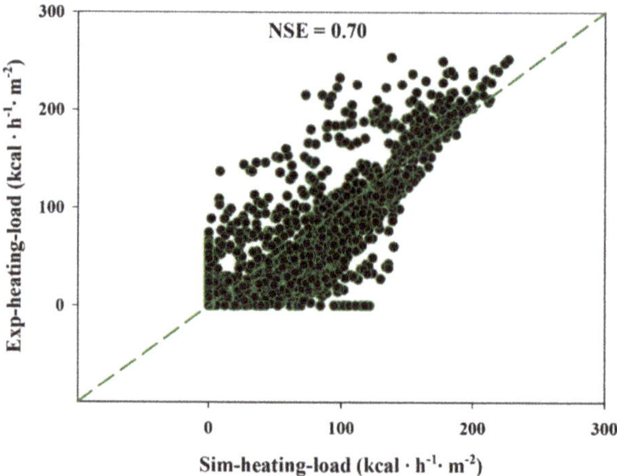

Figure 12. Statistical analysis of measured vs. computed heating supplied from water storage tank to greenhouse.

After validating the greenhouse's internal air temperature and heating load, the experiments were further extended to validate the AWHP results. The output (hot water temperature) of the experimental heat pumps was compared with the simulated hot water temperature. Because the heat pump's function was the same throughout the season, two days' results were compared. The simulated heat pump used ambient air temperature and relative humidity as an input, and a 10 min time step, the same as the time of the data logger, was used for the simulations. The results correlated well with the experimental and simulated results (Figure 13), with a small RMSE of 0.4 °C. Three heat pumps were used in this study to heat the water and store it in the water storage tank. The storage tank temperature controlled the heat pumps' ON/OFF setting. Figure 14a–c shows the seven-day data of the experimental and simulated heat pumps, 1, 2, and 3, respectively, and the electrical consumption shows the switching ON/OFF of the heat pumps according to the need. Furthermore, the validation of both experimental and simulated results correlated well.

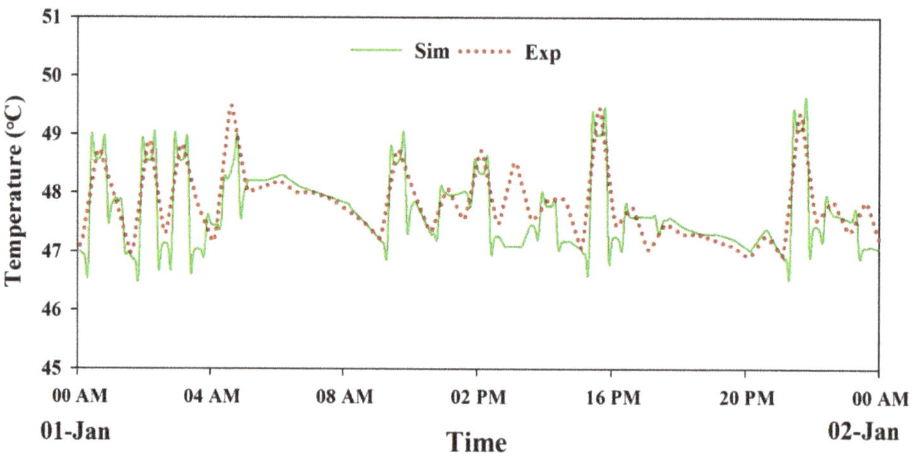

Figure 13. Heat pump output water temperature validation.

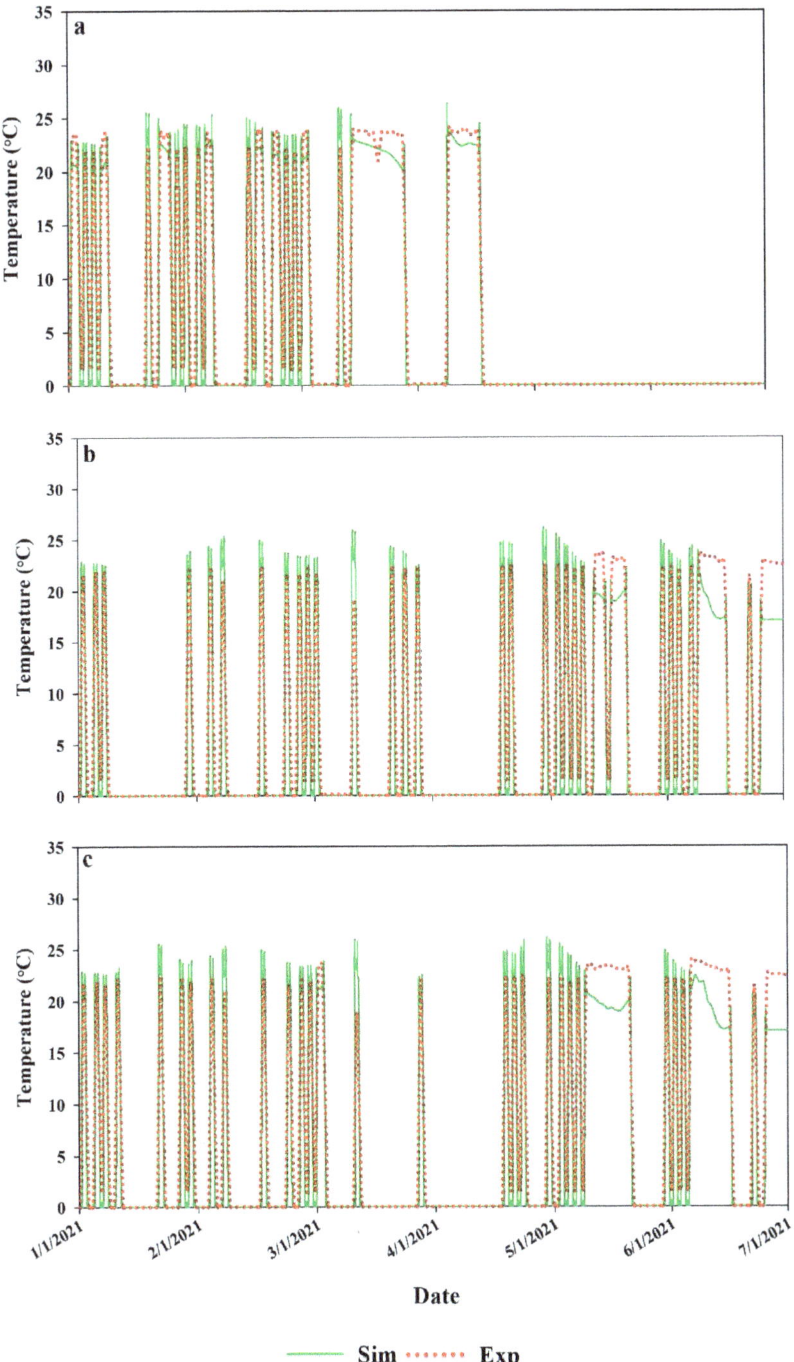

Figure 14. Experimental and simulated power usages of (**a**) heat pump 1, (**b**) heat pump 2, and (**c**) heat pump 3.

Figure 15 shows the COP of the heat pumps with greenhouse internal and ambient temperatures. The heat pumps' heating performance was evaluated during the maximum heating requirement period. The results shown are from 7–8 January 2021, when the ambient temperature was at its lowest (-13 °C). The results show that when the ambient temperature starts reducing to -13 °C from 1 to 5 a.m., the average COP of the heat pump reduced from 2.0 to 1.7 in heating mode. The calculated average COP value of 2.2 shows the same value as the manufacturer's recommended COP for this type of AWHP.

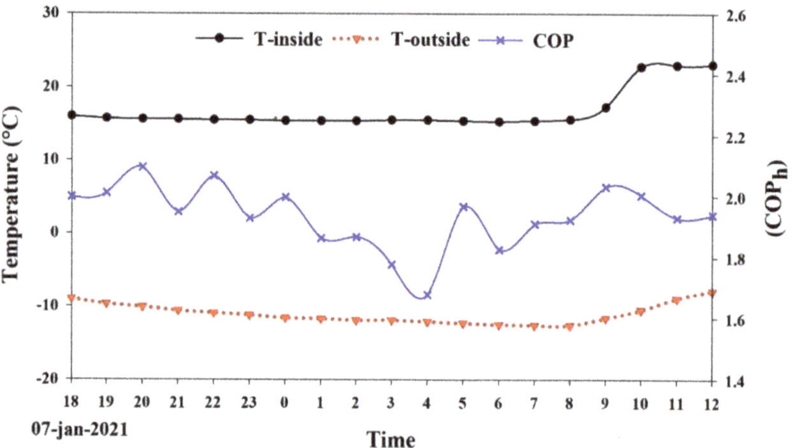

Figure 15. COP of heat pump.

Like other analyses, the normal operation of the water storage tank's charging and discharging was the same; therefore, only some days' results were shown for validation analysis (1–21 January 2021). Figure 16 shows the experimental and simulated storage tank temperatures. The validation results correlated well with an RMSE of 0.5 °C. The validation results show that the storage tank model is fair enough to be adapted. Only one water temperature sensor was installed on the top of the tank; therefore, the average water temperature results are not shown in the study.

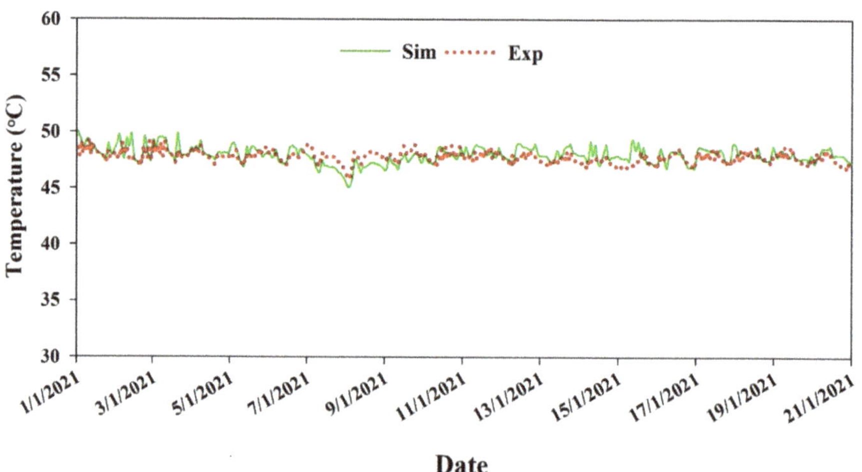

Figure 16. Water storage tank temperature validation.

4. Conclusions

This study validates the proposed BES model of an AWHP system integrated with a multi-span greenhouse using the TRNSYS-18 program. The model was validated by comparing experimental and simulated results, namely, internal air temperature of greenhouse, the heating load of the greenhouse, heat supply from the water storage tank to the greenhouse, heat pumps' output water temperature, power used by heat pumps, COP of heat pumps, and water storage tank temperature. The BES model's performance was evaluated by calculating the RMSE and NSE coefficient of validation results. The specific statistical analyses of all validation results are as follows:

- The RMSE values for the internal greenhouse air temperatures were 1.9, 1.8, and 2.0 °C, indicating the maximum temperature difference between the predicted and experimental results. The NSE values were 0.71, 0.70, and 0.65, respectively.
- The RMSE values of the energy load results for greenhouse compartments 1, 2, and 3 were 5140, 3674, and 5141 Kcal·h^{-1}, respectively, and the NSE values of greenhouse compartments 1, 2, and 3 were, 0.73, 0.81, and 0.67, respectively.
- The validation results of the energy supplied from the water storage tank to the greenhouse showed an RMSE value of 20 Kcal·h^{-1}·m^{-2} and an NSE value of 0.70.
- The heat pump output water temperature validation results showed an RMSE of 0.4 °C, and the COP of the heat pump was 2.2.
- The validation results for the water storage tank temperature show an RMSE value of 0.5 °C.
- The maximum heating energy demand for the studied greenhouse was found to be 250 kcal·h^{-1}·m^{-2}.

The overall results correlate well with experimental and simulated results and encourage adopting the BES model. The model implemented in the TRNSYS-18 program can run year-round simulations. The proposed BES model of an AWHP integrated with a multi-span greenhouse was designed simply, and detailed information on each step is provided to make it easy to use for engineers, researchers, and consultants. The presented model is developed for being used as decision-making tool for the dynamic thermal simulations of specific greenhouse design and control requirements according to the local weather conditions. All the validation results put in evidence that the proposed model is capable of evaluating such a system. Moreover, AWHP analysis can help to find a feasible solution to increase the COP. Future work will consider validating the model in the cooling mode. The proposed model can be used to optimize the control strategies and improvements to systems.

Author Contributions: Conceptualization, A.R., H.W.L., and J.W.L.; methodology, A.R. and H.W.L.; software, A.R.; validation, J.W.L. and H.T.K.; investigation, W.H.N. and A.R.; resources, H.W.L.; writing—original draft preparation, A.R.; writing—review and editing, A.R., H.T.K., W.H.N., and H.W.L.; supervision, H.W.L. All authors have read and agreed to the published version of the manuscript.

Funding: This work was supported by the Korea Institute of Planning and Evaluation for Technology in Food, Agriculture, and Forestry (IPET) through the Agricultural Energy Self-Sufficient Industrial Model Development Program, funded by the Ministry of Agriculture, Food, and Rural Affairs (MAFRA) (120096-3). This work was supported by the Korea Institute of Planning and Evaluation for Technology in Food, Agriculture, Forestry (IPET) through the Agriculture, Food, and Rural Affairs Convergence Technologies Program for Educating Creative Global Leaders, funded by the Ministry of Agriculture, Food, and Rural Affairs (MAFRA) (717001-7). This research was supported by a Basic Science Research Program through the National Research Foundation of Korea (NRF) funded by the Ministry of Education (NRF-2019R1I1A3A01051739).

Institutional Review Board Statement: Not applicable.

Informed Consent Statement: Not applicable.

Data Availability Statement: Not applicable.

Conflicts of Interest: There is no conflict of interest regarding the publication of this research.

Appendix A

The AWHP model component in the TRNSYS requires a heating performance data file as an input. Table A1 shows the heating performance data used in the study, as provided by the manufacturer. Table A2 presents the data sheet provided by the manufacturer and the testing conditions of the heat pump.

Table A1. Heating performance data of AWHP.

25	30	35	40	45	50	T_water_in
2.2	7.2	12.2	15	20		T_air_in
0.759	0.787	!Fraction capacity and power at T_air = 2.2				deg. C and T_water_in = 25
1.08	0.868	!Fraction capacity and power at T_air = 7.2				deg. C and T_water_in = 25
1.137	0.843	!Fraction capacity and power at T_air = 12.2				deg. C and T_water_in = 25
1.233	0.843	!Fraction capacity and power at T_air = 15				deg. C and T_water_in = 25
1.403	0.844	!Fraction capacity and power at T_air = 20				deg. C and T_water_in = 25
0.737	0.86	!Fraction capacity and power at T_air = 2.2				deg. C and T_water_in = 30
1.048	0.938	!Fraction capacity and power at T_air = 7.2				deg. C and T_water_in = 30
1.106	0.923	!Fraction capacity and power at T_air = 12.2				deg. C and T_water_in = 30
1.199	0.924	!Fraction capacity and power at T_air = 15				deg. C and T_water_in = 30
1.359	0.924	!Fraction capacity and power at T_air = 20				deg. C and T_water_in = 30
0.714	0.944	!Fraction capacity and power at T_air = 2.2				deg. C and T_water_in = 35
1.017	1.044	!Fraction capacity and power at T_air = 7.2				deg. C and T_water_in = 35
1.075	1.016	!Fraction capacity and power at T_air = 12.2				deg. C and T_water_in = 35
1.165	1.017	!Fraction capacity and power at T_air = 15				deg. C and T_water_in = 35
1.314	1.018	!Fraction capacity and power at T_air = 20				deg. C and T_water_in = 35
0.692	1.027	!Fraction capacity and power at T_air = 2.2				deg. C and T_water_in = 40
0.986	1.136	!Fraction capacity and power at T_air = 7.2				deg. C and T_water_in = 40
1.043	1.108	!Fraction capacity and power at T_air = 12.2				deg. C and T_water_in = 40
1.131	1.109	!Fraction capacity and power at T_air = 15				deg. C and T_water_in = 40
1.269	1.112	!Fraction capacity and power at T_air = 20				deg. C and T_water_in = 40
0.67	1.132	!Fraction capacity and power at T_air = 2.2				deg. C and T_water_in = 45
0.955	1.255	!Fraction capacity and power at T_air = 7.2				deg. C and T_water_in = 45
1.012	1.224	!Fraction capacity and power at T_air = 12.2				deg. C and T_water_in = 45
1.097	1.226	!Fraction capacity and power at T_air = 15				deg. C and T_water_in = 45
1.224	1.229	!Fraction capacity and power at T_air = 20				deg. C and T_water_in = 45
0.648	1.249	!Fraction capacity and power at T_air = 2.2				deg. C and T_water_in = 50
0.923	1.385	!Fraction capacity and power at T_air = 7.2				deg. C and T_water_in = 50
0.981	1.352	!Fraction capacity and power at T_air = 12.2				deg. C and T_water_in = 50
1.062	1.355	!Fraction capacity and power at T_air = 15				deg. C and T_water_in = 50
1.18	1.359	!Fraction capacity and power at T_air = 20				deg. C and T_water_in = 50

Table A2. Manufacturer's data sheet of the studied AWHP.

Hot Water Outlet Temperature	Ambient Temperature °C													
	−10		−6		−2		2		7		10		13	
	Capacity	Power	Capacity	Power	Capacity	Power	Capacity	Power	Capacity	Power	Capacity	Power	Capacity	Power
°C	kW	kW	kW	kW	kW	kW	kW	kW	kW	kW	kW	kW	kW	kW
40	43.51	13.70	54.39	15.57	63.98	17.30	71.09	18.81	77.28	19.80	86.55	20.98	99.53	22.66
41	42.05	13.98	52.63	15.89	61.98	17.65	68.95	19.19	75.03	20.20	83.88	21.41	96.29	23.12
42	40.83	14.27	51.17	16.21	60.34	18.01	67.19	19.58	73.20	20.61	81.69	21.85	93.61	23.60
43	39.85	14.56	50.0	16.54	59.03	18.38	65.80	19.98	71.76	21.03	79.94	22.29	91.45	24.08
44	39.08	14.86	49.09	16.88	58.03	18.76	64.76	20.39	70.70	21.46	78.62	22.75	89.78	24.57
45	38.51	15.16	48.44	17.23	57.32	19.14	64.05	20.81	70.00	21.90	77.70	23.21	88.58	25.07
46	37.76	15.31	47.55	17.40	56.34	19.33	63.02	21.01	68.95	22.12	76.40	23.45	86.94	25.32
47	36.64	15.62	46.20	17.75	54.81	19.72	31.38	21.43	67.23	22.56	74.35	23.92	84.46	25.83
48	35.19	16.09	44.44	18.28	52.77	20.31	59.16	22.08	64.87	23.24	71.62	24.63	81.22	26.60
49	33.28	16.73	42.07	19.01	50.02	21.12	56.14	22.96	61.63	24.17	67.92	25.62	76.88	27.67
50	31.14	17.57	39.41	19.96	46.92	22.18	52.72	24.11	57.93	25.38	63.73	26.90	72.01	29.05

Capacity calculation standard for heating: inlet/outlet water temperature: 40 °C/45 °C, ambient temperature: 7 °C. Nominal data are shown with a gray background.

References

1. Rasheed, A.; Lee, J.; Lee, H. Development and optimization of a building energy simulation model to study the effect of greenhouse design parameters. *Energies* **2018**, *11*, 2001. [CrossRef]
2. Bartzanas, T.; Tchamitchian, M.; Kittas, C. Influence of the heating method on greenhouse microclimate and energy consumption. *Biosyst. Eng.* **2005**, *91*, 487–499. [CrossRef]
3. Yang, S.H.; Lee, C.G.; Lee, W.K.; Ashtiani, A.A.; Kim, J.Y.; Lee, S.D.; Rhee, J.Y. Heating and cooling system for utilization of surplus air thermal energy in greenhouse and its control logic. *J. Biosyst. Eng.* **2012**, *37*, 19–27. [CrossRef]
4. Lee, J.-Y. Current status of ground source heat pumps in korea. *J. Renew. Sustain. Energy Rev.* **2009**, *13*, 1560–1568. [CrossRef]
5. Dongellini, M.; Morini, G.L. On-off cycling losses of reversible air-to-water heat pump systems as a function of the unit power modulation capacity. *Energy Convers. Manag.* **2019**, *196*, 966–978. [CrossRef]
6. Zhang, Q.; Zhang, L.; Nie, J.; Li, Y. Techno-economic analysis of air source heat pump applied for space heating in northern china. *Appl. Energy* **2017**, *207*, 533–542. [CrossRef]
7. Carlini, M.; Monarca, D.; Biondi, P.; Honorati, T.; Castellucci, S. A simulation model for the exploitation of geothermal energy for a greenhouse in the viterbo province. In *Work Safety and Risk Prevention in Agro-Food and Forest Systems, Proceedings of the International Conference Ragusa SHWA2010, Ragusa Ibla, Italy, 16–18 September 2010*; ElleDue Editore: Ragusa Ibla, Italy, 2010; pp. 16–18.
8. Kozai, T. Thermal performance of an oil engine driven heat pump for greenhouse heating. *J. Agric. Eng. Res.* **1986**, *35*, 25–37. [CrossRef]
9. Tong, Y.; Kozai, T.; Nishioka, N.; Ohyama, K. Greenhouse heating using heat pumps with a high coefficient of performance (cop). *Biosyst. Eng.* **2010**, *106*, 405–411. [CrossRef]
10. Chargui, R.; Sammouda, H.; Farhat, A. Geothermal heat pump in heating mode: Modeling and simulation on trnsys. *Int. J. Refrig* **2012**, *35*, 1824–1832. [CrossRef]
11. Rasheed, A.; Lee, J.W.; Lee, H.W. A review of greenhouse energy management by using building energy simulation. *Prot. Hortic. Plant. Fact.* **2015**, *24*, 317–325. [CrossRef]
12. Benli, H.; Durmuş, A. Evaluation of ground-source heat pump combined latent heat storage system performance in greenhouse heating. *Energy Build.* **2009**, *41*, 220–228. [CrossRef]
13. Yang, S.-H.; Lee, S.-D.; Kim, Y.J.R. Greenhouse heating and cooling with a heat pump system using surplus air and underground water thermal energy. *Eng. Agric. Environ. Food* **2013**, *6*, 86–91. [CrossRef]
14. Kim, M.-H.; Lee, D.-W.; Yun, R.; Heo, J. Operational energy saving potential of thermal effluent source heat pump system for greenhouse heating in jeju. *Int. J. Air Cond. Refrig.* **2017**, *25*, 01–12. [CrossRef]
15. Yildirim, N.; Bilir, L. Evaluation of a hybrid system for a nearly zero energy greenhouse. *Energy Convers. Manag.* **2017**, *148*, 1278–1290. [CrossRef]
16. Boughanmi, H.; Lazaar, M.; Guizani, A. A performance of a heat pump system connected a new conic helicoidal geothermal heat exchanger for a greenhouse heating in the north of tunisia. *Sol. Energy* **2018**, *171*, 343–353. [CrossRef]
17. Hassanien, R.H.E.; Li, M.; Tang, Y. The evacuated tube solar collector assisted heat pump for heating greenhouses. *Energy Build.* **2018**, *169*, 305–318. [CrossRef]
18. Ozgener, O.; Hepbasli, A. Experimental investigation of the performance of a solar-assisted ground-source heat pump system for greenhouse heating. *Int. J. Energy Res.* **2005**, *29*, 217–231. [CrossRef]
19. Ozgener, O.; Hepbasli, A. Experimental performance analysis of a solar assisted ground-source heat pump greenhouse heating system. *Energy Build.* **2005**, *37*, 101–110. [CrossRef]
20. Ozgener, O.; Hepbasli, A. A parametrical study on the energetic and exergetic assessment of a solar-assisted vertical ground-source heat pump system used for heating a greenhouse. *Build. Environ.* **2007**, *42*, 11–24. [CrossRef]
21. Ozgener, O. Use of solar assisted geothermal heat pump and small wind turbine systems for heating agricultural and residential buildings. *J. Energy* **2010**, *35*, 262–268. [CrossRef]
22. Aye, L.; Fuller, R.J.; Canal, A. Evaluation of a heat pump system for greenhouse heating. *Int. J. Therm. Sci.* **2010**, *49*, 202–208. [CrossRef]
23. Le, K.; Shah, N.; Huang, M.; Hewitt, N. High temperature air-water heat pump and energy storage: Validation of trnsys models. In Proceedings of the World Congress on Engineering and Computer Science WCECS 2017, San Francisco, CA, USA, 25–27 October 2017; Volume II.
24. Moon, J.; Kwon, J.; Kim, S.; Kang, Y.; Park, S.; Lee, J. Effect on yield increase and energy saving in partial heating for high-bed strawberries by using an air to water heat pump. In Proceedings of the 2019 ASABE Annual International Meeting, Boston, MA, USA, 7–10 July 2019; American Society of Agricultural and Biological Engineers: Boston, MA, USA, 2019; p. 1.
25. Lim, T.; Baik, Y.-K.; Kim, D.D. Heating performance analysis of an air-to-water heat pump using underground air for greenhouse farming. *Energies* **2020**, *13*, 3863. [CrossRef]
26. Nemś, A., Nemś, M.; Świder, K. Analysis of the possibilities of using a heat pump for greenhouse heating in polish climatic conditions—A case study. *Sustainability* **2018**, *10*, 3483. [CrossRef]
27. Klein, S.A. Trnsys, a transient system simulation program. In *Solar Energy Laborataory*; University of Wisconsin: Madison, WI, USA, 2012.

28. Casetta, D. Implementation and Validation of a Ground Source Heat Pump Model in Matlab. Master's Thesis, Chalmers University of Technology, Gothenburg, Sweden, 2012.
29. Rasheed, A.; Na, W.H.; Lee, J.W.; Kim, H.T.; Lee, H.W. Optimization of greenhouse thermal screens for maximized energy conservation. *Energies* **2019**, *12*, 3592. [CrossRef]
30. Rasheed, A.; Kwak, C.S.; Na, W.H.; Lee, J.W.; Kim, H.T.; Lee, H.W. Development of a building energy simulation model for control of multi-span greenhouse microclimate. *Energies* **2020**, *10*, 1236. [CrossRef]
31. Lee, S.-N.; Park, S.-J.; Lee, I.-B.; Ha, T.-H.; Kwon, K.-S.; Kim, R.-W.; Yeo, U.-H.; Lee, S.-Y. Design of energy model of greenhouse including plant and estimation of heating and cooling loads for a multi-span plastic-film greenhouse by building energy simulation. *Prot. Hortic. Plant. Fact.* **2016**, *25*, 123–132. [CrossRef]
32. Seo, I.-H.; Lee, I.-B.; Kwon, K.-S.; Park, S.-J.J.A.H. Bes computation for periodical energy load of greenhouse with geothermal heating system. *Acta Hortic* **2014**, *1037*, 113–118.
33. Ahamed, M.S.; Guo, H.; Tanino, K. Modeling heating demands in a chinese-style solar greenhouse using the transient building energy simulation model trnsys. *J. Build. Eng.* **2020**, *29*, 101114. [CrossRef]
34. Baglivo, C.; Mazzeo, D.; Panico, S.; Bonuso, S.; Matera, N.; Congedo, P.M.; Oliveti, G. Complete greenhouse dynamic simulation tool to assess the crop thermal well-being and energy needs. *Appl. Therm. Eng.* **2020**, *179*, 115698. [CrossRef]
35. Ritter, A.; Munoz-Carpena, R. Performance evaluation of hydrological models: Statistical significance for reducing subjectivity in goodness-of-fit assessments. *J. Hydrol.* **2013**, *480*, 33–45. [CrossRef]

 *energies*

Article

Assessing Crop Water Requirements and a Case for Renewable-Energy-Powered Pumping System for Wheat, Cotton, and Sorghum Crops in Sudan

Zafar A. Khan [1,2,*], Muhammad Imran [2], Jamal Umer [3], Saeed Ahmed [1], Ogheneruona E. Diemuodeke [4] and Amged Osman Abdelatif [5]

1 Department of Electrical Engineering, Mirpur University of Science and Technology, Mirpur 10250, Pakistan; saeed.ahmed@must.edu.pk
2 Mechanical, Biomedical and Design Engineering, College of Engineering and Physical Sciences Aston University, Birmingham B4 7ET, UK; m.imran12@aston.ac.uk
3 Department of Mechanical Engineering, University of Engineering and Technology Lahore, Lahore 54890, Pakistan; jamalumer@uet.edu.pk
4 Energy and Thermofluid Research Group, Department of Mechanical Engineering, Faculty of Engineering, University of Port Harcourt, Port Harcourt 500102, Nigeria; ogheneruona.diemuodeke@uniport.edu.ng
5 Department of Civil Engineering, Faculty of Engineering, University of Khartoum, Khartoum 51111, Sudan; Amged.Abdelatif@uofk.edu
* Correspondence: zafarakhan@ieee.org

Citation: Khan, Z.A.; Imran, M.; Umer, J.; Ahmed, S.; Diemuodeke, O.E.; Abdelatif, A.O. Assessing Crop Water Requirements and a Case for Renewable-Energy-Powered Pumping System for Wheat, Cotton, and Sorghum Crops in Sudan. *Energies* **2021**, *14*, 8133. https://doi.org/10.3390/en14238133

Academic Editor: Donato Morea

Received: 4 November 2021
Accepted: 29 November 2021
Published: 4 December 2021

Publisher's Note: MDPI stays neutral with regard to jurisdictional claims in published maps and institutional affiliations.

Abstract: Climate change is changing global weather patterns, with an increase in droughts expected to impact crop yields due to water scarcity. Crops can be provided with water via underground pumping systems to mitigate water shortages. However, the energy required to pump water tends to be expensive and hazardous to the environment. This paper explores different sites in Sudan to assess the crop water requirements as the first stage of developing renewable energy sources based on water pumping systems. The crop water requirements are calculated for different crops using the CROPWAT and CLIMWAT simulation tools from the Food and Agriculture Organization (FAO) of the United Nations. Further, the crop water requirements are translated into electrical energy requirements. Accurate calculations of the energy needed will help in developing cost-effective energy systems that can help in improving yields and reducing carbon emissions. The results suggest that the northern regions tend to have higher energy demands and that the potential for renewable energy should be explored in these regions, which are more susceptible to drought and where crops tend to be under higher stress due to adverse climate conditions.

Keywords: agriculture; CROPWAT; irrigation management; crop water requirement

1. Introduction

Agriculture is considered to be the backbone of Sudan's economy, with an approximate contribution of up to one-third of the total GDP and with an estimated 39% of workers employed in the sector [1]. Sudan was known as an Arab breadbasket in the 1970s, having high agricultural exports. However, a lack of resources and climate change exacerbated stresses for Sudanese people already struggling with poverty and environmental degradation, which impact those in rural areas more severely. The population of Sudan increased to 45 million in 2021, up from 31 million in 2008 [1]. This increasing population requires more food, while the effects of climate change are imminent, involving changing weather patterns and droughts that increase uncertainty, resulting in a high risk to food security. Moreover, the use of fossil fuels to meet the energy demands exacerbates the impacts of climate change.

According to the Köppen–Geiger climate classification map of Sudan [2], as can be seen in Figures 1 and 2, most of the northern part of the country comes under the arid or desert climate classification (BWh), with very low rainfall that mainly occurs in July and August. The southern part has a semi-arid or semi-desert (BSh) climate, with maximum rainfall from June to September [3–5]. A small southern portion of the country has a tropical savanna climate (Aw) and experiences frequent rainfall from May to October. The rainfall data were collected from the CRU TS 3.21 dataset, which contains monthly average rainfall rates from 1901 to 2012 [4]. The lack of rain in large areas renders agriculture practices unproductive, since a large area of the country is totally dependent on rain.

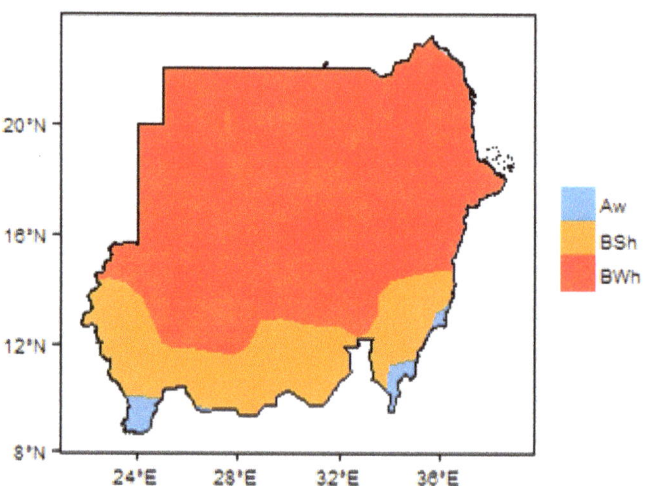

Figure 1. Köppen–Geiger climate classification map of Sudan [2].

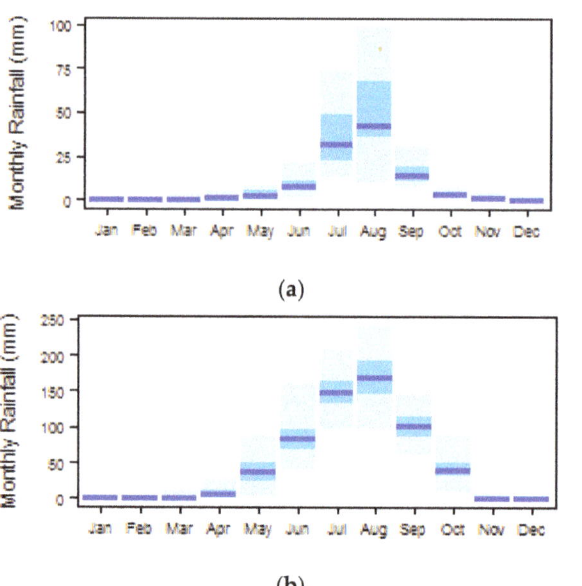

Figure 2. *Cont.*

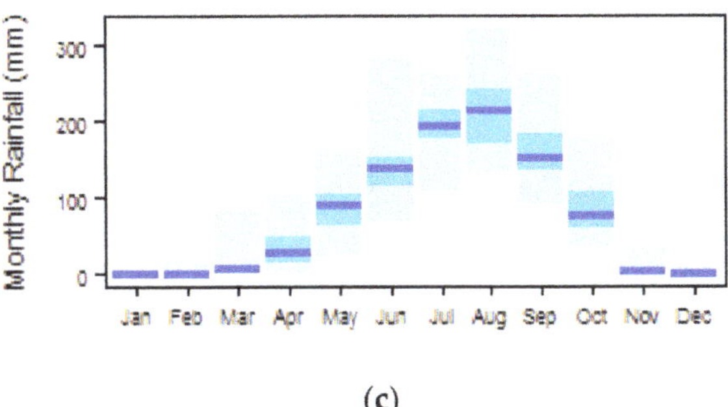

(c)

Figure 2. Average rainfall amounts in different climatic regions of Sudan [5] (*x*-axis shows month of year and *y*-axis show monthly rainfall in mm): (**a**) rainfall in arid/desert (BWh) regions; (**b**) rainfall in semi-desert (BSh) regions; (**c**) rainfall in tropical savanna (Aw) regions.

Agriculture in countries with mostly arid climates tends to struggle. Moreover, global warming and climate change are also suppressing crop yields. Droughts are expected to continue to impose immense pressure on food security in fragile ecosystems [6]. Global leaders have come together to reduce greenhouse gas emissions to control climate change, although the crops of today should be ready to endure new stresses, such as increasing droughts due to climate change, or these stresses should be managed via alternate options. To endure these stresses, the role of symbiotic microorganisms in plant phenotypic adjustments to environmental stresses has been explored in different studies [7–10]. According to the authors of [11], fungal endophytes, an important constituent of the plant microbiome, may be key to the ability of plants to adapt to climatic stressors. Although these are interesting solutions that must be explored and adopted, immediate solutions to reduce the stressors are required to mitigate the water needs of crops.

Sudan contains 175 million feddans (73.5 million hectares) of suitable agriculture land, although only 26 million hectares is currently being sown [1]. However, only 1 million hectares of the total arable land is irrigated, with 6.7 million hectares being managed using semi-mechanized rainfed agriculture methods and a whopping 9 million hectares being completely dependent on traditional rainfed agriculture methods. It is important to understand the climatic conditions in Sudan to fully understand the importance of irrigation for agriculture. However, in addition to Sudan, an overview of irrigation water requirements in Africa as whole shows that countries in Northern Africa tend to have higher crop water requirements, as shown by the in–out values for 53 countries (Figure 3), with 15 countries having irrigation water requirement rates greater than 1000 mm. It is evident that these countries are in Northern Africa, with Sudan being one such country with high water demands. After the separation of South Sudan, Sudan's reliance on agriculture increased, meaning the irrigation water requirements consequently increased.

Despite the fact that the irrigation water requirements in Sudan are very high, the country has a higher percentage of cropland equipped with irrigation as compared to most other African countries. This is clear in Figure 4, showing that although the percentage of irrigation in Sudan is high, the country suffers from food insecurity.

The areas that are irrigated use the Nile River as the key source of water. Sudan has one of the largest irrigation schemes in the world, called the Jazirah scheme, which irrigates approximately 1 million hectares of the total 1.6 million hectares of irrigated land in Sudan [14]. Some studies have pointed out the irrigation management challenges in the Gezira scheme [15,16]; however, better planning could always be utilized to improve the irrigation management [17]. Apart from the river Nile, spate flows from seasonal rivers in the Gash and Tokar Deltas also contribute to irrigated agriculture [1]. Irrigated agriculture also benefits from rain in summer, e.g., the Suki irrigation scheme sources 40% of its water from rain. However, from the hydrogeology map of Sudan [18] presented in Figure 5, it can be observed that the country has abundant resources of underground water [18]. At the same time, from Figure 6 it is clear that the Nile is the main perennial river, while the rest of the rivers are non-perennial.

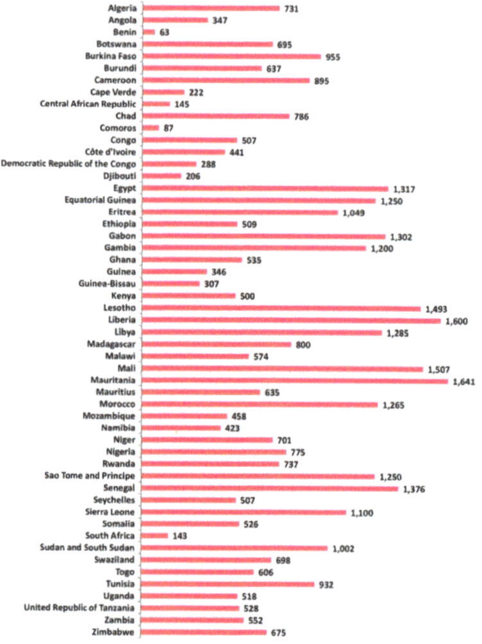

Figure 3. Irrigation water requirements (mm/year) in 2011 (in several cases the index values for the previous years are reported due to a lack of data) [12].

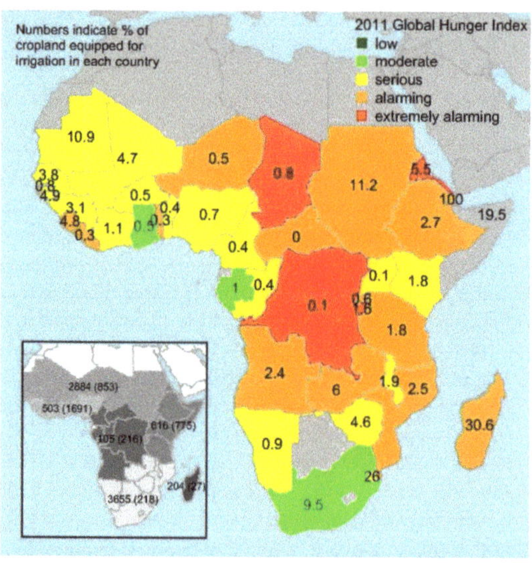

Figure 4. 2011 Global Hunger Index and irrigation coverage [13].

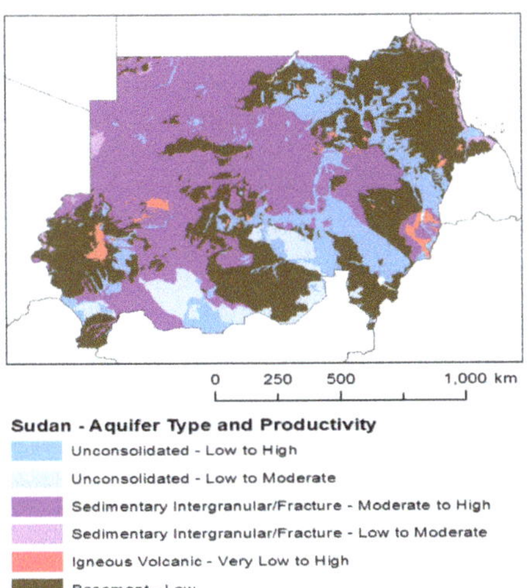

Figure 5. Hydrogeology of Sudan [18].

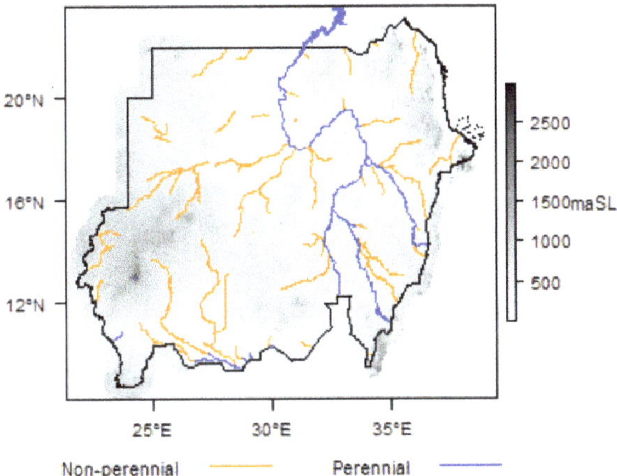

Figure 6. Major surface water features in Sudan [18].

Groundwater is the major source of water for locations away from the Nile basin and other non-Nelotic river wells. It is estimated that Sudan contains more than 900 billion cubic meters (BCM) of water reserves, with 1.563 BCM of annual recharge [19,20]. Despite having such an abundant amount of water to recharge aquifers, agriculture is still dependent on rain-fed conditions.

The utilization of the groundwater presents a multitude of challenges. The primary challenge in the utilization of groundwater is the lack of electricity in remote areas for water pumping [15,16]. Increasing fuel prices combined with increasing inflation make the utilization of diesel generators costly, hazardous for the environment, and financially unsustainable. Recent developments in renewable energy systems have increase their potential for utilization in irrigation. However, the costs of renewable energy sources such as solar photovoltaic (PV) systems are considered to be high due to the high initial capital costs [21,22]. Although there are many financing opportunities being provided by Sudan's government and international organizations such as the United Nations Development Program (UNDP) to promote renewable energy for sustainable agriculture, the accurate determination of water requirements dictates the cost and viability of renewable energy options. This cost is often overestimated due to a lack of knowledge regarding the crop water requirements. Oversizing incurs extra costs, while undersizing can lead to insufficient performance. This is why there is a need for scientific studies that consider irrigation holistically starting from the crop water requirements to determine the amount of energy required and to size the water pumping system.

Some studies have made efforts to determine the crop water requirements, e.g., Schumacher et al. [23] studied the water requirements for urban agriculture in Khartoum. Similarly, Elhag and Ahmed [14] analyzed the irrigation water requirements for the Gezira scheme, while Jabow et al. [24] explored the crop water requirements for tomato, common bean, and chickpea crops in Hudeiba, Sudan. However, a study that considers crop water requirements for different areas in Sudan is urgently required to enable meaningful water planning and the development of cost-effective sustainable irrigation systems across Sudan to alleviate food insecurity. This study seeks to fill this gap by establishing crop water requirements for important crops in Sudan, namely wheat, cotton, and sorghum crops. This will not only improve the knowledge regarding crop water requirements and help in determining the resources required but will also improve the planning process and water management for irrigation. Moreover, the limitation of the data required to calculate the average size of the water pumps is addressed by exploring crop water requirements for different areas and by calculating the amounts of energy required at all sites for each crop. The rest of the paper is organized as below.

Section 2 present the methodology adopted to calculate the crop water requirements and pump size. Section 3 delineates the case studies considered in this paper. The results of the case studies are presented in Section 4. Section 5 concludes the paper.

2. Methodology

Sudan is divided into 18 states for administrative purposes, as can be seen in Figure 7 [25]. However, as the aim of this study is to compare crop water requirements for different crops in different regions, two factors are considered herein for site selection. Firstly, the classification of Sudan according to climatic conditions (Figure 1) is considered as the crop water requirements are highly dependent on climate. Lastly, crop water requirements depend on the type of soil; thus, the soil map of Sudan given in Figure 8 is used to determine the soil types at selected sites for which soil data were not available. Thus, the selected areas are based on the different climatic conditions and dominant soil types.

Figure 7. Different states of Sudan [25].

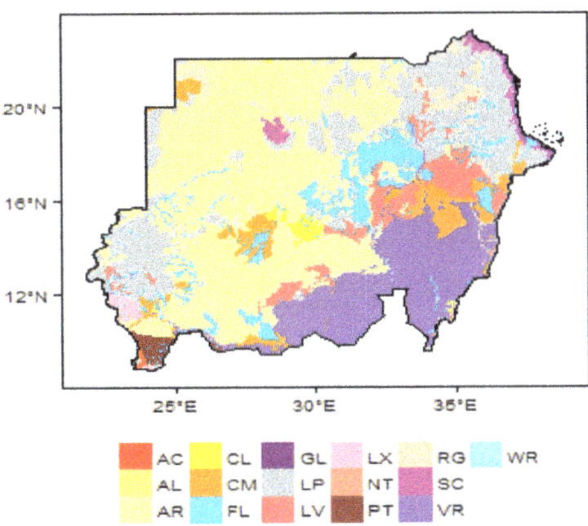

Figure 8. Soil map of Sudan [18]. Acrisols (AC), Alisols (AL), Calcisols (CL), Cambisols (CM), Fluvisols (FL), Gleysols (GL), Arenosols (AR), Leptosols (LP), Lixisols (LX), Luvisols (LV), Plinthosols (PT), Regosols (RG), Solonchaks (SC), Nitisols (NT), Vertisols (VR), Water Body (WR).

Once the study sites are selected, the process for the calculation of the crop water requirements is initiated. Here, the crop water requirements are assessed using CROPWAT software, which is a decision support system developed by the FAO of UN [15]. It considers a number of inputs, such as the crop type, soil type, and climatic data to calculate reference evapotranspiration (ET_0), crop evapotranspiration (ET_c), net irrigation water requirement (NIWR), and gross irrigation water requirement (GIWR) values. To calculate the crop water requirements, the methodology presented in Figure 9 is adopted herein.

The calculation of the crop water requirements can be divided into three distinctive stages, i.e., data acquisition, computation of the ET_c, and computation of the effective rainfall. Soil data were obtained from the FAO database, which classifies soil types into different categories. The literature was used to correctly classify the soils at the selected sites [26]. The climatic data from the FAO [15] for long-term monthly averages for 15 years covering different periods from 1971 to 2000 are available in the CLIMWAT 2 module, which can be directly used in CROPWAT [15]. Although the dataset from CLIMWAT contains averaged data for a long period collected from over 5000 weather stations globally, considering the ongoing climate changes and carbon emissions, this dataset needs to be updated. However, due to a lack of data, the existing CLIMWAT dataset will be used. Apart from the rainfall and temperature data, the most important value is evapotranspiration. Transpiration and evaporation occur at the same time, and when combined with each other are often referred to as evapotranspiration (ET). The rate of ET from a hypothetical crop is called the reference evapotranspiration (ET_0). The rate of ET from a hypothetical crop with a height of 0.12 m, albedo of 0.23, and fixed canopy resistance of 70 sm^{-1} is called the reference evapotranspiration [27].

The FAO CROPWAT model uses the FAO Penman–Monteith equation for the calculation of ET_0 based on weather data [28]. The net irrigation water requirement is calculated using the crop evapotranspiration (ET_c) and effective rainfall (R_{eff}) values. The ET_c is calculated using (1) [27]:

$$ET_c = K_c + ET_0 \tag{1}$$

where K_c is the crop coefficient. K_c represents an integration of the effects of four essential qualities that differentiate the crops from the reference grass, which covers the albedo (reflectance) of the crop–soil surface, crop height, canopy resistance, and evaporation from the soil [29]. K_c varies with the crop development and is divided into four different stages, i.e., initial, development, mid-season, and late-season stages.

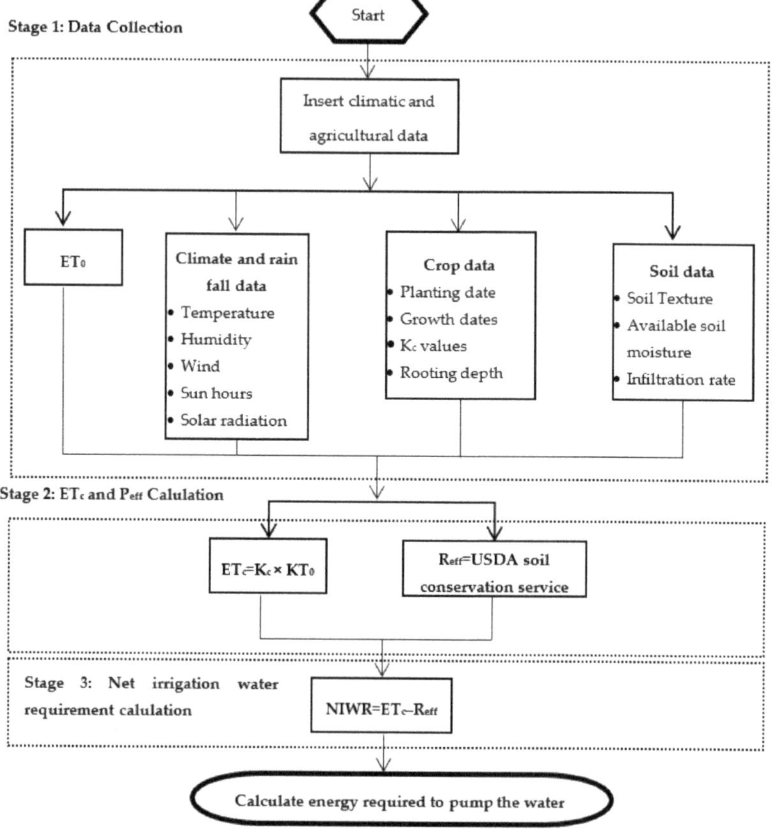

Figure 9. Flow chart of the crop water requirements and energy calculation.

The crop water requirement is defined as the amount of water needed to maintain the moisture level by identifying the water lost through evapotranspiration (ET_c) for a disease free crop. Therefore, the crop water requirement essentially represents the ET_c values obtained over different seasons. However, when calculating the irrigation requirements other sources of water also need to be considered. If there is no rainfall, the ET_c is considered to be the water required for the crop, whereas for rain or deep seepage, the net irrigation water requirement (NIWR) will be lower, as given in (2) [27]:

$$NIWR = ET_0 - R_{eff} \qquad (2)$$

Once the crop water requirement is calculated, the total energy requirement to meet the crop water requirement is calculated. To calculate the power requirement, the total water discharge calculated using CROPWAT is used as in (3) [30]:

$$P = QgH\rho/3.6 \times 10^6 \qquad (3)$$

where Q is the flow of water (m^3/s), ρ is the density of water (kg/m^3), g is the acceleration due to gravity (m/s^2), and H is the differential head (m). Using (3), the power required to lift water is calculated in kW, which can be converted into daily energy.

3. Case Study

3.1. Site Selection

The site selection process requires two major inputs, rainfall and climatic data, as well as soil data. FAO CLIMWAT 2 provides the weather data from 66 weather stations across Sudan; however, soil data are not commonly available. For the purposes of site selection, a study was carried out by Shallah and Ahmed [31], which considered 10 different sites in different climatic areas, as shown in Figure 10. Apart from these sites, two more sites were selected, which are highlighted in red (sites 11 and 12) in Figure 10 [31].

Figure 10. Selected sites used for case study.

The details of the locations and types of soil for the selected sites are given in Table 1. Although the purpose of the soil testing was not agricultural, shallow samples of up to 1.5 m can provide the information that can be used for soil classification. For CROPWAT, soil classification was adopted according to the FAO guidelines for soil description documents [26]. The location name for site 7 could not be found on the map, so an approximate location from the map presented in Figure 10 was used to allocate the nearest weather station.

Table 1. Site locations and soil types (adopted from [31]).

No.	Sample Location	Climate Zone	No. and Name of Nearest Weather Station on CLIMEWAT	Coordinates of Weather Station Long–Lat.	Soil Type
1	North East of Karima	Arid	54-Karima	31.85–18.55	Poorly graded sand
2	South East of Abuhamad	Arid	63-Abu-Hamed	33.31–19.53	Poorly graded silty sand
3	Osaif	Arid	8-Dongonab	37.13–21.1	Silty sandy Gravel
4	South west of Toker	Arid	59-Tokar	37.73–18.43	Silty Gravelly sand
5	West of Almatama	Arid	38-Khartoum	32.55–15.6	Gravelly Silty sand
6	Kassala	Semi Arid	41-Kassala	36.4–15.46	Silty Clay
7	Elgurashi	Semi Arid	27-Ed-Dueim	32.33–14	Silty Clay
8	Kadugli	Semi Arid	9-Kadugli	29.71–11	Sandy Silty Clay
9	Almujlad	Semi Arid	7-Babanusa	27.81–11.33	Silty Clay
10	Algadarif	Semi Arid	34-Gadaref	35.4–14.03	Silty Clay (Expansive soil)
11	Kutum North of Darfur	Semi Arid	33-Kutum	24.66–14.2	Silty Clay
12	Wadi-Halfa	Arid	3-Wadi-Halfa	31.48–21.01	Poorly graded sand

3.2. Climatic Data

As discussed in the methodology, the climatic data were acquired using CLIMWAT 2. It was not possible to include dataset for all 12 sites; therefore, climatic data for a more central site, i.e., site 5, are presented in Table 2. Climatic data for all 12 sites are presented in Appendix A.

Table 2. CROPWAT 8 model climatic rainfall data and ETo values of Khartoum Station (site 5), Sudan.

Station: Khartoum		Altitude: 380 m			Latitude: 15.6°N		Longitude: 32.55° E		
Month	**Min Temp**	**Max Temp**	**Humidity**	**Wind**	**Sun**	**Rad**	**ETo**	**Rain**	**Eff Rain**
	°C	°C	%	km/day	hours	MJ/m^2/day	mm/day	mm	mm
January	15.6	30.8	34	346	9.7	20	6.94	0	0
February	17	33	25	389	10.7	23.1	8.64	0	0
March	20.5	36.8	18	389	10.5	24.7	10.23	0	0
April	23.6	40.1	16	346	10.9	26.2	10.76	0.4	0.4
May	27.1	41.9	19	311	10.4	25.5	10.53	4	4
June	27.3	41.3	26	346	9.8	24.2	10.6	5.4	5.4
July	25.9	38.4	47	346	9	23.1	8.65	46.3	42.9
August	25.3	37.3	55	346	8.7	22.7	7.77	75.2	66.2
September	26	39.1	43	311	9.2	22.9	8.45	25.4	24.4
October	25.5	39.3	32	268	9.2	21.4	8.2	4.8	4.8
November	21	35.2	30	346	9.7	20.3	8.24	0.7	0.7
December	17.1	31.8	35	346	9.9	19.6	7.06	0	0
Average	22.7	37.1	32	341	9.8	22.8	8.84	162.2	148.6

3.3. Crop Data

Sudan produces a variety of crops; however, wheat and sorghum are the most widely planted irrigated crops, as can be seen in Figure 11 [32]. Recently, it has been observed that cotton is being planted more due to the financial gains associated with it. Therefore, cotton, wheat, and sorghum were selected for this study. The crop data for all three crops under study are presented in Table 3.

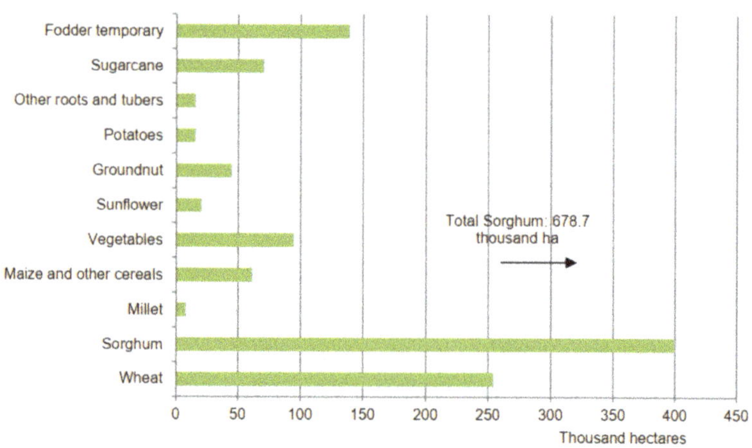

Figure 11. Main irrigated crops in Sudan [32].

Table 3. Crop data.

Crop	Planting and Harvesting Dates	Critical Depletion Fraction	Rooting Depths (m)	Crop Growth Periods (Days)			
				Initial	Crop Development	Mid-Season	Late Season
Wheat	15 Nov–24 Mar	0.55	0.30	30	30	40	30
Cotton	1 Jul–11 Jan	0.65	0.30	30	50	60	55
Sorghum	1 Jun–1 Oct	0.60	0.30	20	35	40	30

The sowing season for cotton starts in July and continues until the middle of September, while it takes four to six months for a cotton seed to grow and mature into a plant with bolls ripe for picking [33]. Sorghum is sown between the beginning of May and the end of June and is harvested from October to December [32]. Wheat is sown in October and harvested in March [32]. Thus, the crop dates were taken following the dates provided by the FAO.

3.4. Soil Data

The types of soil differed according to site, as presented in Table 1. Using Table 1 and the FAO guidelines for soil description documents [26], the classification of each soil type was carried out and the FAO soil data files for soil type and class were used for each site.

4. Results and Discussions

Once all data were available, the crop water requirement was calculated, i.e., ET_c. Irrigation management can be improved with precise knowledge of the irrigation water requirements for the crop and irrigation time. Controlling the amount, timing, and rate of irrigation in an efficient and planned manner is the key to successful irrigation management. Therefore, it is imperative to calculate the water requirement and the irrigation schedule. It is important to note that the K_c values of each crop varied with the development stage. Depending on the K_c value, the crop water requirement will vary. The water requirements for all 3 crops at site 5 (Khartoum) are presented in Tables 4–6.

Table 4. Crop water requirements for wheat.

Month	Decade	Stage	K_c	ET_c	ET_c	Eff. Rain	Irr. Req.
			coeff	mm/day	mm/dec	mm/dec	mm/dec
Nov.	2	Init	0.30	2.5	15	0.10	15
Nov.	3	Init	0.30	2.38	23.8	0.10	23.70
Dec.	1	Init	0.30	2.22	22.2	0.10	22.10
Dec.	2	Dev	0.37	2.54	25.4	0	25.40
Dec.	3	Dev	0.67	4.68	51.5	0	51.50
Jan.	1	Dev	1	6.84	68.4	0	68.40
Jan.	2	Mid	1.22	8.25	82.5	0	82.50
Jan.	3	Mid	1.23	9.09	100	0	100
Feb.	1	Mid	1.23	9.95	99.5	0	99.50
Feb.	2	Mid	1.23	10.65	106.5	0	106.5
Feb.	3	Late	1.15	10.55	84.4	0	84.40
Mar.	1	Late	0.87	8.55	85.5	0	85.50
Mar.	2	Late	0.56	5.83	58.3	0	58.30
Mar.	3	Late	0.35	3.63	14.5	0	14.50
		Total			**837.3**	**0.3**	**837**

(Init = initial; Dev = development; Eff. Rain = effective rain; Irr. Req. = irrigation requirements; ET_c = crop evapotranspiration; mm/dec = millimeter per decade).

Table 5. Crop water requirements for cotton.

Month	Decade	Stage	K_c	ET_c	ET_c	Eff. Rain	Irr. Req.
			coeff	mm/day	mm/dec	mm/dec	mm/dec
Jul.	1	Init	0.35	3.23	32.30	10.60	21.70
Jul.	2	Init	0.35	2.99	29.90	14.90	15
Jul.	3	Dev	0.35	2.91	32.10	17.30	14.70
Aug.	1	Dev	0.47	3.73	37.30	21.50	15.80
Aug.	2	Dev	0.64	4.94	49.40	25.10	24.30
Aug.	3	Dev	0.83	6.60	72.60	19.40	53.20
Sep.	1	Dev	1.02	8.46	84.60	12.10	72.50
Sep.	2	Mid	1.19	10.21	102.10	7.10	94.90
Sep.	3	Mid	1.24	10.49	104.90	5.30	99.60
Oct.	1	Mid	1.24	10.3	103	3.30	99.70
Oct.	2	Mid	1.24	10.19	101.90	0.90	101
Oct.	3	Mid	1.24	10.21	112.30	0.70	111.60
Nov.	1	Mid	1.24	10.31	103.10	0.60	102.50
Nov.	2	Late	1.24	10.32	103.20	0.10	103.10
Nov.	3	Late	1.15	9.14	91.40	0.10	91.30
Dec.	1	Late	1.05	7.77	77.70	0.10	77.60
Dec.	2	Late	0.95	6.6	66	0	66
Dec.	3	Late	0.84	5.83	64.20	0	64.20
Jan.	1	Late	0.73	5	50	0	50
Jan.	2	Late	0.67	4.53	4.5	0	4.50
		Total			**1422.4**	**139.1**	**1283.2**

(Init = initial; Dev = development; Eff. Rain = effective rain; Irr. Req. = irrigation requirements; ETc = crop evapotranspiration; mm/dec = millimeter per decade).

Table 6. Crop water requirements for sorghum.

Month	Decade	Stage	K_c	ET_c	ET_c	Eff. Rain	Irr. Req.
			coeff	mm/day	mm/dec	mm/dec	mm/dec
Jul.	1	Init	0.30	2.77	27.70	10.60	17.10
Jul.	2	Init	0.30	2.56	25.60	14.90	10.70
Jul.	3	Dev	0.43	3.54	39	17.30	21.60
Aug.	1	Dev	0.65	5.20	52	21.50	30.50
Aug.	2	Dev	0.86	6.61	66.10	25.10	41
Aug.	3	Mid	1.03	8.18	90	19.40	70.50
Sep.	1	Mid	1.04	8.65	86.50	12.10	74.40
Sep.	2	Mid	1.04	8.92	89.20	7.10	82.10
Sep.	3	Mid	1.04	8.80	88	5.30	82.70
Oct.	1	Late	1	8.28	82.80	3.30	79.50
Oct.	2	Late	0.85	6.99	69.90	0.90	69
Oct.	3	Late	0.69	5.68	62.50	0.70	61.80
Nov.	1	Late	0.59	4.92	9.80	0.10	9.80
		Total			**789**	**138.3**	**650.8**

(Init = initial; Dev = development; Eff. Rain = effective rain; Irr. Req. = irrigation requirements; ET_c = crop evapotranspiration; mm/dec = millimeter per decade).

Tables 4–6 show that the cotton crop requires more water than wheat and sorghum. The selected site experiences little rain and the overall impact of the rain is not significant during the examined months. This highlights the need for a robust irrigation system for crops in the area to improve the yield and ensure high productivity. Figure 12 shows a comparison of ET_c, effective rain, and net irrigation requirement values for site 5 for 3 different crops during their lifetimes. This shows the lack of rain and higher water requirements for cotton and sorghum.

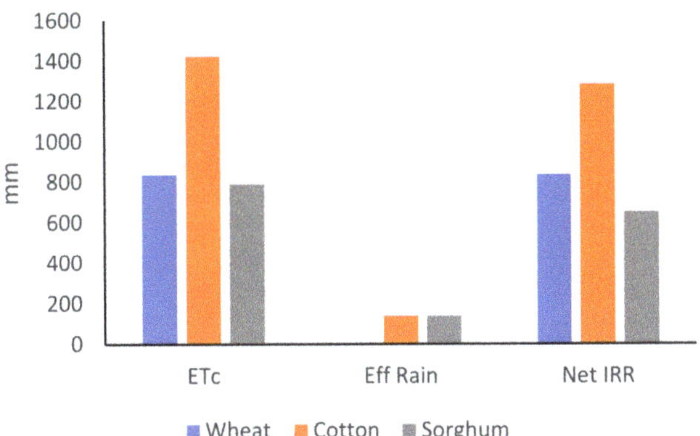

Figure 12. Comparison of ET$_c$ (crop evapotranspiration), effective rain (Eff. Rain), and net irrigation (Net IRR.) values for selected crops at site 5.

Further to the knowledge of irrigation requirements, the irrigation schedule plays an important role in the crop yield. The crop yield can be increased with appropriate irrigation at the appropriate time. Tables 7–9 provide detailed irrigation water requirement data, including the net irrigation, gross irrigation, and flow rates required. The schedule is given in Figures 13–15. The figures show that wheat requires 21 irrigation cycles with a total of 837 mm of water, cotton requires 22 irrigation cycles with 1283 mm of water, and sorghum requires 14 irrigation cycles with 650.8 mm of water.

Table 7. Irrigation schedule for wheat.

Date	Day	Stage	Rain	Ks	Eta	Depl	Net Irr	Deficit	Loss	Gr. Irr	Flow
			mm	fact	%	%	mm	mm	mm	mm	l/s/ha
19-Nov	5	Init	0	1	100	55	12.4	0	0	17.8	0.41
26-Nov	12	Init	0	1	100	58	16.7	0	0	23.9	0.39
5-Dec	21	Init	0	1	100	56	20.5	0	0	29.4	0.38
16-Dec	32	Dev	0	1	100	56	26.3	0	0	37.6	0.40
25-Dec	41	Dev	0	1	100	61	33.6	0	0	48	0.62
1-Jan	48	Dev	0	1	100	57	34.9	0	0	49.9	0.82
07-Jan	54	Dev	0	1	100	62	41.1	0	0	58.7	1.13
13-Jan	60	Dev	0	1	100	63	45.3	0	0	64.7	1.25
18-Jan	65	Mid	0	1	100	57	41.2	0	0	58.9	1.36
23-Jan	70	Mid	0	1	100	61	43.8	0	0	62.5	1.45
28-Jan	75	Mid	0	1	100	63	45.4	0	0	64.9	1.50
2-Feb	80	Mid	0	1	100	65	47.2	0	0	67.4	1.56
6-Feb	84	Mid	0	1	100	55	39.8	0	0	56.8	1.64
10-Feb	88	Mid	0	1	100	55	39.8	0	0	56.8	1.64
14-Feb	92	Mid	0	1	100	59	42.6	0	0	60.8	1.76
18-Feb	96	Mid	0	1	100	59	42.6	0	0	60.8	1.76
22-Feb	100	Mid	0	1	100	59	42.4	0	0	60.6	1.75
26-Feb	104	End	0	1	100	59	42.2	0	0	60.3	1.74
3-Mar	109	End	0	1	100	65	46.7	0	0	66.8	1.55
09-Mar	115	End	0	1	100	71	51.3	0	0	73.2	1.41
18-Mar	124	End	0	1	100	77	55.2	0	0	78.8	1.01

(Init = initial; Dev = development; Ks = crop factor; Eta = actual crop evapotranspiration; Dep = Depletion; Net Irr = net irrigation; Gr. Irr = gross irrigation).

Table 8. Irrigation schedule for cotton.

Date	Day	Stage	Rain	Ks	Eta	Depl	Net Irr	Deficit	Loss	Gr. Irr	Flow
			mm	fract.	%	%	mm	mm	mm	mm	l/s/ha
9-Jul	9	Init	0	1	100	70	17.8	0	0	25.5	0.33
31-Jul	31	Dev	0	1	100	69	30.2	0	0	43.1	0.23
21-Aug	52	Dev	0	1	100	74	44.9	0	0	64.2	0.35
31-Aug	62	Dev	0	1	100	70	48.4	0	0	69.2	0.80
8-Sep	70	Dev	0	1	100	72	54.6	0	0	77.9	1.13
14-Sep	76	Dev	0	1	100	67	54.2	0	0	77.4	1.49
20-Sep	82	Mid	0	1	100	69	57.7	0	0	82.4	1.59
26-Sep	88	Mid	0	1	100	72	60.3	0	0	86.1	1.66
2-Oct	94	Mid	0	1	100	74	62.5	0	0	89.3	1.72
08-Oct	100	Mid	0	1	100	72	60.1	0	0	85.8	1.66
14-Oct	106	Mid	0	1	100	73	60.9	0	0	87	1.68
20-Oct	112	Mid	0	1	100	72	60.7	0	0	86.7	1.67
26-Oct	118	Mid	0	1	100	73	60.9	0	0	87.1	1.68
1-Nov	124	Mid	0	1	100	73	61.4	0	0	87.7	1.69
7-Nov	130	Mid	0.3	1	100	73	61.3	0	0	87.5	1.69
13-Nov	136	Mid	0.1	1	100	74	61.8	0	0	88.3	1.70
19-Nov	142	End	0	1	100	74	61.9	0	0	88.4	1.70
26-Nov	149	End	0	1	100	78	65.1	0	0	93.1	1.54
4-Dec	157	End	0	1	100	80	67.6	0	0	96.6	1.40
13-Dec	166	End	0	1	100	79	66.4	0	0	94.8	1.22
24-Dec	177	End	0	1	100	83	69.5	0	0	99.3	1.04
7-Jan	191	End	0	1	100	90	75.8	0	0	108.3	0.90

(Init = initial; Dev = development; Ks = crop factor; Eta = actual crop evapotranspiration; Dep = Depletion; Net Irr = net irrigation; Gr. Irr = gross irrigation).

Table 9. Irrigation schedule for sorghum.

Date	Day	Stage	Rain	Ks	Eta	Depl	Net Irr	Deficit	Loss	Gr. Irr	Flow
			mm	fract.	%	%	mm	mm	mm	mm	L/s/ha
11-Jul	11	Init	0	1	100	61	19.1	0	0	27.3	0.29
31-Jul	31	Dev	0	1	100	59	32.5	0	0	46.4	0.27
12-Aug	43	Dev	0	1	100	61	42.7	0	0	60.9	0.59
21-Aug	52	Dev	0	1	100	58	46.7	0	0	66.7	0.86
29-Aug	60	Mid	0	1	100	55	46.2	0	0	66.1	0.96
4-Sep	66	Mid	0	1	100	53	44.4	0	0	63.4	1.22
10-Sep	72	Mid	0	1	100	54	45.3	0	0	64.8	1.25
16-Sep	78	Mid	0	1	100	59	50	0	0	71.4	1.38
21-Sep	83	Mid	0	1	100	53	44.5	0	0	63.5	1.47
27-Sep	89	Mid	2.6	1	100	57	47.5	0	0	67.9	1.31
2-Oct	94	Mid	0	1	100	51	43	0	0	61.4	1.42
8-Oct	100	End	0	1	100	57	48	0	0	68.6	1.32
16-Oct	108	End	0	1	100	69	58.1	0	0	82.9	1.20
26-Oct	118	End	0	1	100	73	61.7	0	0	88.2	1.02

(Init = initial; Dev = development; Ks = crop factor; Eta = actual crop evapotranspiration; Dep = Depletion; Net Irr = net irrigation; Gr. Irr = gross irrigation).

Figure 16 shows that in all cases cotton requires the highest level of water, although the differences between the irrigation water rates for wheat and cotton vary significantly according to site. This is due to a multitude of factors, particularly the effective rain rate, crop characteristics, and soil characteristics. These factors play vital roles, although the ability of soil to retain moisture and effective rain plays a significant role in this variation. It is interesting that the areas in the northern region require significantly higher levels of water, except for site 3. Site 3 is located close to the red sea and the climatic conditions are comparatively less harsh. Site 4, which is also located near the red sea, tends to require a higher amount of water due to higher temperatures. This points out the importance of temperature, which plays a vital role in evapotranspiration. A comparison with the reference evapotranspiration rate is shown in Figure 17, which reveals that the northern regions have higher evapotranspiration as compared to the southern regions.

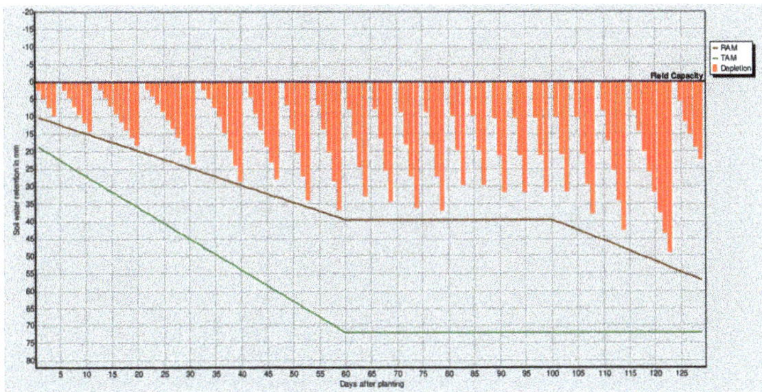

Figure 13. Irrigation schedule for wheat at site 5 (*x*-axis shows days after planting and *y*-axis shows soil water retention in mm). TAM = total available moisture; RAM = readily available moisture.

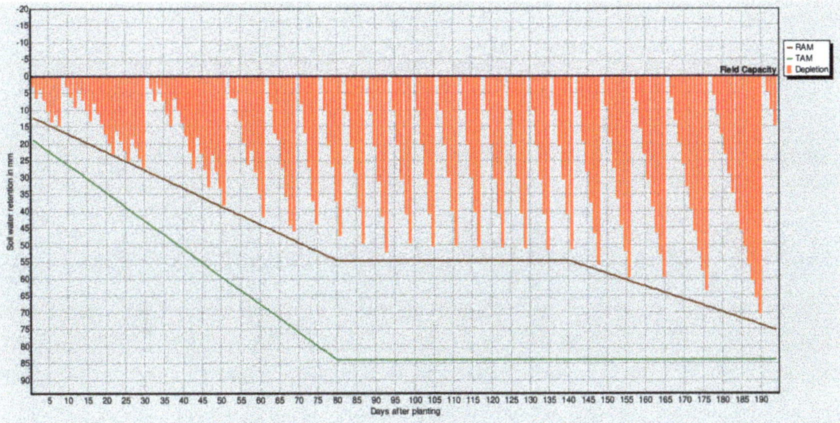

Figure 14. Irrigation schedule for cotton at site 5 (*x*-axis shows days after planting and *y*-axis shows soil water retention in mm). TAM = total available moisture; RAM = readily available moisture.

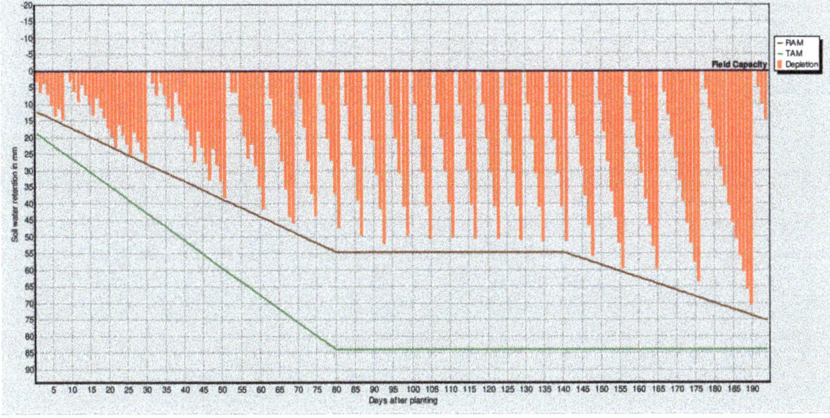

Figure 15. Irrigation schedule for sorghum at site 5 (*x*-axis shows days after planting and *y*-axis shows soil water retention in mm). TAM = total available moisture; RAM = readily available moisture.

Figure 16 presents irrigation water requirements for the three crops at all 12 sites.

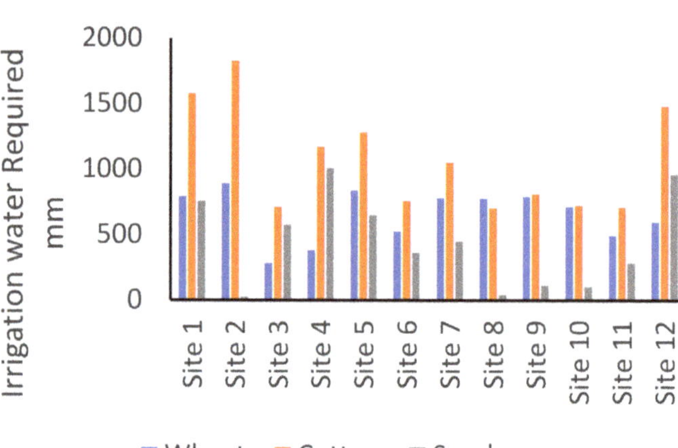

Figure 16. Irrigation water amounts required at all 12 sites.

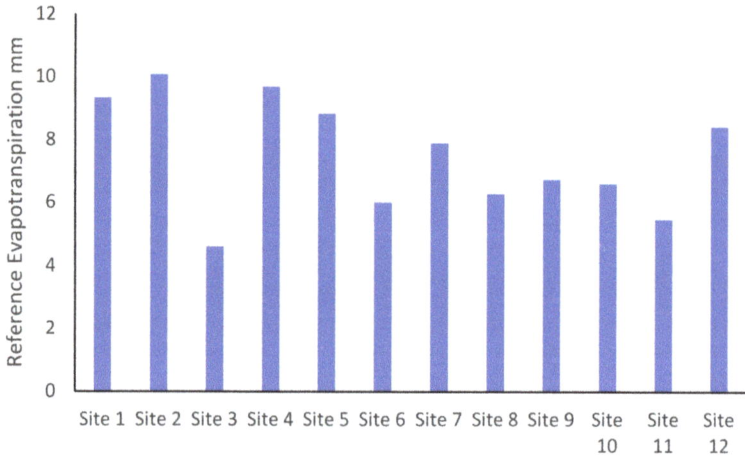

Figure 17. ET$_0$ values for selected sites.

The comparison of the number of irrigation cycles required at different sites with different soil types presents interesting results. With more sandy soil types, the number of irrigation cycles tends to be higher, whereas soil types with higher amounts of clay tend to require lower numbers of irrigation cycles. This can be observed from Figure 18, where the southern sites tend to require lower numbers of irrigation cycles (as low as 2). It is interesting to note that although site 3 has the lowest irrigation water requirement, this does not necessarily mean the number of irrigation cycles will be the lowest. Therefore, a comprehensive understanding based on scientific studies such as this is essential for successful irrigation management.

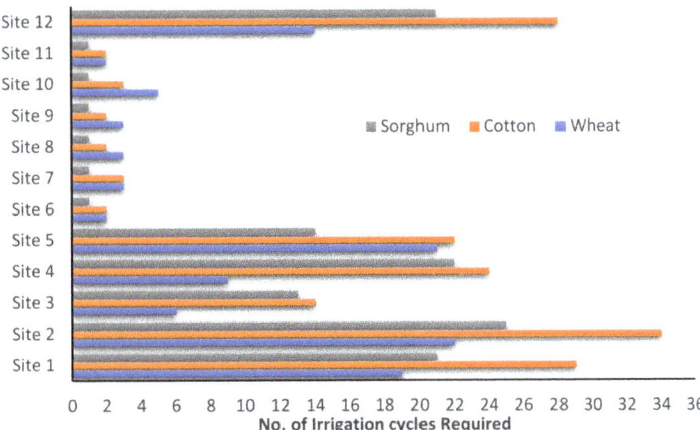

Figure 18. Number of irrigation cycles required at different sites for different crops.

The higher crop water requirements and numbers of irrigation cycles tend to mean that higher amounts of energy will be required to pump the water and more frequent operation of the pump will be required. The crop water requirement data for all 12 sites were used to calculate the amount of energy required by each crop at each site and are given in Figure 19. It was assumed that the water will need to be lifted to a height of 40 m, with a pump efficiency rate of 80%.

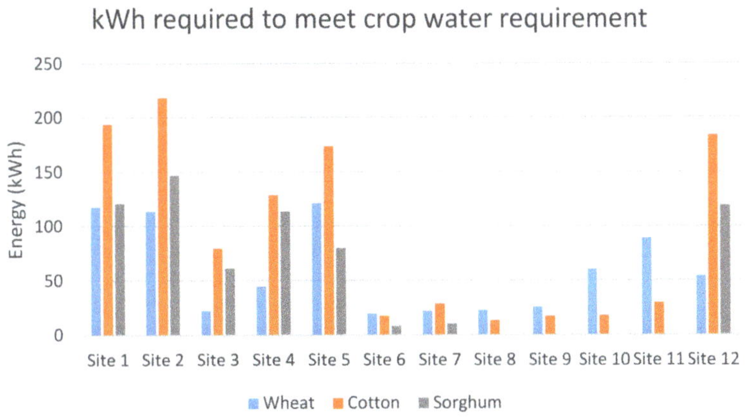

Figure 19. Energy required at each site to meet crop water requirements.

This study has analyzed a number of different sites in terms of the water required for different crops and calculated the amount of energy required to pump this water. The results can be used as a benchmark to compare per hectare crop water requirements in different areas of Sudan and the energy required to aid in the design of the pump size. Overall, arid areas tend to have higher water requirements for the same crop as compared to semi-arid areas. However, variations in crop water requirements in different regions for different crops were observed due to rain patterns. Cotton, being a crop that requires high amount of water, did not require more water at sites 6 to 11. This was mainly due to the weather and rain patterns. Sorghum is a weather-tolerant crop, although at some sites it required higher amounts of water as compared to wheat. These data are important to determine the impacts of climate change and particularly to develop resilient crops that are resistant to stresses resulting from climate change and other factors.

The crop water requirement results were used to calculate the electricity needed (kWh) by water pumps in different areas. The energy requirements at different sites show that the sites located further north tend to have higher energy demands and are potential candidates for renewable energy development. Higher energy demands warrant the use of hybrid energy, which can be beneficial in agricultural applications, helping to reduce energy poverty and improve yields, and consequently helping to reduce poverty and improve food security. Some studies have developed renewable energy systems for irrigation purposes although these studies did not adopt the detailed approach of determining the energy demands according to the crop water requirements [34,35]. A systematic approach to determining the energy demands of crops has been presented herein. The results from this study will be used as a stepping stone to develop feasible renewable energy system options for the different sites in Sudan. This study will pave the way for the integration of future renewable energy sources into the agricultural sector, which will help in reducing pollution by cutting carbon emissions and increasing savings for farmers by reducing their expenses for fuels.

5. Conclusions

This study has presented a comparison of crop water requirements for three different crops at 12 different sites in Sudan. Moreover, the amounts of energy required for all sites to meet the water requirement have been calculated. The comparison suggested that the areas that are arid and located in the north require higher amounts of water and have less effective rainfall. However, the significance of temperature is higher for the sites located near the red sea. The results suggest that depending on the climatic conditions, the crops requiring high amounts of water in arid regions might not necessarily be the ones requiring the highest amounts of energy. The depth of the borehole and other climatic conditions play bigger roles in the energy requirements. The data for crop water requirements have been translated into energy requirements according to the different borehole depths at each site. The northern regions tend to require higher amounts of energy to meet the crop water requirements but at the same time they tend to have longer daylight hours and higher winds, making them suitable candidates for hybrid energy generation.

Thus, the use of tools such as CLIMWAT and CROPWAT can be beneficial in planning and managing irrigation systems with a significant degree of confidence. The use of these tools will not only be beneficial by providing information regarding the crop water requirements but will also provide information about when and how much water is needed for the optimum development of the crops. These tools can be beneficial not only in terms of increasing crop yields but also in planning the use of resources, such as the energy required to supply the water required for crops. Future studies will include the development of hybrid renewable energy systems for the different areas in Sudan using the energy requirements calculated in this study. Specifically, the use of hybrid renewable energy (e.g., solar PV and wind) is particularly advantageous because there are no operational GHG emissions, which have been mainly attributed to climate change and have produced unprecedented consequences in recent time. Additionally, the costs of renewable energy conversion technologies have decreased rapidly in recent years. Other future studies will include an exploration of the techno-economic feasibility of different renewable energy systems to meet energy demands and crop water requirements for the sites and crops presented in this study.

Author Contributions: Conceptualization, M.I. and Z.A.K.; methodology, M.I., Z.A.K. and A.O.A.; software, Z.A.K.; validation, M.I., Z.A.K., J.U., S.A. and O.E.D.; formal analysis, Z.A.K. and A.O.A.; investigation, Z.A.K., J.U. and S.A.; resources, M.I., O.E.D. and J.U.; data curation, Z.A.K. and S.A.; writing—original draft preparation, Z.A.K.; writing—review and editing, M.I., A.O.A., O.E.D., J.U. and Z.A.K.; visualization, Z.A.K.; supervision, M.I.; project administration, M.I.; funding acquisition, M.I., A.O.A. and O.E.D. All authors have read and agreed to the published version of the manuscript

Funding: This research was funded by Innovate UK and Academy of Medical Sciences, grant numbers 60558 and 62327, under the Global Challenge research fund. The APC was funded by grant number 60558.

Institutional Review Board Statement: Not applicable.

Informed Consent Statement: Not applicable.

Data Availability Statement: Additional results and data are presented in Appendix A.

Conflicts of Interest: The authors declare no conflict of interest. The funders had no role in the design of the study; in the collection, analyses, or interpretation of data; in the writing of the manuscript, or in the decision to publish the results.

Appendix A

Table A1. Climate data for 12 selected sites.

Site 1

Month	Min Temp	Max Temp	Humidity	Wind	Sun	Rad	ETo	Rain	Eff Rain
	°C	°C	%	km/day	hours	MJ/m²/day	mm/day	mm	mm
January	11.9	28	34	389	8.9	18.1	6.55	0	0
February	13.5	30.5	26	423	9.5	20.8	8.09	0	0
March	17.3	34.7	20	389	9.5	22.7	9.39	0	0
April	21.5	38.8	19	389	9.9	24.6	10.81	0	0
May	25.1	42	17	389	9.9	24.8	11.8	0.1	0.1
June	26.9	43.4	19	346	9.4	23.9	11.33	0.1	0.1
July	27	41.9	27	311	9.7	24.4	10.2	7.7	7.6
August	27.1	41.4	38	311	9	23.2	9.27	11.4	11.2
September	27.1	42.1	24	346	8.9	22.2	10.49	1.1	1.1
October	23.8	39.2	26	346	9.3	21	9.54	0.3	0.3
November	18	33.2	34	389	9.4	19.2	7.89	0	0
December	13.8	29.4	35	389	9.3	17.9	6.75	0	0
Average	21.1	37	27	368	9.4	21.9	9.34	20.7	20.4

Site 2

Month	Min Temp	Max Temp	Humidity	Wind	Sun	Rad	ETo	Rain	Eff Rain
	°C	°C	%	km/day	hours	MJ/m²/day	mm/day	mm	mm
January	12.5	28.1	33	501	8.9	17.9	7.37	0.2	0.2
February	13.8	30.4	25	501	9.3	20.3	8.75	0	0
March	17.3	34.6	20	467	9.8	23.1	10.27	0	0
April	21	38.7	18	423	10.5	25.4	11.33	0.1	0.1
May	25	41.9	17	346	10.5	25.7	11.17	0.2	0.2
June	27.2	43.5	16	346	10.7	25.9	11.7	0.1	0.1
July	26.9	42.1	23	346	9.9	24.6	10.97	5.1	5.1
August	27.2	41.7	30	346	9.5	23.9	10.34	6	5.9
September	27.3	42.4	21	423	9.3	22.7	11.98	0.7	0.7
October	23.7	39.2	24	423	9.8	21.4	10.68	0.1	0.1
November	18.9	33.3	33	501	9.2	18.6	8.94	0.1	0.1
December	14.4	29.4	36	501	8.9	17.2	7.45	0	0
Average	21.3	37.1	25	427	9.7	22.2	10.08	12.6	12.5

Site 3

Month	Min Temp	Max Temp	Humidity	Wind	Sun	Rad	ETo	Rain	Eff Rain
	°C	°C	%	km/day	hours	MJ/m²/day	mm/day	mm	mm
January	17.2	26.1	87	259	6.2	14.1	2.63	1	1
February	17.2	26.7	86	259	7.4	17.3	3.18	2	2
March	17.8	28.3	84	242	8.4	20.8	3.96	0	0
April	18.9	31.1	81	242	9.3	23.5	4.84	0	0
May	21.7	34.4	78	216	9.7	24.6	5.6	2	2
June	22.8	37.8	73	190	10	25	6.24	0	0

Table A1. *Cont.*

Month	Min Temp	Max Temp	Humidity	Wind	Sun	Rad	ETo	Rain	Eff Rain
July	25	38.9	75	216	10.1	25	6.5	0	0
August	26.1	38.9	70	216	9.7	24.1	6.64	0	0
September	24.4	37.2	76	190	9	22	5.64	0	0
October	22.2	33.3	92	190	8.2	18.9	3.96	9	8.9
November	21.1	30.5	85	242	7	15.4	3.36	12	11.8
December	18.9	27.2	82	242	6.5	13.9	2.91	11	10.8
Average	21.1	32.5	81	225	8.5	20.4	4.62	37	36.4

Site 4

Month	Min Temp	Max Temp	Humidity	Wind	Sun	Rad	ETo	Rain	Eff Rain
	°C	°C	%	km/day	hours	MJ/m^2/day	mm/day	mm	mm
January	20	28.2	71	346	5.3	13.6	3.97	15	14.6
February	19.7	28.7	68	268	5.2	14.9	4.08	3	3
March	20.9	30.9	66	268	7.1	19.3	5.04	0	0
April	22	33.7	62	268	8.8	22.9	6.21	1	1
May	23.1	37.7	54	268	9.6	24.4	7.42	2	2
June	25.1	42.3	41	311	10	24.8	9.45	0	0
July	28	42.9	36	467	9.7	24.3	11.87	6	5.9
August	28.3	42.4	39	423	9.3	23.7	11.01	6	5.9
September	26.2	41.3	45	268	9	22.3	8.24	0	0
October	25.3	36.8	60	346	8.2	19.5	6.8	6	5.9
November	23.6	33.1	65	311	7.5	16.8	5.24	19	18.4
December	21.2	29.8	69	311	6.3	14.4	4.23	15	14.6
Average	23.6	35.6	56	321	8	20.1	6.96	73	71.5

Site 5

Month	Min Temp	Max Temp	Humidity	Wind	Sun	Rad	ETo	Rain	Eff Rain
	°C	°C	%	km/day	hours	MJ/m^2/day	mm/day	mm	mm
January	15.6	30.8	34	346	9.7	20	6.94	0	0
February	17	33	25	389	10.7	23.1	8.64	0	0
March	20.5	36.8	18	389	10.5	24.7	10.23	0	0
April	23.6	40.1	16	346	10.9	26.2	10.76	0.4	0.4
May	27.1	41.9	19	311	10.4	25.5	10.53	4	4
June	27.3	41.3	26	346	9.8	24.2	10.6	5.4	5.4
July	25.9	38.4	47	346	9	23.1	8.65	46.3	42.9
August	25.3	37.3	55	346	8.7	22.7	7.77	75.2	66.2
September	26	39.1	43	311	9.2	22.9	8.45	25.4	24.4
October	25.5	39.3	32	268	9.2	21.4	8.2	4.8	4.8
November	21	35.2	30	346	9.7	20.3	8.24	0.7	0.7
December	17.1	31.8	35	346	9.9	19.6	7.06	0	0
Average	22.7	37.1	32	341	9.8	22.8	8.84	162.2	148.6

Site 6

Month	Min Temp	Max Temp	Humidity	Wind	Sun	Rad	ETo	Rain	Eff Rain
	°C	°C	%	km/day	hours	MJ/m^2/day	mm/day	mm	mm
January	16.5	33.7	48	156	8.8	18.8	4.81	0	0
February	17.2	35.2	43	156	9.4	21.3	5.48	0	0
March	20.1	38.3	40	156	9.3	22.9	6.27	0.3	0.3
April	23	40.8	37	156	9.6	24.3	7	2	2
May	25.8	41.6	39	156	9.4	23.9	7.18	8.8	8.7
June	25.7	39.8	42	190	8.8	22.8	7.31	27.9	26.7
July	23.9	36.1	57	233	7.5	20.8	6.41	76	66.8
August	23.4	34.9	63	233	7.5	20.8	5.9	85	73.4
September	24	36.8	55	156	8.9	22.5	5.95	40.5	37.9
October	24.3	38.7	44	112	9.3	21.6	5.6	9.9	9.7
November	21.4	37	47	156	9.1	19.5	5.46	0.8	0.8
December	17.9	34.4	52	156	8.8	18.3	4.78	0	0
Average	21.9	37.3	47	168	8.9	21.5	6.01	251.2	226.2

Table A1. *Cont.*

Site 7

Month	Min Temp	Max Temp	Humidity	Wind	Sun	Rad	ETo	Rain	Eff Rain
	°C	°C	%	km/day	hours	MJ/m^2/day	mm/day	mm	mm
January	16.3	31.3	34	311	8.9	19.4	6.73	0	0
February	17.7	33.2	28	346	9.3	21.5	8	0	0
March	20	37.3	24	346	9.5	23.4	9.32	0.1	0.1
April	23.4	40.1	23	346	9.7	24.4	10.26	0.1	0.1
May	24.3	41.2	28	311	9.6	24.1	9.77	5.8	5.7
June	25.3	39.5	39	311	9	22.9	8.88	22.1	21.3
July	25	36.2	53	311	7.7	21	7.27	65.8	58.9
August	23.6	34.6	61	268	7.4	20.7	6.21	99.6	83.7
September	23.9	36.5	53	268	8.3	21.7	6.89	38.1	35.8
October	24.3	38.1	42	233	8.8	21.1	7.12	5.2	5.2
November	21	35.4	32	311	9.2	20.1	7.75	0.4	0.4
December	17.9	32.2	37	311	9.1	19.1	6.75	0	0
Average	21.9	36.3	38	306	8.9	21.6	7.91	237.2	211.2

Site 8

Month	Min Temp	Max Temp	Humidity	Wind	Sun	Rad	ETo	Rain	Eff Rain
	°C	°C	%	km/day	hours	MJ/m^2/day	mm/day	mm	mm
January	17.4	34.4	23	268	9.4	20.8	7.32	0	0
February	19.3	36.1	20	268	9.6	22.6	8.01	0	0
March	22	38.9	20	233	9.6	23.9	8.24	3.3	3.3
April	23.5	40	27	190	9.5	24.2	7.71	10.9	10.7
May	23.8	38.3	47	156	8.9	22.9	6.52	60.6	54.7
June	22.5	35.1	65	156	8	21	5.45	99.7	83.8
July	21.7	32.1	78	156	6.7	19.2	4.47	134.9	105.8
August	21.2	31.4	82	156	6.6	19.4	4.27	163.1	120.5
September	20.7	32.8	83	112	7.3	20.4	4.39	98.7	83.1
October	20.2	35.2	68	156	8.5	21.2	5.14	61	55
November	18.7	36.4	37	233	9.4	21.1	6.79	0.9	0.9
December	17.9	34.9	30	268	9.4	20.2	7.08	0	0
Average	20.7	35.5	48	196	8.6	21.4	6.28	633.1	517.9

Site 9

Month	Min Temp	Max Temp	Humidity	Wind	Sun	Rad	ETo	Rain	Eff Rain
	°C	°C	%	km/day	hours	MJ/m^2/day	mm/day	mm	mm
January	16.1	32.3	26	268	10.2	21.9	6.99	0	0
February	19.1	35.1	21	311	10	23.1	8.44	0	0
March	22.6	38.1	19	268	9.1	23.1	8.67	0.9	0.9
April	24.5	40.1	21	190	9.3	23.9	7.81	6.2	6.1
May	25.7	39.3	38	190	9.1	23.2	7.4	21.7	20.9
June	19.5	36.3	54	190	8.1	21.3	6.16	88	75.6
July	22.9	32.7	66	190	6.5	19	5.01	125.2	100.1
August	22.7	32.3	71	190	6.8	19.8	4.89	125.3	100.2
September	22.6	33.6	66	190	7.6	20.8	5.31	107	88.7
October	23	35.9	45	190	8.6	21.3	6.23	22.9	22.1
November	20.7	35.7	28	233	10.3	22.2	7.17	0.1	0.1
December	17.5	34.7	25	233	10.1	21.1	6.87	0	0
Average	21.4	35.5	40	220	8.8	21.7	6.75	497.3	414.7

Site 10

Month	Min Temp	Max Temp	Humidity	Wind	Sun	Rad	ETo	Rain	Eff Rain
	°C	°C	%	km/day	hours	MJ/m^2/day	mm/day	mm	mm
January	17.2	34.7	39	268	8.7	19.2	6.48	0	0
February	18.3	36.4	31	268	9.1	21.3	7.39	0	0
March	21.6	39.2	28	268	8.7	22.3	8.28	0.5	0.5
April	23.8	41	25	233	8.9	23.2	8.42	3.4	3.4

Table A1. *Cont.*

Month	Min Temp	Max Temp	Humidity	Wind	Sun	Rad	ETo	Rain	Eff Rain
May	25.2	40.6	36	233	9	23.3	8.16	21.2	20.5
June	23.4	37.5	53	268	8.4	22	7.22	90.9	77.7
July	21.6	33.4	70	268	6.7	19.5	5.42	183.4	129.6
August	21.2	32.2	76	268	6.9	19.9	4.95	184.4	130
September	21.4	34.1	70	233	7.7	20.7	5.34	85.5	73.8
October	22.1	36.8	55	156	8.3	20.3	5.5	31.4	29.8
November	21	37.1	38	190	8.9	19.7	6.12	3	3
December	18.1	35.2	40	233	9.1	19.1	6.13	0	0
Average	21.2	36.5	47	240	8.4	20.9	6.62	603.7	468.2

Site 11

Month	Min Temp	Max Temp	Humidity	Wind	Sun	Rad	ETo	Rain	Eff Rain
	$^\circ$C	$^\circ$C	%	km/day	hours	MJ/m^2/day	mm/day	mm	mm
January	6.8	28.5	32	164	9.2	19.7	4.7	0	0
February	7.2	29.1	30	138	9.7	22	4.82	0	0
March	11.5	33.1	28	164	9.7	23.7	5.94	0	0
April	16.2	35.5	26	164	10	25	6.62	1	1
May	18.6	36.3	26	156	10.2	25	6.7	5	5
June	20.5	36	33	164	9.5	23.7	6.63	15	14.6
July	20	33	48	138	7.7	21	5.48	71	62.9
August	19	30.1	63	138	7.5	21	4.83	118	95.7
September	18	33	48	138	8.5	22	5.41	25	24
October	15.8	33.2	35	138	9.3	21.8	5.48	4	4
November	11.1	30.3	33	138	9.2	20	4.78	0	0
December	7.6	28.7	32	138	9.2	19.1	4.36	0	0
Average	14.4	32.2	36	148	9.1	22	5.48	239	207.2

Site 12

Month	Min Temp	Max Temp	Humidity	Wind	Sun	Rad	ETo	Rain	Eff Rain
	$^\circ$C	$^\circ$C	%	km/day	hours	MJ/m^2/day	mm/day	mm	mm
January	9.5	23.3	41	311	10.5	19.1	4.8	0.1	0.1
February	10.5	25.7	31	346	10.9	21.8	6.2	0	0
March	14.6	30.5	20	346	11	24.5	8.1	0	0
April	19.4	36.1	17	346	11.4	26.7	9.89	0	0
May	23.3	39.6	18	346	11.9	27.9	10.94	0	0
June	25	41.1	15	311	11.6	27.4	10.85	0	0
July	24.9	40.7	21	233	11.2	26.7	9.21	0	0
August	24.9	40.1	25	268	10.6	25.4	9.34	0.5	0.5
September	24.4	39.2	25	423	10.6	24.2	11.13	0	0
October	21.5	36.1	28	389	10.4	21.7	9.29	0	0
November	15.7	29.2	39	346	10.1	19	6.4	0	0
December	11.3	24.8	41	311	10	17.8	4.98	0	0
Average	18.8	33.9	27	331	10.9	23.5	8.43	0.6	0.6

References

1. *Special Report–2020/21 FAO Crop and Food Supply Assessment Mission (CFSAM) to the Republic of the Sudan*; Food and Agriculture Organization of the United Nations: Rome, Italy, 2021. [CrossRef]
2. Peel, M.C.; Finlayson, B.L.; McMahon, T.A. Updated world map of the Köppen-Geiger climate classification. *Hydrol. Earth Syst. Sci.* **2007**, *11*, 1633–1644. [CrossRef]
3. University of East Anglia Climatic Research Unit. Available online: https://lr1.uea.ac.uk/cru/data (accessed on 13 June 2021).
4. Harris, I.; Jones, P.D.; Osborn, T.J.; Lister, D.H. Updated high-resolution grids of monthly climatic observations–the CRU TS3. 10 Dataset. *Int. J. Climatol.* **2014**, *34*, 623–642. [CrossRef]
5. British Geological Survey Climate of Sudan. Available online: http://earthwise.bgs.ac.uk/index.php/Climate_of_Sudan (accessed on 13 June 2021).
6. Ahmed, S.M. Impacts of drought, food security policy and climate change on performance of irrigation schemes in Sub-saharan Africa: The case of Sudan. *Agric. Water Manag.* **2020**, *232*, 106064. [CrossRef]
7. Shankar Naik, B. Functional roles of fungal endophytes in host fitness during stress conditions. *Symbiosis* **2019**, *79*, 99–115 [CrossRef]

8. Redman, R.S.; Kim, Y.O.; Woodward, C.J.D.A.; Greer, C.; Espino, L.; Doty, S.L.; Rodriguez, R.J. Increased fitness of rice plants to abiotic stress via habitat adapted symbiosis: A strategy for mitigating impacts of climate change. *PLoS ONE* **2011**, *6*, e14823. [CrossRef] [PubMed]

9. Acuña-Rodríguez, I.S.; Newsham, K.K.; Gundel, P.E.; Torres-Díaz, C.; Molina-Montenegro, M.A. Functional roles of microbial symbionts in plant cold tolerance. *Ecol. Lett.* **2020**, *23*, 1034–1048. [CrossRef]

10. Decunta, F.A.; Pérez, L.I.; Malinowski, D.P.; Molina-Montenegro, M.A.; Gundel, P.E. A Systematic Review on the Effects of Epichloë Fungal Endophytes on Drought Tolerance in Cool-Season Grasses. *Front. Plant Sci.* **2021**, *12*, 380. [CrossRef]

11. Suryanarayanan, T.S.; Shaanker, R.U. Can fungal endophytes fast-track plant adaptations to climate change? *Fungal Ecol.* **2021**, *50*, 101039. [CrossRef]

12. Valipour, M. Land use policy and agricultural water management of the previous half of century in Africa. *Appl. Water Sci.* **2015**, *5*, 367–395. [CrossRef]

13. Burney, J.A.; Naylor, R.L.; Postel, S.L. The case for distributed irrigation as a development priority in sub-Saharan Africa. *Proc. Natl. Acad. Sci. USA* **2013**, *110*, 12513–12517. [CrossRef]

14. Elhag, A.E.; Abdelkarim, A.A. Analysis of Irrigation Water Requirements in Gezira Scheme Using Geographic Information Systems: Case Study Block Number 26 (Dolga). *J. Eng. Comput. Sci.* **2021**, *22*, 81–91.

15. Food and Agriculture Organization of the United Nations, Cropwat. Available online: https://www.fao.org/land-water/databases-and-software/cropwat/en/ (accessed on 20 April 2021).

16. Barnett, T. Why are bureaucrats slow adopters? The case of water management in the Gezira scheme. *Sociol. Rural.* **1979**, *19*, 60–70. [CrossRef]

17. Elshaikh, A.E.; Yang, S.; Jiao, X.; Elbashier, M.M. Impacts of Legal and Institutional Changes on Irrigation Management Performance: A Case of the Gezira Irrigation Scheme, Sudan. *Water* **2018**, *10*, 1579. [CrossRef]

18. British Geological Survey Hydrology of Sudan. Available online: http://earthwise.bgs.ac.uk/index.php/Africa_Groundwater_Atlas_Hydrogeology_Maps (accessed on 15 May 2021).

19. MacDonald, A.M.; Bonsor, H.C.; Dochartaigh, B.É.Ó.; Taylor, R.G. Quantitative maps of groundwater resources in Africa. *Environ. Res. Lett.* **2012**, *7*, 24009. [CrossRef]

20. MacDonald, A.M.; Davies, J. *A Brief Review of Groundwater for Rural Water Supply in Sub-Saharan Africa*; BGS Technical Report WC/00/33 2000; BGS: Nottinghamshire, UK, 2000.

21. Saeed, T.M. Sustainable energy potential in Sudan. *J. Eng. Comput. Sci.* **2020**, *20*, 1–10.

22. Fadlallah, S.O.; Serradj, D.E.B. Determination of the optimal solar photovoltaic (PV) system for Sudan. *Sol. Energy* **2020**, *208*, 800–813. [CrossRef] [PubMed]

23. Schumacher, J.; Luedeling, E.; Gebauer, J.; Saied, A.; El-Siddig, K.; Buerkert, A. Spatial expansion and water requirements of urban agriculture in Khartoum, Sudan. *J. Arid Environ.* **2009**, *73*, 399–406. [CrossRef]

24. Jabow, M.; Salih, A.; Abdelhadi, A.; Ahmed, B. Crop water requirements for tomato, common bean and chick pea in Hudeiba, River Nile State, Sudan. *Sudan J. Agric. Res.* **2013**, *22*, 11–22.

25. Mappr States of Sudan. Available online: https://www.mappr.co/counties/states-of-sudan/ (accessed on 22 August 2021).

26. *Guidelines for Soil Description*; Food and Agriculture Organization of the United Nations: Rome, Italy, 2006; ISBN 9251055211.

27. Allen, R.G.; Pereira, L.S.; Raes, D.; Smith, M. Crop evapotranspiration-Guidelines for computing crop water requirements-FAO Irrigation and drainage paper 56. *Fao Rome* **1998**, *300*, D05109.

28. Penman, H.L. Evaporation: An introductory survey. *Neth. J. Agric. Sci.* **1956**, *4*, 9–29. [CrossRef]

29. Ewaid, S.H.; Abed, S.A.; Al-Ansari, N. Crop water requirements and irrigation schedules for some major crops in southern Iraq. *Water* **2019**, *11*, 756. [CrossRef]

30. Protogeropoulos, C.; Pearce, S. Laboratory evaluation and system sizing charts for a 'second generation' direct PV-powered, low cost submersible solar pump. *Sol. energy* **2000**, *68*, 453–474. [CrossRef]

31. Mahmoud, S.D.M. Sudan Subgrade Soils Characteristics. *IOSR J. Eng.* **2014**, *4*, 48–56. [CrossRef]

32. *Country Profile–Sudan*; Food and Agriculture Organization of the United Nations: Rome, Italy, 2015.

33. Abdallah, N.A. The Story behind Bt Cotton: Where does Sudan stand? *GM Crop. Food* **2014**, *5*, 241–243. [CrossRef] [PubMed]

34. Kose, F.; Aksoy, M.H.; Ozgoren, M. Experimental investigation of solar/wind hybrid system for irrigation in Konya, Turkey. *Therm. Sci.* **2019**, *23*, 4129–4139. [CrossRef]

35. Elkadeem, M.R.; Wang, S.; Sharshir, S.W.; Atia, E.G. Feasibility analysis and techno-economic design of grid-isolated hybrid renewable energy system for electrification of agriculture and irrigation area: A case study in Dongola, Sudan. *Energy Convers. Manag.* **2019**, *196*, 1453–1478. [CrossRef]

 *energies*

Article

Assessment of Wind and Solar Hybrid Energy for Agricultural Applications in Sudan

Zafar A. Khan [1,2,*], Muhammad Imran [2], Abdullah Altamimi [3], Ogheneruona E. Diemuodeke [4] and Amged Osman Abdelatif [5]

1 Department of Electrical Engineering, Mirpur University of Science and Technology, Mirpur A.K. 10250, Pakistan
2 College of Engineering and Physical Sciences, Mechanical, Biomedical and Design Engineering, Aston University, Birmingham B4 7ET, UK; m.imran12@aston.ac.uk
3 Department of Electrical Engineering, College of Engineering, Majmaah University, Al-Majmaah 11952, Saudi Arabia; a.altmimi@mu.edu.sa
4 Energy and Thermofluid Research Group, Department of Mechanical Engineering, Faculty of Engineering, University of Port Harcourt, Port Harcourt 500102, Nigeria; ogheneruona.diemuodeke@uniport.edu.ng
5 Department of Civil Engineering, Faculty of Engineering, University of Khartoum, P.O. Box 321, Khartoum 51111, Sudan; Amged.Abdelatif@uofk.edu
* Correspondence: zafarakhan@ieee.org

Citation: Khan, Z.A.; Imran, M.; Altamimi, A.; Diemuodeke, O.E.; Abdelatif, A.O. Assessment of Wind and Solar Hybrid Energy for Agricultural Applications in Sudan. *energies* **2022**, *15*, 5. https://doi.org/10.3390/en15010005

Academic Editors: Md Shamim Ahamed, Muhammad Sultan, Redmond R. Shamshiri, Muhammad Wakil Shahzad and Gianpiero Colangelo

Received: 10 November 2021
Accepted: 17 December 2021
Published: 21 December 2021

Publisher's Note: MDPI stays neutral with regard to jurisdictional claims in published maps and institutional affiliations.

Abstract: In addition to zero-carbon generation, the plummeting cost of renewable energy sources (RES) is enabling the increased use of distributed-generation sources. Although the RES appear to be a cheaper source of energy, without the appropriate design of the RES with a true understanding of the nature of the load, they can be an unreliable and expensive source of energy. Limited research has been aimed at designing small-scale hybrid energy systems for irrigation pumping systems, and these studies did not quantify the water requirement, or in turn the energy required to supply the irrigation water. This paper provides a comprehensive feasibility analysis of an off-grid hybrid renewable energy system for the design of a water-pumping system for irrigation applications in Sudan. A systematic and holistic framework combined with a techno-economic optimization analysis for the planning and design of hybrid renewable energy systems for small-scale irrigation water-pumping systems is presented. Different hybridization cases of solar photovoltaic, wind turbine and battery storage at 12 different sites in Sudan are simulated, evaluated, and compared, considering the crop water requirement for different crops, the borehole depth, and the stochasticity of renewable energy resources. Soil, weather, and climatic data from 12 different sites in Sudan were used for the case studies, with the key aim to find the most robust and reliable solution with the lowest system cost. The results of the case studies suggest that the selection of the system is highly dependent on the cost, the volatility of the wind speed, solar radiation, and the size of the system; at present, hybridization is not the primary option at most of sites, with the exception of two. However, with the reduction in price of wind technology, the possibility of hybrid generation will rise.

Keywords: hybrid renewable energy; techno-economic optimization; net present cost; HOMER Pro®

1. Introduction

With the technological developments, the drive for renewable energy has intensified the use of renewable energy sources (RES), particularly wind and solar photovoltaic (PV) generation. The use of renewable energy and hybridization is taking new shapes, but PV and wind energy tend to be the center point of renewable energy hybridization [1,2]. This is mainly due to the aim of the reduction of greenhouse gas emissions, but the increased penetration of RES is also due to the dropping prices of the renewable energy sources. The technological advancements in the PV [3] and wind generation systems have been formidable. The global cumulative installed capacity of all solar PV (utility scale and rooftop) increased from 42 GW in 2010 to 714 GW in 2020 [4]. Similarly, the on-shore wind

cumulative installed capacity grew from 178 GW to 699 GW between 2010 and 2020 [4]. Th
increased generation capacity of wind and PV generation is a result of the decline in price
from USD 4731/kW to USD 883/kW for PV, and from USD 1971/kW to USD 1355/kV
between 2010 and 2020 [4].

Although there has been a precipitous decline in the cost of RES, many countrie
are struggling to adopt these technologies due to a number of reasons, ranging from th
financial resources to the technical capability for design, installation, and maintenanc
This is particularly common in developing countries, and Sudan is one such country. Th
bulk of Sudan's energy needs are met by using biofuels as key sources of energy, which i
supported by oil and hydroelectric generation. The bulk of electricity in Sudan is generatec
by hydroelectric generation, and this is mainly due to the hydroelectric project completec
after 2009, where hydroelectric power generation overtook oil-based generation [5]. Despit
having abundant sunlight and high winds, almost 46% of the population do not have acces
to electricity [6]. The international community have provided support to Sudan during th
years 2010 to 2012 to promote renewable energy, but the support could not be sustainec
for long. In year 2020, 90% of renewable electricity was produced from hydro-electri
generation, 9% from biofuels, and only 1% from solar PV [7].

Sudan has a very high potential for solar and wind energy, as can be seen from
Figure 1 [8] and Figure 2 [9]. The wind and solar generation capacity rise from the soutl
to the north. The northern regions tend to have higher solar irradiance and wind speec
whereas higher sunlight is also linked with the arid climate. Therefore, despite having a
huge area available for crop cultivation, the arid nature of the terrain limits the potentia
farming. At the same time, there are huge reserves of underground water which car
potentially help to convert the arid land into fertile pastures. The conventional rain
fed irrigation can be converted into controlled irrigation by providing the farmers witl
opportunities to use underground water when and how they need it. The key challenges ir
pumping the water out are the limitations of the grid supply in these regions, increasing
prices of fuel for diesel generators, the lack of technical capability for the maintenance o:
diesel generator, and concerns around carbon emissions. The solution to these problems is
the use of RES, which can be cheaper, easy to maintain, and emissions free.

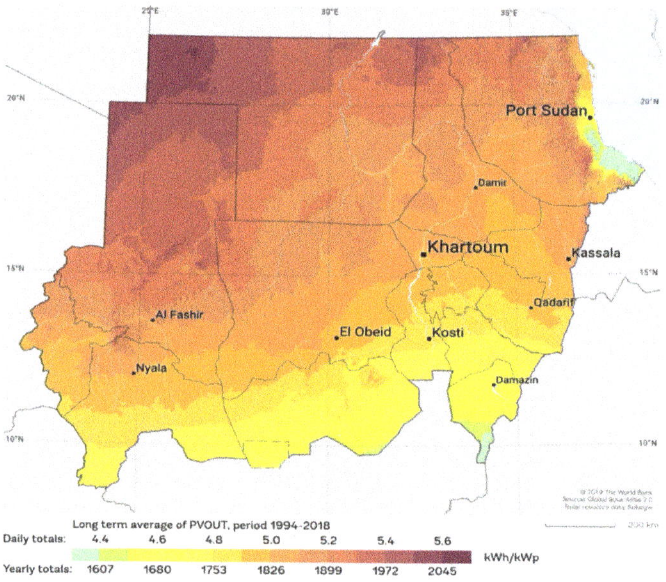

Figure 1. Solar capacity map of Sudan [8].

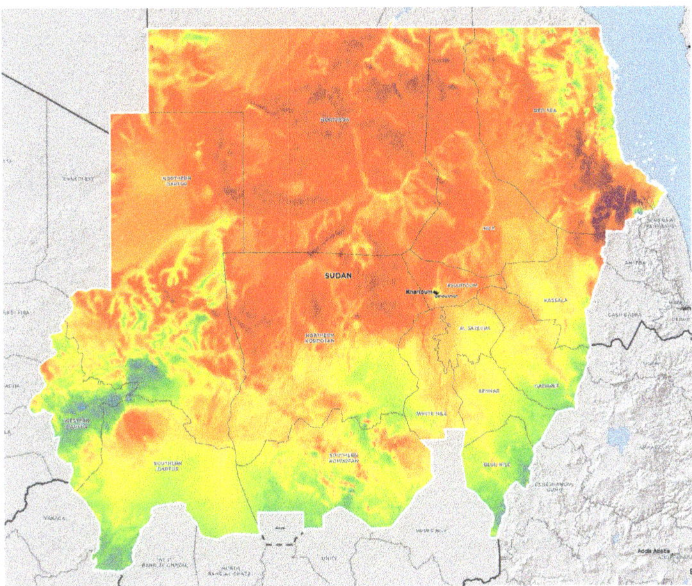

Figure 2. Wind capacity map of Sudan [9].

As discussed above, the agriculture sector in Sudan can potentially benefit from RES. However, very limited research has been conducted in the use of RES, and in particular the hybridization of RES in the agriculture sector in Sudan. A number of studies simulated the use of renewable energy in Sudan, but only a few considered specific applications to irrigation [10]. Moreover, the studies considering the irrigation application were also aimed at a higher level, and the technical feasibility of the hybrid renewable energy system was not explored at a smaller level [10]. Moreover, all of the previous studies considered the irrigation load as standard, and did not consider the climatic effects and the soil data to evaluate the water requirement of different crops. Therefore, a study is needed to optimize the size of the renewable energy source according to the real crop water requirement.

Considering the above, this paper endeavored to bridge the research gap by presenting a holistic technical feasibility approach to explore the potential of a hybrid renewable energy system for small-farm irrigation applications. The paper systematically develops the approach by considering crop evapotranspiration, crop coefficients, soil types and hydrogeology to determine the load values according to the individual crop water requirement and climatic conditions. The paper is expected to serve as a benchmark study for the design of hybrid renewable energy systems for small-scale irrigation water pumps by considering the individual crop water requirement.

The rest of the paper is organized as follows. Section 2 presents the methodology adopted to develop the approach. Section 3 delineates the case studies, and Section 4 gives the results and discussions of the case studies. Finally, Section 5 gives the conclusion of the paper.

2. Methodology

The optimal design of a hybrid energy system for irrigation requires the development of an integrated framework which can lead to the optimal design solutions. As the aim of the paper is to explore the techno-economic feasibility of the hybrid renewable energy system for irrigation applications, the objective function used for the optimization is the total Net Present Cost (NPC). Using the optimization function, the optimal system configuration is determined whilst considering the design constraints. Figure 3 gives the flow chart for the approach adopted to design the hybrid renewable energy system.

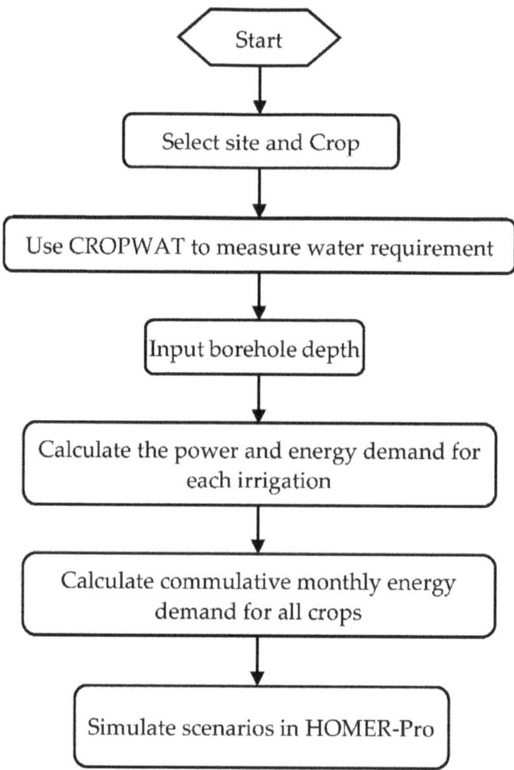

Figure 3. Approach for the design of a hybrid RE system for agriculture.

When determining the irrigation load, the decisive factor is the crop water requirement. The crop water requirement depends on a multitude of factors, including the crop type, soil type, and climate parameters, etc. Therefore, prior to the consideration of the development of the renewable energy system, the type of crop and site needs to be selected. Once the study sites are selected, the process for the calculation of the crop water requirements is initiated. The crop water requirement evaluation is carried out using CROPWAT software, which is a decision support system developed by the Food and Agriculture Organization (FAO) of the United Nations (UN) [11]. It considers a number of inputs, such as the crop, soil type, climatic data to calculate reference evapotranspiration (ET_0), crop evapotranspiration (ET_c), net irrigation water requirement (NIWR), and gross irrigation water requirement (GIWR). In order to calculate the crop water requirement, the methodology presented in Figure 4 was adopted.

Once the site and crop are selected, the climatic and soil data can be collected and used to evaluate the crop water requirement. Apart from the input parameters related to the crop and soil—i.e., crop evapotranspiration, root depth, and type of soil, etc.—climatic data including the temperature and rain is also required to finalize the net irrigation requirement for the crop.

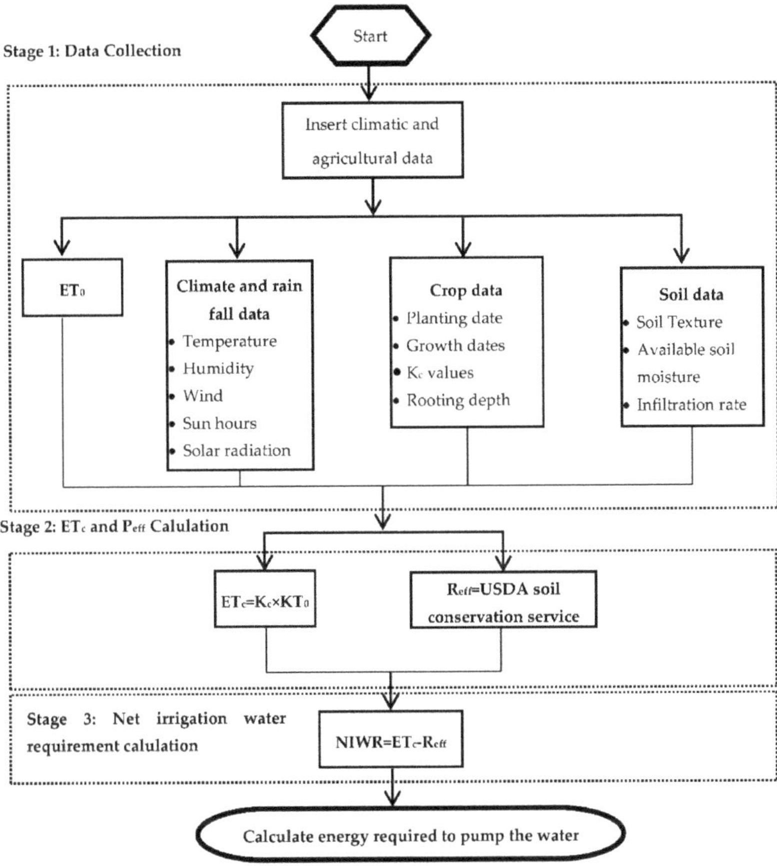

Figure 4. Flow chart for the crop water requirement and energy calculation [12].

The crop water requirement is converted into an electrical load by considering the borehole depth to design the water-pumping system. The load values of the water-pumping system are converted into energy values by considering the amount of irrigation required for each crop. Finally, monthly energy data are used to design the optimum renewable energy system for the selected site using HOMER software. A detailed study on the crop water requirement is presented in Part I of this paper [12].

The software commonly known as HOMER stands for Hybrid Optimization of Multiple Energy Resources (HOMER), and is used to optimize the use of multiple energy sources. HOMER was developed by the National Renewable Energy Laboratory (RNEL), USA for the technoeconomic modeling, simulation, and optimization of HRESs [13]. HOMER can support the user in designing a hybrid renewable energy system with economic and technical optimization, with optimal sizing to determine the optimal system configuration. HOMER can also carry out a sensitivity analysis to ascertain the sensitivity of a solution to variations in certain parameters.

A number of studies have used HOMER, and it is considered to be the global standard for the optimal planning and design of HRES in all energy sectors [10]. Its application can be found in different sectors, such as radio telecommunication stations [14], desalination plants [15], rural households [16], urban cities [17], and large residential communities [18]. Considering the wider applications of HOMER, it is used in this study to evaluate the techno-economic feasibility of a renewable hybrid energy system for small-scale irrigation purposes in Sudan.

In this paper, 12 sites in Sudan were selected to evaluate the feasibility of the hybrid renewable energy system, and the optimal energy system configuration was simulated for each site. A number of studies using HOMER have considered the application of renewable energy to the irrigation load [10,19–22]; however, most of these studies consider the irrigation load as a standard periodic load. This practice can potentially result in the over-sizing of the system. On the other hand, the irrigation is calculated either on a daily basis or on a decade basis. This necessarily means that the overall amount of water required in 10 days does not necessarily need to be supplied on the same hour of the day as it is calculated. Thus, the irrigation load is a deferrable load, and has the flexibility to shift the load in time. Considering the above, this paper models the irrigation load as a deferrable load.

Figure 5 shows the flow chart of the process adopted by HOMER to optimize the design of RES. Firstly, all of the input parameters and design constraints—such as meteorological data and system constraints, as well as the design and economic parameters of the system components—are fed to HOMER. Then, the monthly energy required by the water-pumping system at each site is provided as a deferrable load. The energy resources, i.e., wind and solar energy, along with battery storage, are selected, and finally, after selecting the final configuration of the system, the model is simulated with the constraints to find the optimum solution, i.e., the solution with the minimum net present cost.

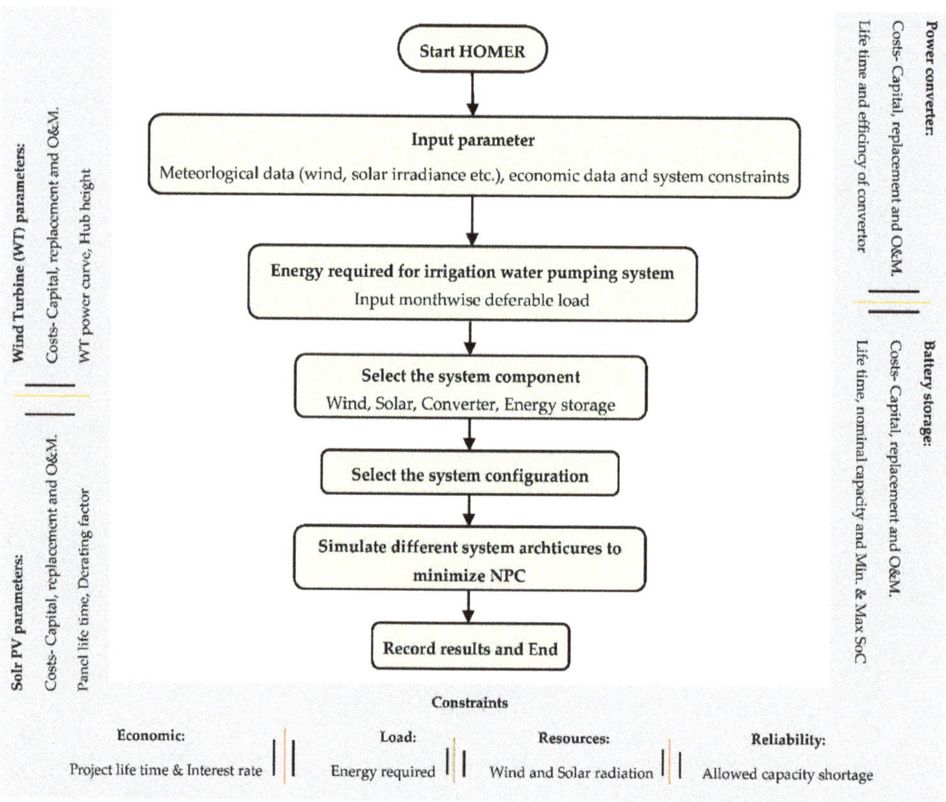

Figure 5. Homer process flow for the optimization of the hybrid renewable system.

The process is repeated for each site, and the optimum system architecture is obtained with the economic analysis.

3. Case Studies

This paper aims to explore the techno-economic feasibility of a wind–solar hybrid energy system for small-scale irrigation applications in Sudan. Considering the aim, 12 different sites were selected across Sudan. The selected sites were used to evaluate the crop water requirement for three crops, i.e., wheat, cotton and sorghum. According to a special report of the FAO on food in Sudan, most farmers use traditional rain-fed irrigation to irrigate 9 million hectares of land, and this sector serves the largest number of farmers with land between 2 and 50 hectares [23]. As the system was intended for small-scale application, it was assumed that each crop is irrigated in 1 hectare of land, and a total of 3 hectares of land was considered for irrigation purposes at each site. The location of the selected sites is given in Table 1, and the nearest weather station data available in UN FAO CLIMWAT for CROPWAT is also given in the table.

Table 1. Site locations and soil types (Adopted from [24]).

No.	Sample Location	Climate Zone	No. and Name of Nearest Weather Station on CLIMEWAT	Co-Ordinates of Weather Station Long.–Lat.	Soil Type
1	North East of Karima	Arid	54-Karima	31.85–18.55	Poorly graded sand
2	South East of Abuhamad	Arid	63-Abu-Hamed	33.31–19.53	Poorly graded silty sand
3	Osaif	Arid	8-Dongonab	37.13–21.1	Silty sandy Gravel
4	South west of Toker	Arid	59-Tokar	37.73–18.43	Silty Gravelly sand
5	West of Almatama	Arid	38-Khartoum	32.55–15.6	Gravelly Silty sand
6	Kassala	Semi Arid	41-Kassala	36.4–15.46	Silty Clay
7	Elgurashi	Semi Arid	27-Ed-Dueim	32.33–14	Silty Clay
8	Kadugli	Semi Arid	9-Kadugli	29.71–11	Sandy Silty Clay
9	Almujlad	Semi Arid	7-Babanusa	27.81–11.33	Silty Clay
10	Algadarif	Semi Arid	34-Gadaref	35.4–14.03	Silty Clay
11	Kutum North of Darfur	Semi Arid	33-Kutum	24.66–14.2	Silty Clay
12	Wadi-Halfa	Arid	3-Wadi-Halfa	31.48–21.01	Poorly graded sand

The climate data, soil data and other variables available in the CLIMWAT and CROPWAT were used to calculate the crop water requirement and irrigation requirements. Each site has a different bore hole depth, and the bore hole depth data were used from the British Geological Survey, which holds hydrogeology data [25]. The bore hole depth varied between 40 m and 80 m. An overall analysis of the irrigation requirements at all of the sites showed that the sites located further north require more water due to their arid climate, and thus more energy is required. However, at the same time, the northern sites have more wind and solar energy potential, as is apparent from the wind and solar maps for Sudan.

Figure 6 shows the peak load for all 12 sites for each crop individually. It is evident that cotton has a higher peak load, as cotton is a high water-requiring crop. The climate and rain patterns tend to reduce the need for water at different sites for some crops. However, as discussed earlier, the water-pumping system is designed by considering three crops cultivated in a 3 hectare area, and thus the combined load of all three crops is used to design the water-pumping system, which is given in Figure 7. It is not necessary that the peaks of the load coincide, and thus careful consideration should be given to calculate the combined peak in order to avoid the over- or under-sizing of the system. The peak load at each site represents the size of the water pump required at each site in kW.

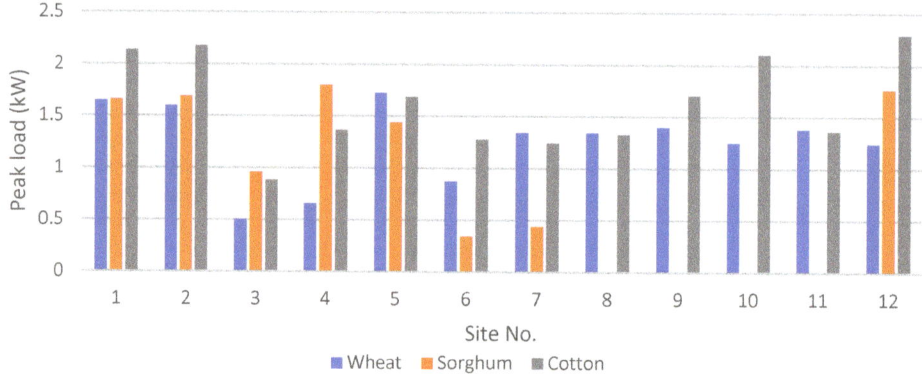

Figure 6. Peak load for each crop at each site.

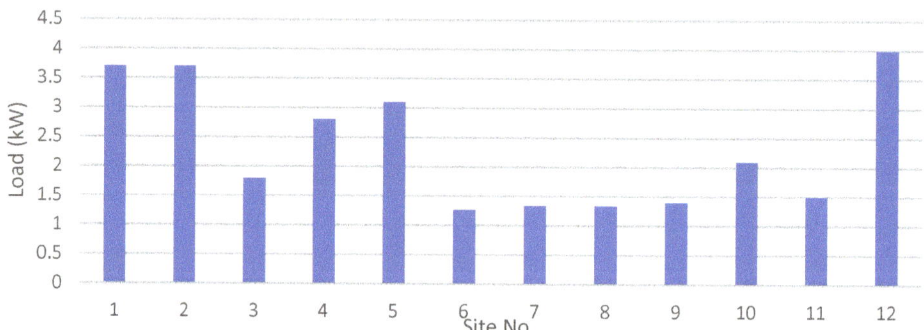

Figure 7. Cumulative peak load for all of the sites [12].

Apart from the peak load, in order to design a hybrid renewable energy system, the energy required is the key parameter. The energy required at different sites in kWh for the entire year is given in Figure 8. It should be noted that the peak loads and energy consumption do not necessarily have a linear relationship. Often, in everyday loads the peak of the load is correlated with the energy usage, but in the case of agricultural applications it is not compulsory. There is a multitude of factors that affect the energy use of the water-pumping system. These factors primarily govern the crop water requirement and are discussed above.

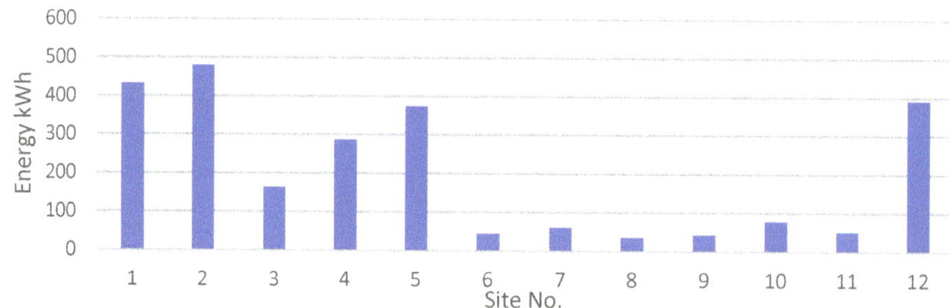

Figure 8. Total energy consumption at each site.

The energy input required in HOMER should be the monthly energy demand in order to simulate the scenarios for a deferrable load. The monthly energy demand is given in

Figure 9, where September is the month with the highest energy demand. Figure 9 presents the energy consumption at each site throughout the year, with the individual contribution of each site in each month. The cumulative effect of the energy consumption is shown by stacking all of the sites. For example, all of the sites show a very high energy demand in September, and thus the overall energy consumption from all of the sites combined is more than 400 kWh in September. A detailed discussion on the cropping patterns and seasons is given in Part I of the paper, which was published separately [12]. A sensitivity analysis of the area for agricultural use and load was carried out, and a direct correlation between the energy use, peak load and the area of agriculture was observed.

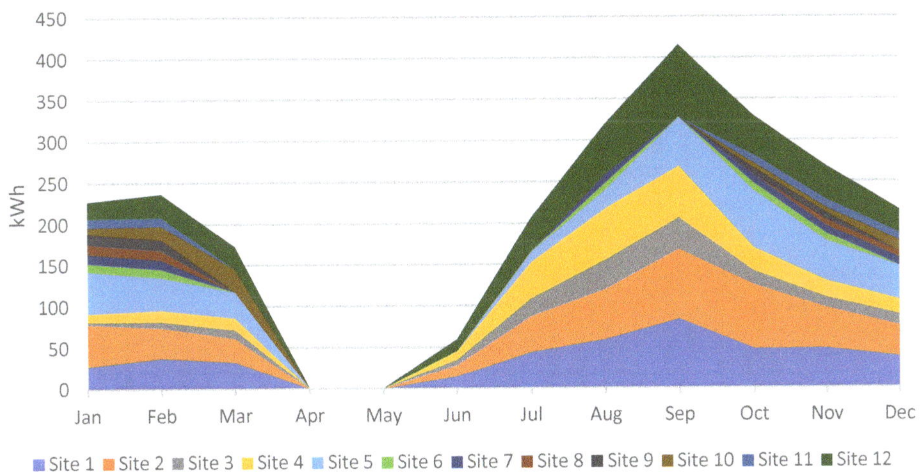

Figure 9. Monthly energy consumption at each site throughout the year.

The energy required is spread across the entire day, and thus the calculation considers the maximum load based on the assumption that the pump can run any time. Spreading the load delivery over 24 h can potentially benefit the system by reducing the renewable generation size by spreading the load over a longer period of time, and the potential for the hybridization of renewable energy sources is also promoted. It is pertinent to mention that the water pump was designed at peak load, which can only be for a few days during the crop season; the rest of the days, the pump might not need to run all day. However, due to the intermittent nature of the wind and solar PV energy, the water-pumping load was taken as a deferable load.

As the overall load and energy calculations were performed, each site was analyzed for wind and solar potential. As is evident from the solar and wind maps of Sudan, a high potential of wind and solar energy is present is Sudan. Figures 10 and 11 show the solar and wind generation potential at each site, as solar irradiance and wind speed. The variation in the wind speed is significant, and during high load periods the wind speed is low, except for sites 11 and 12. Because September is a developing stage of the cotton crop, the energy demand is high, as more water is needed at this stage. The reduction of the wind speed can potentially result in wind being less favorable compared to solar PV, which is not as volatile as wind.

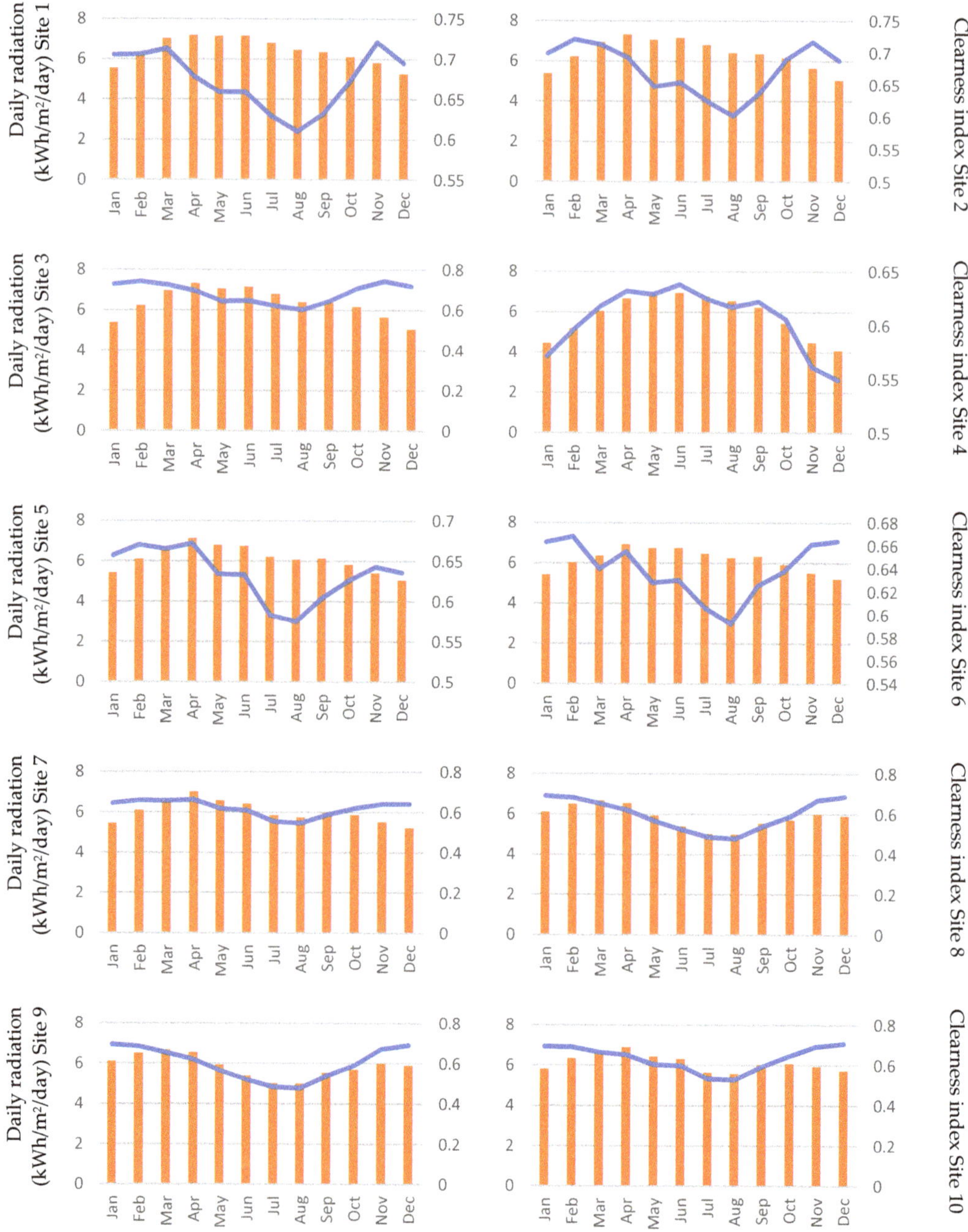

Figure 10. *Cont.*

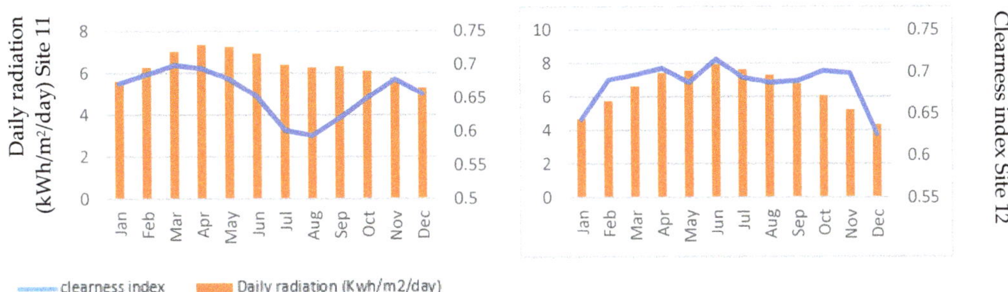

Figure 10. Daily solar radiation and clearness index for the 12 sites (the *x*-axis represents the month of the year, and the primary *y*-axis represents the daily radiation in kWh/m^2/day; the secondary *y*-axis represents the clearness index).

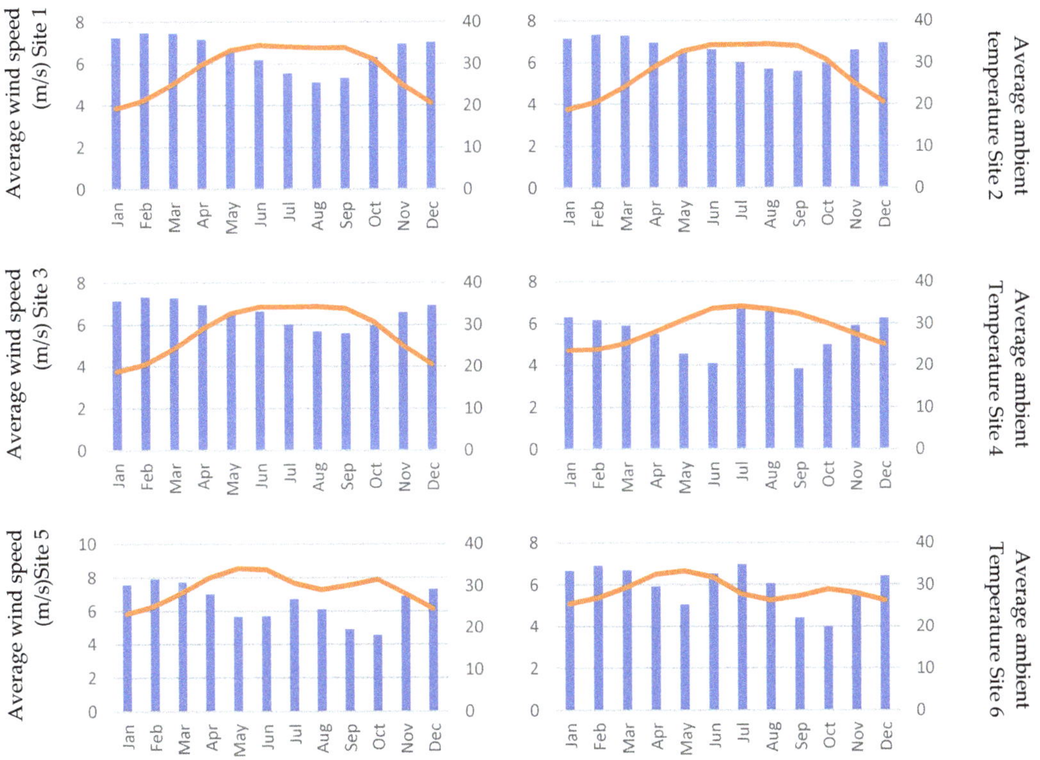

Figure 11. *Cont.*

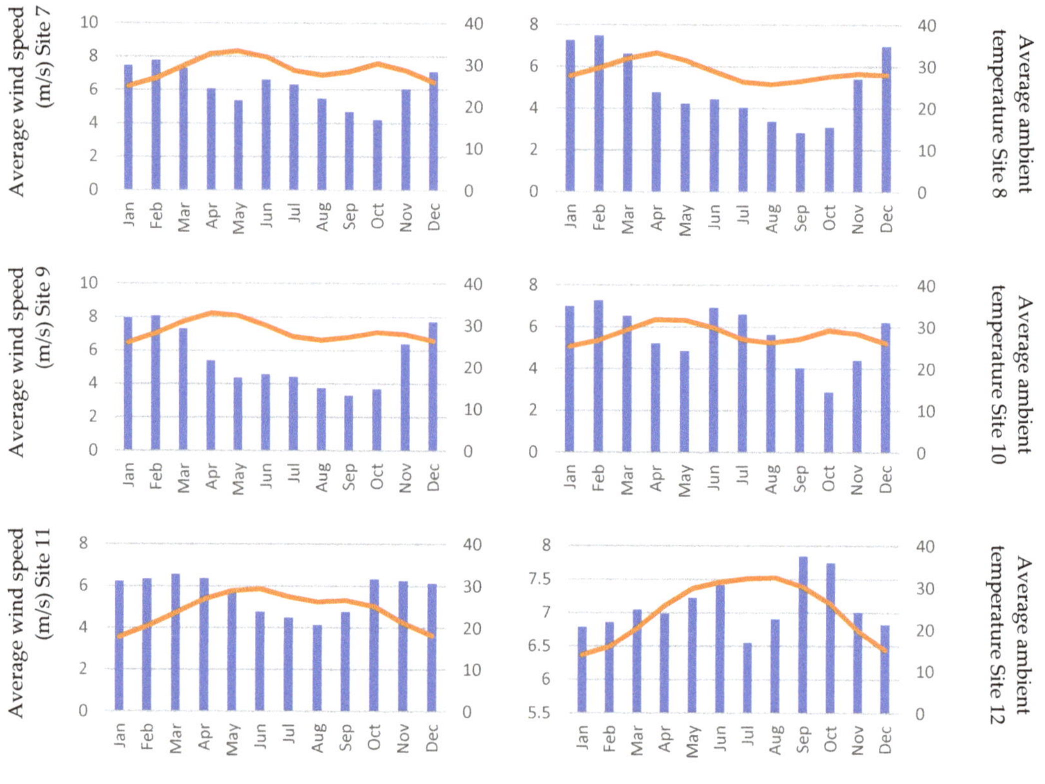

Figure 11. Average wind speed and ambient temperature for the 12 sites (the *x*-axis represents the month of the year, and the primary *y*-axis represents the speed in m/s; the secondary *y*-axis represents the average ambient temperature).

As given in Figure 5, a number of constraints are used when designing a hybrid renewable energy system using HOMER Pro. The average wind speed and solar irradiance at each site were the primary constraints. The project life time was set as 25 years, which is a commonly adopted approach considering the fact that such projects are usually government led, and governments tend to have long-term projects. A typical discount rate of 8% was considered. The reliability was set at 100%, i.e., no capacity shortage was allowed, which necessarily means that all of the load demand is met. The price for solar PV was set as $883 per kW, and for a wind generator it was set as $1355, which are global averages [4]. The operation and maintenance (OM) cost for PV was taken as 14 $/kW/year, whereas for wind it was taken as 15 $/kW/year [10]. The Surrette 4KS25P type was used and connected as a central storage system, and the capital cost of one battery was taken as 1250 $, while the OM cost was assumed to be 15 $/year [10]. Once the system constraints and the resources, i.e., wind solar generation resources, battery storage and load, were determined, a simulation setup was developed in HOMER Pro to simulate the case studies for each site, as given in Figure 12.

The simulation setup used in the case studies included wind and solar generation, a battery storage system, and a converter. A schematic diagram of the simulation setup is shown in Figure 12.

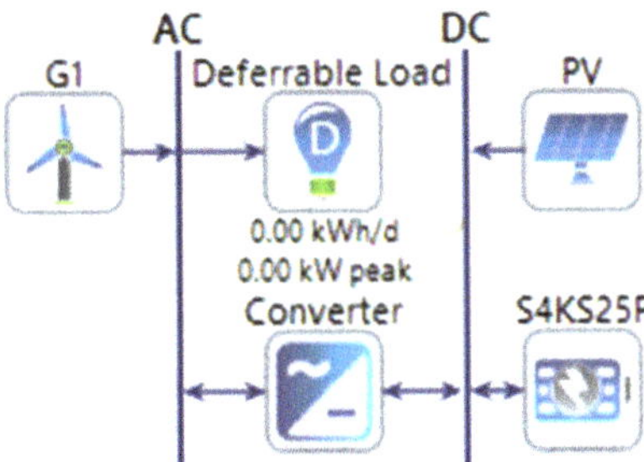

Figure 12. HOMER Pro simulation set up.

4. Results and Discussion

The case studies for all 12 sites were simulated using the load data for each site, and the resources constraints determined by climatic data were adopted. As is clearly evident from Figures 10 and 11, almost all of the sites have an abundant amount of wind and solar energy. With the decreasing prices of wind and solar generation technologies, it is expected that due to the intermittent nature of both energy sources and the availability of wind even at night, the hybrid system can potentially perform better both technically and economically. The load calculated considering the crop water requirement was used to determine the optimal system architecture, and the net present cost (NPC) and cost of electricity (COE) were calculated to compare the feasibility of a wind–solar hybrid generation system at each site. Table 2 shows the results of all 12 sites, and the two best architectures are given for each site. Moreover, the excess electricity generated is also given for each site and case.

Table 2. Results of HOMER Pro for 12 Sites.

Site No.	Best Case	Architecture	PV (kW)	WT (kW)	Batt (Nos.)	Conv (kW)	NPC ($)	COE ($)	Excess Electricity (%)
1	i	PV-Battery/Converter	31.4	0	21	16.4	70,987	0.417	75.1
1	ii	PV-WT-Battery/Converter	31.2	1	21	15.9	73,010	0.429	75.8
2	i	PV-Battery/Converter	32.9	0	20	9.35	68,183	0.362	73.6
2	ii	PV-WT-Battery/Converter	32.1	1	19	9.43	68,302	0.362	73.9
3	i	PV-Battery/Converter	14.3	0	10	4.88	31,960	0.496	73.6
3	ii	PV-WT-Battery/Converter	13.8	1	9	4.99	32,450	0.504	73.9
4	i	PV-Battery/Converter	32.3	0	14	7.62	57,485	0.508	82.3
4	ii	PV-WT-Battery/Converter	28.5	1	15	7.71	57,729	0.510	80.6
5	i	PV-Battery/Converter	24.7	0	17	10.1	55,516	0.377	71
5	ii	PV-WT-Battery/Converter	25.2	1	16	14.7	58,803	0.399	72.8
6	i	PV-Battery/Converter	9.11	0	1	0.130	10,832	0.609	91.8
6	ii	PV-WT-Battery/Converter	8.85	1	1	0.260	11,411	0.641	92.3
7	i	PV-Battery/Converter	14.8	0	1	0.130	16,595	0.687	93
7	ii	PV-WT-Battery/Converter	15.1	1	1	0.260	19,396	0.803	93
8	i	PV-Battery/Converter	4.03	0	2	15	13,234	0.939	83
8	ii	PV-WT-Battery/Converter	2.79	1	2	18.3	15,750	1.12	79.5
9	i	PV-Battery/Converter	11.9	0	1	0.680	13,612	0.814	93.8
9	ii	PV-WT-Battery/Converter	11.5	1	1	0.521	15,811	0.945	93.9

Table 2. *Cont.*

Site No.	Best Case	Architecture	PV (kW)	WT (kW)	Batt (Nos.)	Conv (kW)	NPC ($)	COE ($)	Excess Electricity (%)
10	i	PV-WT-Battery/Converter	6.48	1	6	4.46	20,125	0.658	81
10	ii	PV-Battery/Converter	7.53	0	7	4.68	20,386	0.666	81.9
11	i	PV-Battery/Converter	21.4	0	1	0.13	23,222	1.15	96.2
11	ii	PV-WT-Battery/Converter	21.4	1	1	0.26	25,723	1.28	96.3
12	i	PV-WT-Battery/Converter	22.8	2	17	9.71	58,312	0.379	72.5
12	ii	PV-Battery/Converter	29.3	0	17	11.1	61,656	0.400	76.1

Only two sites were identified, which resulted in a preference of a hybrid of wind and solar for the purposes of supporting the irrigation load. Despite the optimality of hybrid energy, solar energy was the dominant source of electricity. The hybridization scenario at site 12 showed interesting results compared to all of the other sites, where removing 2 kW wind generation resulted in a 6.5 kW addition in solar generation. This shows the significance of the wind generation, and is an indicator of the fact that despite being almost double in the price, the hybridization of these sources can potentially provide significant benefits.

As the overall cost of a wind turbine tends to be higher than solar PV, the NPC tends to be higher than the solar PV generation at most sites, except site 10 and site 12. The lowest cost of electricity was $0.362, and highest was $1.15. The high costs are the results of the excess electricity generated. The lowest excess electricity was 73.6%, and the highest was 96.2%.

The power generated by the wind and PV generators, and the total electrical load served at site 10 and site 12 is shown in Figures 13–16. Figures 13 and 15 show the generation by all of the sources during a single year. It is pertinent to mention that these loads are not operational throughout the year, and the energy demand varies every day and month. However, in order to ensure that all of the load demand is met, the energy system's size was optimized to provide 100% load serving reliability. Therefore, a large amount of excess energy was observed at all of the sites. As can be seen from the wind profiles, the power production using a wind turbine at site 12 tends to be better than site 10. Moreover, the amount of water and the number of irrigations required at site 12 also contribute to the need of additional power which is provided by the wind generation. As the pump rating is calculated considering the maximum load requirement, Site 12 tends to have a higher demand of electricity and to meet that demand using the rated pump; wind energy provides the extra energy, which can serve the base load to meet the energy needs. The weather patterns, in combination with the soil type, reduce the load significantly for site 10, which is evident from Figure 14. It is evident from Figures 13 and 15 that a great amount of excess energy is produced, which is wasted. Apart from the water-pumping load, the storage system's annual throughput for site 10 is 440 kWh/year, and for site 12 it is 5383 kWh/year. Moreover, the converter losses for site 10 are 19.6 kWh/year, and for site 12 they are 241 kWh/year. While considering the overall generation, the only energy consumed is through the energy used by the water pump, energy storage throughput and converter losses; a large amount of the energy is surplus, which can potentially be used for other purposes. The energy required at each site for the water-pumping system is given in Figure 8. The excess amount of energy in each case is given in Table 2.

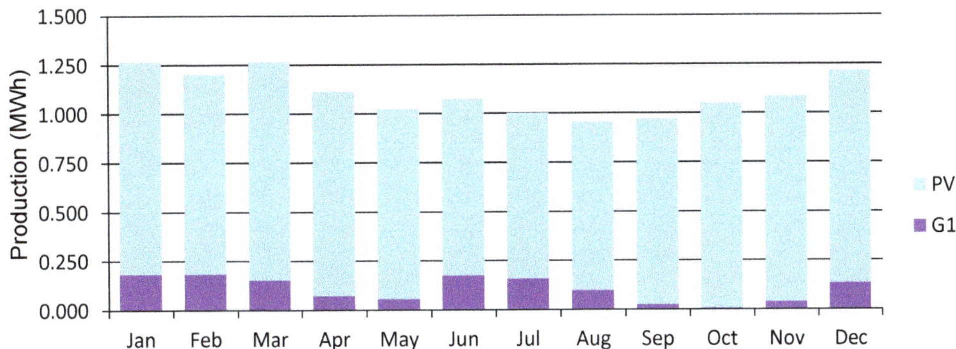

Figure 13. Power generation share at Site 10 (PV represents solar PV generation, and G1 represents wind turbine generation).

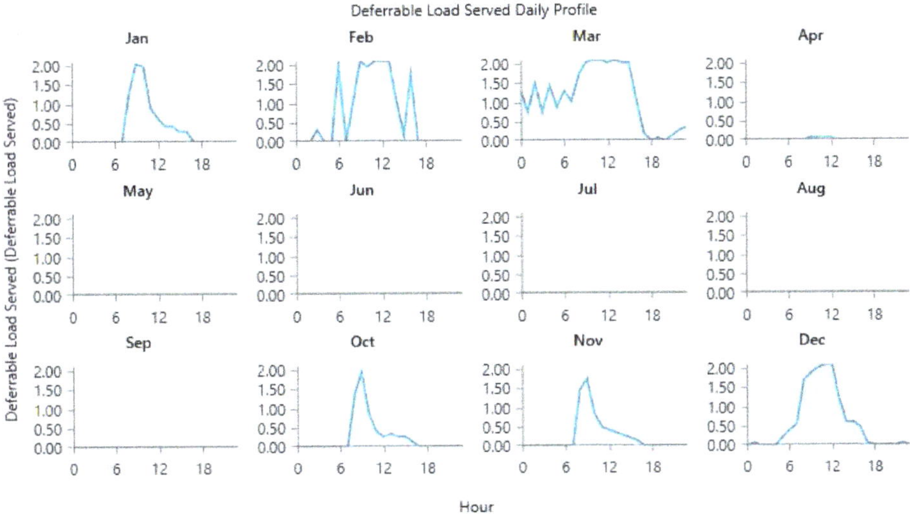

Figure 14. Electric load served (daily profile) at Site 10.

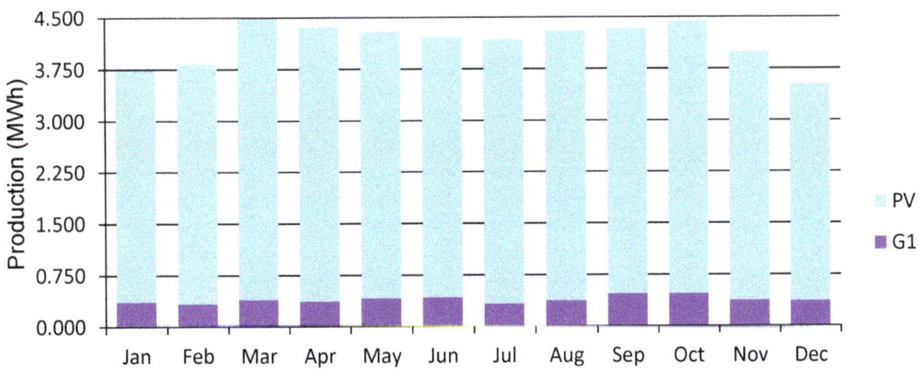

Figure 15. Power generation share at Site 12 (PV represents solar PV generation, and G1 represents wind turbine generation).

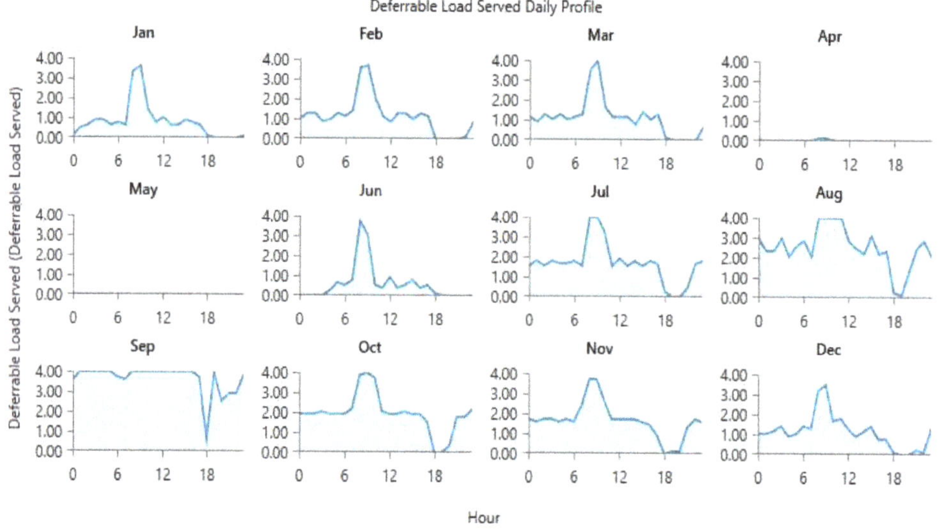

Figure 16. Electric load served (daily profile) at Site 12.

The overall results suggest that the hybridization of wind and solar energy can potentially benefit the system by reducing the need for solar PV multifold. The availability of wind over 24 h can spread the load and reduce the system size; thus, a smaller wind system might be able to replace a bigger solar system; however, the cost of a wind generator compared to solar PV is as yet at a higher level for applications on a small scale. Although the present reduction in cost of on-shore wind generation is promising, wind energy has yet to take the long journey to become feasible for small-scale applications. The application of hybrid energy systems in small scale applications, and particularly in irrigation applications, is not yet feasible in Sudan.

5. Conclusions

This paper presents a techno-economic assessment of wind–solar hybrid generation for a water-pumping system for small-scale irrigation projects. A comprehensive framework is presented which systematically develops the energy system using the crop and site data to measure the energy required. The data from 12 sites in Sudan were used to calculate the crop water requirement using FAO CROPWAT software, and the energy need was evaluated using the crop water requirement. HOMER Pro was used for the techno-economic assessment of different architectures, and to evaluate the feasibility of the hybrid energy systems for small irrigation applications. The results suggest that wind energy is suitable at two sites, i.e., site 10 and site 12. Site 12 tends to be more suitable for the hybridization of energy generation compared to all of the other sites. From the case studies, it is apparent that hybrid energy generation is yet not economically feasible due to the higher prices of on-shore wind generation. Therefore, due to the higher costs and technical challenges of wind turbines, and the ease of installation and maintenance of solar PV, solar PV tends to be more suitable for such applications as opposed to a hybrid of wind and solar PV. In conclusion, the wind–solar hybrid energy system for small-scale irrigation applications in Sudan is not yet feasible in most regions.

Author Contributions: Conceptualization, M.I., Z.A.K. and A.O.A.; methodology, Z.A.K. and M.I.; software, Z.A.K. and A.A.; validation, Z.A.K., M.I. and O.E.D.; formal analysis, Z.A.K.; investigation, Z.A.K. and A.A.; resources, M.I. and A.O.A.; data curation, Z.A.K.; writing—original draft preparation, Z.A.K. and A.A.; writing—review and editing, M.I. and O.E.D.; visualization, Z.A.K.; supervision, M.I. and A.O.A.; project administration, M.I.; funding acquisition, M.I. and A.O.A. All authors have read and agreed to the published version of the manuscript.

Funding: This research was funded by Innovate UK and the Academy of Medical Sciences, grant number 60558 and 62327, under the global challenge research fund.

Conflicts of Interest: The authors declare no conflict of interest.

References

1. Wu, Y.; Qu, J.; Chu, P.K.; Shin, D.-M.; Luo, Y.; Feng, S.-P. Hybrid photovoltaic-triboelectric nanogenerators for simultaneously harvesting solar and mechanical energies. *Nano Energy* **2021**, *89*, 106376. [CrossRef]
2. Alotaibi, M.A.; Eltamaly, A.M. A Smart Strategy for Sizing of Hybrid Renewable Energy System to Supply Remote Loads in Saudi Arabia. *Energies* **2021**, *14*, 7069. [CrossRef]
3. Eltamaly, A.M.; Mohamed, M.A.; Abo-Khalil, A.G. Design and Comprehensive Analysis of Maximum Power Point Tracking Techniques in Photovoltaic Systems. In *Advanced Technologies for Solar Photovoltaics Energy Systems*; Springer: Berlin/Heidelberg, Germany, 2021; pp. 253–284.
4. IRENA Renewable Cost Database. Renewable Power Generation Costs in 2020. 2020. Available online: https://www.irena.org/-/media/Files/IRENA/Agency/Publication/2018/Jan/IRENA_2017_Power_Costs_2018.pdf (accessed on 15 June 2021).
5. International Energy Agency (IEA). Sudan, Key Energy Statistics. 2019. Available online: https://www.iea.org/countries/sudan (accessed on 10 October 2021).
6. Energy Sector Management Assistatance Program. Sudan Tracking SDG7, the Energy Progress Report. 2021. Available online: https://trackingsdg7.esmap.org/country/sudan (accessed on 30 October 2021).
7. Energy Profile. 2018. Available online: https://www.irena.org/IRENADocuments/Statistical_Profiles/Africa/Sudan_Africa_RE_SP.pdf (accessed on 15 June 2021).
8. World Bank Group, Global Solar Atlas. [Data/information/map] Obtained from the "Global Solar Atlas 2.0, a Free, Web-Based Application Is Developed and Operated by the Company Solargis s.r.o. on Behalf of the World Bank Group, Utilizing Solargis Data, with Funding Provided by the Energy Sector Management Assistance Program (ESMAP)". Available online: https://globalsolaratlas.info/download/sudan (accessed on 6 July 2021).
9. World Bank Group, Global Wind Atlas. [Data/information/map] Obtained from the "Global Wind Atlas 3.0, a Free, Web-Based Application Developed, Owned and Operated by the Technical University of Denmark (DTU). The Global Wind Atlas 3.0 Is Released in Partnership with the World Bank Group, Utilizing Data Provided by Vortex, Using Funding Provided by the Energy Sector Management Assistance Program (ESMAP)". Available online: https://globalwindatlas.info/area/Sudan (accessed on 6 July 2021).
10. Elkadeem, M.R.; Wang, S.; Sharshir, S.W.; Atia, E.G. Feasibility analysis and techno-economic design of grid-isolated hybrid renewable energy system for electrification of agriculture and irrigation area: A case study in Dongola, Sudan. *Energy Convers. Manag.* **2019**, *196*, 1453–1478. [CrossRef]
11. Food And Agriculture Organization of the United Nations CropWat. 2021. Available online: http://www.fao.org/land-water/databases-and-software/cropwat/en/ (accessed on 19 August 2021).
12. Khan, Z.A.; Imran, M.; Umer, J.; Ahmed, S.; Diemuodeke, O.E.; Abdelatif, A.O. Assessing Crop Water Requirements and a Case for Renewable-Energy-Powered Pumping System for Wheat, Cotton, and Sorghum Crops in Sudan. *Energies* **2021**, *14*, 8133. [CrossRef]
13. NREL. Homer Pro. 2021. Available online: https://www.homerenergy.com/products/pro/index.html (accessed on 4 May 2021).
14. Kanase-Patil, A.B.; Saini, R.P.; Sharma, M.P. Integrated renewable energy systems for off grid rural electrification of remote area. *Renew. Energy* **2010**, *35*, 1342–1349. [CrossRef]
15. Halabi, L.M.; Mekhilef, S.; Olatomiwa, L.; Hazelton, J. Performance analysis of hybrid PV/diesel/battery system using HOMER: A case study Sabah, Malaysia. *Energy Convers. Manag.* **2017**, *144*, 322–339. [CrossRef]
16. Baghdadi, F.; Mohammedi, K.; Diaf, S.; Behar, O. Feasibility study and energy conversion analysis of stand-alone hybrid renewable energy system. *Energy Convers. Manag.* **2015**, *105*, 471–479. [CrossRef]
17. Baek, S.; Park, E.; Kim, M.-G.; Kwon, S.J.; Kim, K.J.; Ohm, J.Y.; del Pobil, A.P. Optimal renewable power generation systems for Busan metropolitan city in South Korea. *Renew. Energy* **2016**, *88*, 517–525. [CrossRef]
18. Abedi, S.; Alimardani, A.; Gharehpetian, G.B.; Riahy, G.H.; Hosseinian, S.H. A comprehensive method for optimal power management and design of hybrid RES-based autonomous energy systems. *Renew. Sustain. Energy Rev.* **2012**, *16*, 1577–1587. [CrossRef]
19. Ronad, B.F.; Jangamshetti, S.H. Optimal cost analysis of wind-solar hybrid system powered AC and DC irrigation pumps using HOMER. In Proceedings of the 2015 International Conference on Renewable Energy Research and Applications (ICRERA), Palermo, Italy, 22–25 November 2015; pp. 1038–1042.

20. Sarkar, N.I.; Sifat, A.I.; Rahim, N.; Reza, S.M.S. Replacing diesel irrigation pumps with solar photovoltaic pumps for sustainable irrigation in Bangladesh: A feasibility study with HOMER. In Proceedings of the 2015 2nd International Conference on Electrical Information and Communication Technologies (EICT), Khulna, Bangladesh, 10–12 December 2015; pp. 498–503.
21. Abdel-Salam, M.; Ahmed, A.; Ziedan, H.; Sayed, K.; Amery, M.; Swify, M. A solar-wind hybrid power system for irrigation in Toshka area. In Proceedings of the 2011 IEEE Jordan Conference on Applied Electrical Engineering and Computing Technologies (AEECT), Amman, Jordan, 6–8 December 2011; pp. 1–6. [CrossRef]
22. Haffaf, A.; Lakdja, F.; Meziane, R.; Abdeslam, D.O. Study of economic and sustainable energy supply for water irrigation system (WIS). *Sustain. Energy Grids Netw.* **2021**, *25*, 100412. [CrossRef]
23. FAO. Special Report—2020 FAO Crop and Food Supply Assessment Mission (CFSAM) to the Republic of the Sudan. 2021. Available online: https://doi.org/10.4060/cb4159en (accessed on 15 June 2021).
24. Shallal, D.M.M. Sudan Subgrade Soils Characteristics. *IOSR J. Eng.* **2014**, *4*, 48–56. [CrossRef]
25. British Geological Survey. Hydrology of Sudan. Available online: http://earthwise.bgs.ac.uk/index.php/Africa_Groundwater_Atlas_Hydrogeology_Maps (accessed on 13 June 2021).

Article

Techno-Economic Analysis of Fast Pyrolysis of Date Palm Waste for Adoption in Saudi Arabia

Sulaiman Al Yahya [1,*], Tahir Iqbal [2,3], Muhammad Mubashar Omar [4] and Munir Ahmad [5]

1 Department of Mechanical Engineering, Qassim University, Qassim 51452, Saudi Arabia
2 Faculty of Agricultural Engineering & Technology, PMAS-Arid Agriculture University, Rawalpindi 46000, Pakistan; tahir.iqbal@uaar.edu.pk
3 National Centre of Industrial Biotechnology, PMAS-Arid Agriculture University, Rawalpindi 46000, Pakistan
4 Department of Energy Systems Engineering, University of Agriculture Faisalabad, Faisalabad 38000, Pakistan; mubi_mahmood@yahoo.com
5 Pakistan Agricultural Research Council, Islamabad 44000, Pakistan; drmunir.abei@hotmail.com
* Correspondence: syhya@qec.edu.sa; Tel.: +966-56-519-7373

Citation: Yahya, S.A.; Iqbal, T.; Omar, M.M.; Ahmad, M. Techno-Economic Analysis of Fast Pyrolysis of Date Palm Waste for Adoption in Saudi Arabia. *Energies* **2021**, *14*, 6048. https://doi.org/10.3390/en14196048

Academic Editor: Gabriele Di Giacomo

Received: 20 July 2021
Accepted: 8 September 2021
Published: 23 September 2021

Publisher's Note: MDPI stays neutral with regard to jurisdictional claims in published maps and institutional affiliations.

Abstract: Date palm trees, being an important source of nutrition, are grown at a large scale in Saudi Arabia. The biomass waste of date palm, discarded of in a non-environmentally-friendly manner at present, can be used for biofuel generation through the fast pyrolysis technique. This technique is considered viable for thermochemical conversion of solid biomass into biofuels in terms of the initial investment, production cost, and operational cost, as well as power consumption and thermal application cost. In this study, a techno-economic analysis has been performed to assess the feasibility of converting date palm waste into bio-oil, char, and burnable gases by defining the optimum reactor design and thermal profile. Previous studies concluded that at an optimum temperature of 525 °C, the maximum bio-oil, char and gases obtained from pyrolysis of date palm waste contributed 38.8, 37.2 and 24% of the used feed stock material (on weight basis), respectively, while fluidized bed reactor exhibited high suitability for fast pyrolysis. Based on the pyrolysis product percentage, the economic analysis estimated the net saving of USD 556.8 per ton of the date palm waste processed in the pyrolysis unit. It was further estimated that Saudi Arabia could earn USD 44.77 million per annum, approximately, if 50% of the total date palm waste were processed through fast pyrolysis, with a payback time of 2.57 years. Besides that, this intervention will reduce 2029 tons of greenhouse gas emissions annually, contributing towards a lower carbon footprint.

Keywords: pyrolysis; date palm waste; techno-economic; fluidized bed reactor; biofuels

1. Introduction

In addition to abundant oil reserves, Saudi Arabia is known for the production of delicious dates. The major crops grown in this water-scarce country include cereals, vegetables, fruits (date palm, grapes) and forage crops (alfalfa). Date palm trees are mostly grown in arid to semi-arid regions such as MENA. It has been estimated that out of total 120 million date palm trees on the planet, 23 million (19.16%) are present in Saudi Arabia. Date palm waste is generated in the form of dry leaves, stems, pits/ seeds, etc. An average date palm tree can produce 15–20 kg biomass per annum, whereas date pits account for 10% of total fruit production [1]. This indicates that 345,000 tons (23 million trees × 15 kg/tree) per annum date palm waste is produced in Saudi Arabia [2]. The date palm trees waste comprises of cellulose, hemicellulose, and lignin. Generally, this waste is burned directly on the farm or disposed of in the landfills resulting in environmental pollution. Due to highly volatile solids and low moisture contents, the date palm biomass can be transformed into biofuels (bio-oil, biochar and syn gas) using thermochemical and extraction techniques [3], thus reducing the dependence on fossil fuels.

The techniques used for the thermochemical conversion of biomass into biofuel include combustion, gasification, and pyrolysis. Date palm leaves and seeds are considered more suitable for pyrolysis compared to the leaf stem on account their of higher calorific value and volatile contents [4]. Based on the initial weight and chemical composition of the feedstock used, the pyrolysis technique produces 60–70% of liquid biofuels, 15–25% char and 10–20% gases [5]. The obtained byproducts of pyrolysis have heating value ranges of 15–30 MJ/kg (bio-oil), 17–36 MJ/kg (biochar) and 6.4–9.8 MJ/kg (syn gas) respectively [6–8].

Various research studies have indicated the potential of date palm waste pyrolysis for biofuel generation. A study recorded the production of 50%wt bio-oil, 30%wt biochar and 20%wt gas through pyrolysis of date palm biomass in a fixed bed reactor for 2 h at 500 °C. The calorific value of the obtained bio-oil was calculated to be 28.636 MJ/kg [9]. While slow pyrolysis of the date palm waste favors biochar production, the quantity of the bio-oil, biochar and gas are heavily influenced by the process temperature. A sharp decrease in the yield of biochar from 43 to 22% was observed when the thermal profile of the pyrolysis reactor jumped from 350 to 650 °C [10]. The optimum thermal profile for fast pyrolysis of date palm biomass was noted to be 525 °C where the maximum percentage of bio-oil, biochar and gas (38.8, 37.2 and 24, respectively) was achieved with energy conversion of 87% [11]. In another study, pyrolysis of date palm (feeding rate of 0.30 kg/h) in a fixed bed reactor at 500 °C with heating rate of 2000 watts produced bio-oil, char and syngas between the respective ranges of 17.0–25.9%, 31.0–36.66% and 39.0–46.33% [12]. The biochar produced after fast pyrolysis of 2 ton feedstock per batch was sold at a price of US 2.48 $/kg [13], whereas, the price of bio-oil obtained at 330 °C was US 1.04 USD/gal [14]. The cost of pyrolytic bio-oil obtained from 100 tons of biomass per day was 0.94 USD/gal slightly higher compared to the previous reported investigation due to variation in biomass collection and transportation rate. The cost of energy generated from bio-oil and char was calculated to be USD 6.35 MMBTU (USD 0.002 per kilowatt hour) [14].

The upgradation of bio-oil to transportation fuel is also possible after treatment. The cost of bio-oil produced from the fast pyrolysis ranges between USD 12–26/GJ. The cost of electricity and sales of char affects the cost of bio-oil production. It is estimated that around 18% of the obtained bio-oil would be used for meeting the initial electricity requirement of the pyrolysis plant [15], whereas the cost of diesel production by upgradation of bio-oil is around USD 0.56–0.82 per liter [16]. In recent years, some research articles have been published regarding techno-economics of date palm pyrolysis for biochar production to be used for soil amendments (such as neutralizing the acidity) and as a low-cost absorbent for organic and non-organic pollutant removals. However, no noticeable work has been carried out specifically targeting bio-oil production.

The objective of this study is to determine the techno-economic feasibility of thermal pyrolysis of date palm residue for production of bio-oil, biochar and syn gas while considering initial investment, cost of production and operation cost as well as power consumption and thermal applications by using the previous literature as a reference This paper focuses on defining the potential impact of byproduct on the cost of energy production (thermal and power applications). Besides the cost estimates, the feasibility of date palm waste pyrolysis has been evaluated based on mass and energy balance. The pyrolysis processes were assessed using mass balance that controlled the quantity of the obtained products, while energy balance was employed to optimize the operational cost based on energy used or lost. The techno-economic assessment of date palm is useful for the commercialization of pyrolysis techniques as well as for calculating the production cost of biofuels and chemicals in Saudi Arabia.

2. Process Description

2.1. Date Palm Waste Analysis

For this techno-economic study, the ultimate and proximate analysis of date palm biomass waste is taken from the previous literature produced by Hussain et al., 2014 [2]. The analysis is shown below in Table 1 below.

Table 1. Ultimate and Proximate analysis of Date palm biomass.

Ultimate Analysis					
Biomass	C	H	N	S	O
Seed	44.1	6.1	0.9	0.6	48.3
Leaf	50.4	6.3	1.1	0.4	41.8
Stem	38.1	5.2	0.8	0.3	55.6
Proximate Analysis					
Biomass	Moisture	Volatile Matter	Ash contents	Fixed Carbon	
Seed	5.1	75.1	9.8	8.1	
Leaf	5.3	77.5	12.1	6.1	
Stem	18.1	52.1	20.2	8.1	

2.2. Thermochemical Conversion Process

The process of converting biomass into biofuels (gases and liquids) is either performed through biochemical, physio-chemical, or thermochemical approach. Thermochemical approach comprises super heating of biomass in the presence of some gasifying medium (air, steam, oxygen) to produce a mixture of burnable and non-burnable gases called syn gas, bio-oil, char, and tarry products. These obtained biofuels can be utilized in meeting the thermal and power needs of the local as well as industrial community, whereas the obtained char could be used for soil amendment and helps in decreasing the acidity of the soil. However, the obtained tar either went for thermal breakdown to low chain hydrocarbons or, after treatment with suitable additives, it can be used in petrochemical industry [17,18]. Figure 1 below shows three chains of processes that allow biomass to meet power demands in the form of electrical or heat energy.

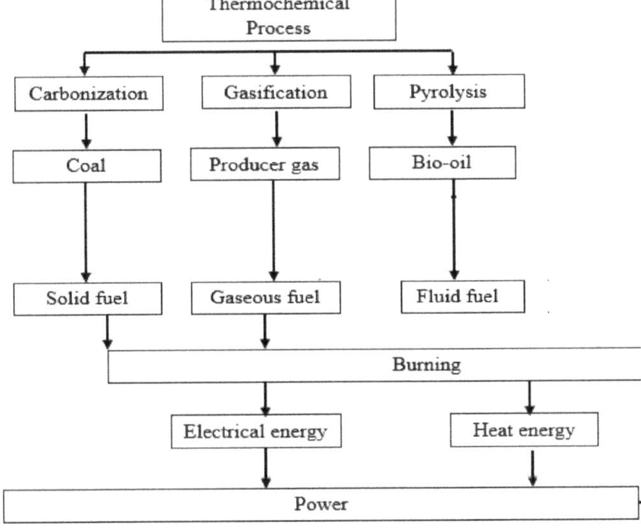

Figure 1. Thermochemical routes to energy.

To carry out biochemical and extraction processes (through physiochemical technique) successfully, several tasks must be convened, such as raw materials' formulation, maintaining ideal thermal profile, and ensuring the allowable ranges of pH and other operational parameters. However, pyrolysis technique makes the system performance efficient and solves the problems of biomass handling by generating energy in a reliable and efficient manner. The biofuel and clean synthetic gas produced through pyrolysis retain 70–80% of the chemical energy.

2.3. The Principle of Fast Pyrolysis

Fast pyrolysis is a process wherein biomass undergoes breakdown into vapors under high temperature in the absence of oxidizing medium. These vapors are condensed in the form of a liquid having an energy value half of the fossil oil. Therefore, a higher yield of biofuel is required to compensate the lower energy value. A higher yield of oils up to 60% has to be ensured through fast pyrolysis at a reaction temperature of 500–550 °C at 1 atm pressure with particle residence time of 30–1500 ms (milliseconds) [8,19,20]. The factors directly controlling the bio-oil yield include feedstock type, reactor internal temperature, vapors residence time, char separation and feedstock ash contents. The maximum yield observed at 500 °C is 75% along with biochar and syn gas [21].

2.4. Pyrolysis Unit Description

The fluidized bed pyrolysis reactor used for this investigation had a capacity of 10 tons per day (dry feed stock with moisture content less than 10%). The reactor was equipped with condenser, cyclone, and gas separator. A shredder was also used for size reduction of biomass waste for a better increase heat transfer during the pyrolysis. The fluidized bed reactor act as both pyrolizer and a combustor. The dried biomass is shredded and milled before feeding into the chamber for efficient transformation into biofuels. The quality of the pyrolysis byproduct is directly affected by the biomass particle size. Therefore, the feed stock was converted into power form of <1 mm in size using the shredder [22]. The power requirement for biomass size reduction is up to 24 kWh/ton [23]. Fast pyrolytic auto thermal technique has been adopted for converting carbonaceous biomass material into oil, biochar, and condensable gases mixture. The schematic of the 10 tpd date palm auto thermal fast pyrolysis unit is shown in Figure 2 below.

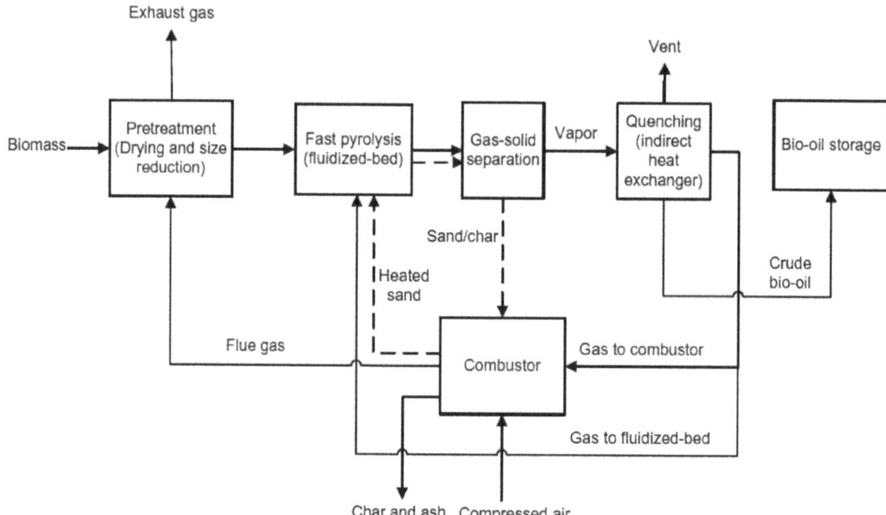

Figure 2. Schematic of auto thermal pyrolysis unit.

The products of pyrolysis, such as char and gas, are separated using cyclone separator or electrostatic separator. The efficiency of the separator is around 99%. Some amount of ash remains in the oil and gas giving them a brown color. The quenching process is carried out to immediately condense the hot pyrolysis vapors (coming out of the auto thermal pyrolysis reactor) to crude bio-oil form. The pyrolytic bio-oil yield is directly affected by the condensation. Equation (1) are used to calculate the percentages of bio-oil or char and syngas, respectively [23].

$$\text{Syngas } (\%) = 100 - (\text{Bio oil})\% - (\text{Biochar})\% \tag{1}$$

3. Techno-Economic Analysis

A techno-economic analysis of biofuels production from date palm waste is required to establish its viability at commercial scale. In this analysis, the technical and economic aspects of date palm waste pyrolysis have been analyzed against each other. First, the mass and energy balances were worked out and then economic analysis was performed to estimate the cost of biofuel production and profit gain [24].

3.1. Technical Analysis

Biochemical or thermochemical techniques can be utilized for converting the date palm biomass into biofuel. The thermochemical transformation can be classified into pyrolysis, gasification, and liquefaction [25]. While pyrolysis is a promising technique to convert biomass such as lignocellulosic into fuel, it has no potential in the case of grain-based biofuels such as corn ethanol [26,27]. At present, flash or fast pyrolysis is one of the areas being most widely researched. It is considered highly promising for producing biofuels from solid biomass [28,29]. Researchers in Saudi Arabia have also described fast pyrolysis a feasible technique for deriving biofuels from date palm waste [2].

The optimum thermal profile range for carrying out biomass pyrolysis is between 450–650 °C under anaerobic conditions. The products obtained in this thermal profile include bio-oil, char, and mixture of burnable and non-burnable gases (burnable and non-burnable). Bio-oil has various advantages over unprocessed biomass given its higher volumetric energy density and economical transportation. However, lower calorific values of biofuels are noted due to the presence of oxygenated compounds and moisture contents. The presence of organic acid is another factor that makes biofuel less preferable over fossil fuels. Mostly, catalytic treatment is carried out during or after pyrolysis to minimize the ratio of oxygenated compounds in pyrolytic bio-oil.

3.1.1. Yield Assessment and Carbon Conversion Efficiency

Yield assessment and biomass conversion are important parameters required for system optimization as well as for economic analysis. The percentage of pyrolysis products can be altered by changing the operational thermal profile of the reactor. The obtained yield of the process can be estimated using Equation (2):

$$\text{Yield } (\%) = \frac{\text{Mass of product (bio oil, biochar and syngas)}}{\text{Mass of the biomass feed}} \tag{2}$$

The rate of biomass conversion into biofuels can be estimated by dividing the energy contents of pyrolysis products to the energy contents of the biomass. The mass of each pyrolysis product and higher calorific values are required for conversion estimation as shown in Equation (3):

$$\text{Conversion efficiency } (\%) = \frac{(m \times \text{HHV})\text{biochar} + (m \times \text{HHV})\text{Bio oil} + (m \times \text{HHV})\text{syngas}}{(m \times \text{HHV}) \text{ biomass}} \tag{3}$$

Makkawi used a bubbling fluidized bed reactor for fast pyrolysis of date palm biomass [11]. The obtained yield of the pyrolysis process comprised 37.2% biochar, 38.8% bio-oil and 24% non-condensable gas (all on dry and ash free basis). However, the bio-oil contained 10% of moisture. One of the factors affecting the yield of bio-oil is the ash content in the biomass. The bio-oil yield has been recorded mostly between 60–70% of the initial feed weight in most of the biomass types [5]. However, theoretical calculations revealed that the efficiency of pyrolysis process using date palm waste would be up to 86.6%. Therefore, bio-oil yield in case of date palm waste was classified as low. The HHV for the bio-oil, char and non-condensable gases was calculated to be 20.88, 19.76 and 8.9 MJ/kg respectively. The HHV of date palm biomass used in this study was taken as 18.83 MJ/kg.

3.1.2. Energy Balance of Pyrolysis Process

Table 2 presents the energy balance of processing one ton of date palm waste using fast pyrolysis unit. It abided by the law of conservation of energy which states that the input energy is equal to the output energy. The higher heating values mentioned in the above paragraph are used for these energy balance calculations.

Table 2. Energy balance for processing one ton of date palm waste from fast pyrolysis unit.

Material	Input Energy (kWh)	Pyrolysis Products from One Ton of Waste (Tons)	Output Energy (kWh)
Date palm waste, 8% MC (wet basis)	5230.5		
Bio-oil		0.388	2250.4
Char		0.372	2041.8
Syngas		0.240	593.3
Ash and heat loss from pyrolysis unit			345.0
Total			5230.5

3.1.3. Energy Requirement of Pyrolysis Process

Table 3 presents energy consumed by the pyrolysis unit for one ton of date palm wastes processed.

Table 3. Energy consumed in pyrolysis process.

S #	Steps in Pyrolysis Process	Energy Consumed (MJ/kg)	Energy Consumed (kWh/ton)
	A. Electrical Energy		
(i)	Electricity consumption by fluidized bed pyrolysis unit	-	200.0 [30]
(ii)	Electricity consumed for biomass handling and pre-processing (chopping and grinding)	-	40 [30]
	Total electric energy consumed per ton		240.00
	B. Thermal energy		
(i)	Drying (removal of residual moisture from 8% to zero%), about 80 kg/ton	0.3	6.66
(ii)	Heat required to raise the biomass to 500 °C and decompose it (920 kg/ton)	1.5	383.3
(iii)	Heat required to raise fluidizing gas up to 500 °C from 50 °C quenching temperature (2.75 kg of fluidizing gas per kg of biomass (Wright et al., 2010)) [31]	0.60 [30]	458.33
(iv)	Radiation, convection, and conduction losses (3% of heat input to pyrolysis reactor)	-	25.45
	Total thermal energy required per ton of date palm waste, kWh		872.74

3.2. Economic Analysis

The economic analysis of date palm fast pyrolysis was performed using the annual cost method. The annual cost is sum of the fixed cost and the variable cost of the complete pyrolysis process.

3.2.1. Fixed Cost

To estimate the per annum fixed cost of the pyrolysis unit, several assumptions were made, as shown in Table 4.

Table 4. Assumption for pyrolysis unit.

Parameters	Assumptions
Financing	100% owned capital
Pyrolysis unit availability	300 days/year
Pyrolysis unit depreciation period	8 years
Interest on investment	10%
Housing and insurance cost	2% of the initial cost of pyrolysis unit
Capacity of pyrolysis unit selected	10 tons per day
Annual capacity of pyrolysis unit for processing date palm waste	3000 tons
Total annual operating hours	7200 h

Table 5 indicates the initial cost of the complete pyrolysis unit. The capacity of pyrolysis unit was adopted to be 10 tons/day. One shredder and two storage tanks were considered for purchase. An amount of USD 0.05 million was earmarked for civil works.

Table 5. The initial cost of pyrolysis unit.

Item Description	Quantity	Price (USD Million)
Pyrolysis unit (10 tons/day capacity)	1	1.70
Shredder	1	0.05
Storage tanks	2	0.05
Civil work		0.05
Miscellaneous equipment		0.15
	Total cost	2.00

3.2.2. Variable Cost

The variable cost consists of repair and maintenance cost, labor cost, date palm wastes purchasing and transportation cost, electricity cost for pyrolysis unit and shredder, nitrogen cost for induction into reactor. Table 6 presents the assumptions made to work out the variable cost of pyrolysis unit, while Table 7 presents the fixed and variable costs per annum as well as per ton of biomass used by the pyrolysis unit.

Table 6. Assumptions to work out variable cost.

Sr #	Parameters	Assumptions
(i)	Repair and maintenance cost	2.0% of the initial cost per annum [24]
(ii)	Labor cost (3 shifts and 3 workers in each shift)	USD 30,000/worker/annum
(iii)	Date palm waste purchasing and transportation cost per ton	USD 50.00
(iv)	Electricity cost (5.4 cents/kWh in US)	0.054 USD/kWh
(v)	Nitrogen cost per ton of biomass [32]	USD 2.66
(vi)	Miscellaneous chemical cost per ton of biomass [32]	USD 4.00

Table 7. Cost analysis of biomass pyrolysis unit having capacity of 10 tons per day.

Sr #	Item	Cost USD/Annum	Cost USD/Ton of Date Palm Waste Processed	Cost USD/h of Date Palm Waste Processed
	A Fixed cost estimates			
(i)	Depreciation cost	225,000	75.0	31.25
(ii)	Interest on investment	110,000	36.0	15.2
(iii)	Housing and insurance cost	40,000	13.3	5.5
	Sum of fixed cost	375,000	124.3	51.95
	B Variable cost estimates			
(i)	Repair and maintenance cost	40,000	13.3	5.5
(ii)	Labor cost	270,000	90.00	37.5
(iii)	Date palm waste purchasing and transportation cost	150,000	50.00	20.83
(iv)	Electricity cost for biomass pyrolysis and pre-processing (240 kWh/ton @ 0.054 USD/kWh)	38,880	12.96	5.4
(v)	Nitrogen induction in the reactor cost	7980	2.66	1.11
(vi)	Miscellaneous chemical cost	12,000	4.00	1.67
	Sum of Variable Cost	518,860	172.92	72.01
	Total Cost (Fixed + Variable)	893,860	297.22	123.96

3.2.3. Net Revenue Generation from Pyrolysis of Date Palm Waste

Based on Makkawi's study about date palm pyrolysis, the pyrolysis products are comprised of 37.2% char, 38.8% bio-oil and 24% non-condensable gas (all on dry and ash free basis) [11]. This means one ton of date palm waste can produce 0.372-, 0.388-, and 0.240-tons char, bio-oil, and non-condensable gas, respectively, after processing through the pyrolysis unit. We may assume the selling price of biochar, bio-oil, and non-condensable gas as 1.2, 0.30 and 0.20 USD/kg, respectively. Makkawi measured the HHV for the bio-oil and char, which was 20.88 MJ/kg, and 19.76 MJ/kg, respectively. They also calculated the HHV for non-condensable gas which was 11.89 MJ/kg. As discussed earlier, the thermal energy required for pyrolysis of one ton of date palm waste was estimated at 872.74 kWh, indicating that 0.18 tons of bio-oil are required to meet the thermal energy demand of pyrolysis of one ton of date palm waste. This becomes around 38.6% of the bio-oil produced through pyrolysis of one ton of date palm waste.

Table 8 indicates that gross income from the sale of pyrolysis products obtained from one ton of date palm waste is in the tune of USD 556.8, whereas the cost of producing these pyrolysis products is in the tune of USD 297.22 (Table 7). This showed the net saving of USD 259.58 for processing one ton of date palm waste through the pyrolysis unit. It is estimated that around 345,000 tons/annum of date palm waste are available in Saudi Arabia. If 50% of this is processed through pyrolysis units, the date growers of the Kingdom can earn USD 44.77 million per annum from date palm waste, whereas the current practices of dealing with this waste are causing disposal and environmental problems for the Kingdom.

Table 8. Revenue generation from pyrolysis of one ton of date palm waste.

Item	Quantity Produced (Tons)	Quantity Consumed in the Process (Tons)	Net Quantity for Selling (Tons)	$/Ton	Income from the Sale of Pyrolysis Products ($)
Bio-oil	0.388	0.180	0.208	300	62.4
Biochar [33]	0.372	-	0.372	1200	446.4
Non-condensable gas	0.240	-	0.240	200	48.0
			Total revenue generated/ton		556.8
			Total revenue generated/h		232.0

3.2.4. Break-Even Analysis

The break-even point (BEP = X) in terms of cost and revenue was found as follows:

$$X = \frac{TFC}{P - V} \tag{4}$$

where:

X = Operating time in years
TFC = Total fixed cost invested in the beginning of the project
V = Total cost (fixed +variable) per annum of pyrolysis unit
P = Gross revenue generated per year through pyrolysis unit
$$X = \frac{2{,}000{,}000}{1{,}670{,}400 - 893{,}860}$$
X = 2.57 years

The payback period for 10 TPD thermal pyrolysis unit was 2.57 years based on assumptions made in this analysis.

3.2.5. Net Present Value of the Pyrolysis Unit

The net present value (NPV) of the project has been calculated to determine the profitability of the project. If the NPV of a project is negative, the project is likely to fail, while a positive value indicates the chances for the project to be a success. In this project, an investment of (USD 2.0 million) was assumed to be made in the beginning of the year. Annual cash flow (annual gross revenue—annual costs) was USD 780,000. The salvage value at the end of project period (8 years) was taken as 10% of the investment (USD 200,000). The discount rate was assumed to be 10%. The NPV of this project was calculated as USD +2254, 543.6 based on the methodology presented by Smith, [34] for calculating the present equivalent of (Revenue-Costs). As the NPV is positive, the project will be beneficial provided that the assumptions made for calculating annual gross income and costs stand valid.

3.2.6. Internal Rate of Return

The internal rate of return is the rate at which an investment project promises to generate a return during its useful life. It is the discount rate at which the present value of a project's net cash inflows becomes equal to the present value of its net cash outflows. In other words, the internal rate of return is the discount rate at which a project's net present value becomes equal to zero. The prospective before-tax internal rate of return on an investment is the interest rate at which present equivalent revenues equal present equivalent costs computed on a before-tax basis. The internal rate of return for this project was calculated as 36.45% based on the methodology devised by Smith [31].

3.2.7. Other Performance Indicators

Table 9 indicates other performance parameters. It revealed that annual bio-oil production excluding bio-oil used in the pyrolysis process was 624 tons. The annual char production was 1116 tons, whereas the annual syngas production was predicted to be around 720 tons. The annual energy yield for the pyrolysis unit was predicted to be 12.12 GWh and the total energy yield calculated for the life of the pyrolysis unit was 96.96 GWh.

Table 9. Results of other performance indicators.

Sr #	Performance Indicator	Amount
1	Annual Bio-oil production	624 tons (Excluding the bio-oil consumed during the process)
2	Annual Char production	1116 tons
3	Annual syngas production	720 tons
4	Annual energy yield	12.12 GWh
5	Total energy yield during the life of the pyrolysis unit	96.96 GWh
6	Levelized cost of energy	0.048 $/kWh
7	Simple payback time	2.57 years
8	Annual GHG emission reduction	2029 tons of CO_2

The projected levelized cost of energy generation was 0.048 $/kWh. This was calculated by discounting net cash flows of the project to the equivalent net present value costs at the first year when the plant commenced operation and dividing the results with total energy yield during the life of the project. The simple payback period for the pyrolysis unit was 2.57 years. The annual capacity of pyrolysis unit for processing date palm waste was estimated 3000 tons, which is expected to produce biochar capacity of 1116 tons per year. The value of carbon content of biochar 0.62 and carbon stability 0.80 has been adopted from literature for medium temperature pyrolysis [32–34].

$$\text{GHG credit from biochar} = \text{biochar (tons)} \times$$
$$\text{carbon content of biochar (\%)} \times \text{carbon stability (\%)} \times \tfrac{44}{12} \tag{5}$$

The annual reduction in GHG (Greenhouse Gases) emissions was anticipated to be 2029 tons of carbon dioxide by using Equation (5). All performance indicators affirm that the project is feasible, technically as well as economically.

4. Conclusions and Recommendations

(i)   Fast pyrolysis is regarded as a promising technology to derive bio-oil, biochar, and non-condensable gas from biomass.

(ii) This techno-economic analysis affirmed the feasibility of fast pyrolysis of date palm waste, based on the experimental data. However, it is necessary to operate pilot plant for a reasonably long period to collect more reliable first-hand data for determining optimal operational conditions and for making this analysis more reliable.

(iii) The economic analysis revealed that a date grower can secure net saving of USD 556.8 by converting one ton of date palm waste into biofuel through fast pyrolysis.

(iv) If 50% of date palm waste produced in Saudi Arabia is converted into biofuel through fast pyrolysis, a net amount of USD 44.77 million per annum can be earned.

(v)   The payback period of a fast pyrolysis unit having 10 tons/day capacity was worked out to be 2.57 years. The net present value of this project was positive, and internal rate of return was calculated as 36.45%.

(vi) It was observed that the most sensitive parameter for the economic analysis was the selling price of the byproducts (bio-oil, biochar, syn gas) of fast pyrolysis unit, particularly, the price variation of biochar is very high. In this analysis, the average selling price of biochar was considered.

(vii) Based on the findings of this techno-economic analysis, it is recommended to frame a research and commercialization project of biomass energy generation in Saudi Arabia. This will help in exploring the potential of biomass energy resources and suggest adoption of appropriate technologies in the Kingdom based on scientific research.

Author Contributions: Conceptualization, S.A.Y. and M.M.O.; methodology, S.A.Y. and M.M.O.; validation, T.I. and S.A.Y.; formal analysis, S.A.Y. and M.A.; investigation, S.A.Y.; resources, S.A.Y.; data curation, S.A.Y. and M.M.O.; writing—original draft preparation, S.A.Y. and M.M.O.; writing—

review and editing, T.I. and M.A.; visualization, S.A.Y.; and M.A.; supervision, S.A.Y.; project administration, S.A.Y.; funding acquisitions, S.A.Y. All authors have read and agreed to the published version of the manuscript.

Funding: This article was realized thanks to the funding of Qassim University, Qassim, Saudi Arabia (the project no. QU-IF-1-3-1).

Acknowledgments: The authors are highly appreciative of and thankful to the Ministry of Education and Qassim University in Saudi Arabia. They are also appreciative of the supervision of the Research Deanship of Qassim University.

Conflicts of Interest: There is no conflict of interest.

Abbreviations or Acronyms

Atm	Atmosphere
MENA	Middle East and North Africa
HHV	Higher Heating Value
$m \times HHV$	Higher Heating Value of the products based on mass
MJ	Mega Joules
wt %	Weight percent
Gj	Gega Joules
Ms	Milliseconds
kWh	Kilo Watt hour
GWh	Giga Watt hour
TEA	Techno-economic Analysis
CV	Calorific Values
Tpd	Tons per day
BEP	Break Even Point
TFC	Total Fixed Cost
X	Operating time in years
V	Total cost (fixed +variable) per annum
P	Gross revenue generated per year
NPV	Net Present Value
IRR	Internal Rate of Return
GHG	Green House Gas

References

1. Al-Farsi, M.A.; Lee, C.Y. Optimization of phenolics and dietary fibre extraction from date seeds. *Food Chem.* **2008**, *108*, 977–985. [CrossRef]
2. Hussain, A.; Farooq, A.; Bassyouni, M.I.; Sait, H.H.; El-Wafa, M.A.; Hasan, S.W.; Ani, F.N. Pyrolysis of Saudi Arabian date palm waste: A viable option for converting waste into wealth. *Life Sci. J.* **2014**, *11*, 12.
3. Elmay, Y.; Jeguirim, M.; Trouvé, G.; Said, R. Kinetic analysis of thermal decomposition of date palm residues using Coats–Redfern method. *Energy Sources Part A Recover. Util. Environ. Eff.* **2016**, *38*, 1117–1124. [CrossRef]
4. Sait, H.H.; Hussain, A.; Salema, A.; Ani, F.N. Pyrolysis and combustion kinetics of date palm biomass using thermogravimetric analysis. *Bioresour. Technol.* **2012**, *118*, 382–389. [CrossRef]
5. Zaman, C.Z.; Pal, K.; Yehye, W.A.; Sagadevan, S.; Shah, S.T.; Adebisi, G.A.; Marliana, E.; Rafique, R.F.; Bin Johan, R. Pyrolysis: A sustainable way to generate energy from waste. In *Pyrolysis*; IntechOpen: Rijeka, Croatia, 2017; Volume 1, p. 316806.
6. Elliott, D.; Beckman, D.; Bridgwater, A.V.; Diebold, J.P.; Gevert, S.B.; Solantausta, Y. Developments in direct thermochemical liquefaction of biomass: 1983–1990. *Energy Fuels* **1991**, *5*, 399–410. [CrossRef]
7. Jung, S.-H.; Kang, B.-S.; Kim, J.-S. Production of bio-oil from rice straw and bamboo sawdust under various reaction conditions in a fast pyrolysis plant equipped with a fluidized bed and a char separation system. *J. Anal. Appl. Pyrolysis* **2008**, *82*, 240–247. [CrossRef]
8. Gust, S. Combustion experiences of flash pyrolysis fuel in intermediate size boilers. In *Developments in Thermochemical Biomass Conversion: Volume 1/Volume 2*; Bridgwater, A.V., Boocock, D.G.B., Eds.; Springer: Dordrecht, The Netherlands, 1997; pp. 481–488.
9. Joardder, M.U.H.; Uddin, S.; Islam, M.N. The Utilization of Waste Date Seed as Bio-Oil and Activated Carbon by Pyrolysis Process. *Adv. Mech. Eng.* **2012**, *4*, 316806. [CrossRef]
10. Mahdi, Z.; El Hanandeh, A.; Yu, Q.J. Date Palm (*Phoenix dactylifera* L.) Seed Characterization for Biochar Preparation. In Proceedings of the 6th International Conference on Engineering, Project, and Production Management (EPPM), Gold Coast, Australia, 2–4 September 2015; Volume 7, pp. 130–138.

11. Makkawi, Y.; El Sayed, Y.; Salih, M.; Nancarrow, P.; Banks, S.; Bridgwater, T. Fast pyrolysis of date palm (Phoenix dactylifera) waste in a bubbling fluidized bed reactor. *Renew. Energy* **2019**, *143*, 719–730. [CrossRef]

12. Bensidhom, G.; Trabelsi, A.B.H.; Alper, K.; Sghairoun, M.; Zaafouri, K.; Trabelsi, I. Pyrolysis of Date palm waste in a fixed-bed reactor: Characterization of pyrolytic products. *Bioresour. Technol.* **2018**, *247*, 363–369. [CrossRef]

13. Jirka, S.; Tomlinson, T. *State of the Biochar Industry 2013—A Survey of Commercial Activity in the Biochar Field*; International Biochar Initiative: Philadeplhia, PA, USA, 2014.

14. Badger, P.; Badger, S.; Puettmann, M.; Steele, P.; Cooper, J. Techno-economic analysis: Preliminary assessment of pyrolysis oil production costs and material energy balance associated with a transportable fast pyrolysis system. *Bioresources* **2010**, *6*, 1 [CrossRef]

15. Bu, Q.; Lei, H.; Zacher, A.H.; Wang, L.; Ren, S.; Liang, J.; Wei, Y.; Liu, Y.; Tang, J.; Zhang, Q.; et al. A review of catalytic hydrodeoxygenation of lignin-derived phenols from biomass pyrolysis. *Bioresour. Technol.* **2012**, *124*, 470–477. [CrossRef]

16. Brown, T.R.; Thilakaratne, R.; Brown, R.C.; Hu, G. Techno-economic analysis of biomass to transportation fuels and electricity via fast pyrolysis and hydroprocessing. *Fuel* **2013**, *106*, 463–469. [CrossRef]

17. Taufiq-Yap, Y.H.; Teo, S.; Rashid, U.; Islam, A.; Hussien, M.Z.; Lee, K.T. Transesterification of Jatropha curcas crude oil to biodiesel on calcium lanthanum mixed oxide catalyst: Effect of stoichiometric composition. *Energy Convers. Manag.* **2014**, *88*, 1290–1296. [CrossRef]

18. Go, A.; Sutanto, S.; Ong, L.K.; Tran-Nguyen, P.L.; Ismadji, S.; Ju, Y.-H. Developments in in-situ (trans) esterification for biodiesel production: A critical review. *Renew. Sustain. Energy Rev.* **2016**, *60*, 284–305. [CrossRef]

19. Czernik, S.; Bridgwater, A.V. Overview of Applications of Biomass Fast Pyrolysis Oil. *Energy Fuels* **2004**, *18*, 590–598. [CrossRef]

20. Bridgwater, A.V.; Meier, D.; Radlein, D. An overview of fast pyrolysis of biomass. *Org. Geochem.* **1999**, *30*, 1479–1493. [CrossRef]

21. Bridgwater, A.V. Review of fast pyrolysis of biomass and product upgrading. *Biomass Bioenergy* **2012**, *38*, 68–94. [CrossRef]

22. Abdullah, N.; Gerhauser, H. Bio-oil derived from empty fruit bunches. *Fuel* **2008**, *87*, 2606–2613. [CrossRef]

23. Do, T.X.; Lim, Y.-I.; Yeo, H. Techno-economic analysis of biooil production process from palm empty fruit bunches. *Energy Convers. Manag.* **2014**, *80*, 525–534. [CrossRef]

24. Swanson, R.M.; Satrio, J.A.; Brown, R.C.; Hsu, D.D. Techno-Economic Analysis of Biofuels Production Based on Gasification Techno-Economic Analysis of Biofuels Production Based on Gasification Alexandru Platon. *Energy* **2010**, *89*. Available online https://www.osti.gov/biblio/994017 (accessed on 1 August 2021).

25. Kole, C.; Joshi, C.P.; Shonnard, D.R. *Handbook of Bioenergy Crop Plants*; CRC Press: Boca Raton, FL, USA, 2012.

26. Naik, S.N.; Goud, V.V.; Rout, P.K.; Dalai, A.K. Production of first and second generation biofuels: A comprehensive review. *Renew. Sustain. Energy Rev.* **2010**, *14*, 578–597. [CrossRef]

27. Shemfe, M.B.; Gu, S.; Ranganathan, P. Techno-economic performance analysis of biofuel production and miniature electric power generation from biomass fast pyrolysis and bio-oil upgrading. *Fuel* **2015**, *143*, 361–372. [CrossRef]

28. Bridgwater, A. Principles and practice of biomass fast pyrolysis processes for liquids. *J. Anal. Appl. Pyrolysis* **1999**, *51*, 3–22 [CrossRef]

29. Lyu, G.; Wu, S.; Zhang, H. Estimation and Comparison of Bio-Oil Components from Different Pyrolysis Conditions. *Front. Energy Res.* **2015**, *3*, 28. [CrossRef]

30. Rogers, J.; Brammer, J. Estimation of the production cost of fast pyrolysis bio-oil. *Biomass Bioenergy* **2012**, *36*, 208–217. [CrossRef]

31. Whittaker, J.D.; Smith, G.W. Engineering Economy, Analysis of Capital Expenditures. *Oper. Res. Q.* **1975**, *26*, 773. [CrossRef]

32. IPCC. App 4. Method for estimating the change in mineral soil organic carbon stocks from biochar amendments: Basis for future methodological development, 2019 Refinement to 2006 IPCC Guidel. *Natl. Greenh. Gas Invent.* **2019**. Available online: https://www.ipcc-tfi.iges.or.jp/public/2019rf/pdf/4_Volume4/19R_V4_Ch02_Ap4_Biochar.pdf (accessed on 1 August 2021).

33. Cheng, F.; Luo, H.; Colosi, L.M. Slow pyrolysis as a platform for negative emissions technology: An integration of machine learning models, life cycle assessment, and economic analysis. *Energy Convers. Manag.* **2020**, *223*, 113258. [CrossRef]

34. Tomczyk, A.; Sokołowska, Z.; Boguta, P. Biochar physicochemical properties: Pyrolysis temperature and feedstock kind effects *Rev. Environ. Sci. Biotechnol.* **2020**, *19*, 1. [CrossRef]

 *energies*

Article

Exergy and Energy Analyses of Microwave Dryer for Cantaloupe Slice and Prediction of Thermodynamic Parameters Using ANN and ANFIS Algorithms

Safoura Zadhossein [1], Yousef Abbaspour-Gilandeh [1,*], Mohammad Kaveh [1], Mariusz Szymanek [2,*], Esmail Khalife [3], Olusegun D. Samuel [4,5], Milad Amiri [6] and Jacek Dziwulski [7]

1   Department of Biosystems Engineering, College of Agriculture and Natural Resources, University of Mohaghegh Ardabili, Ardabil 56199-11367, Iran; safoorazadhossein@yahoo.co.uk (S.Z.); sirwankaweh@gmail.com (M.K.)
2   Department of Agricultural, Forest and Transport Machinery, University of Life Sciences in Lublin, Głęboka 28, 20-612 Lublin, Poland
3   Department of Civil Engineering, Cihan University-Erbil, Kurdistan Region, Erbil 44001, Iraq; esmailkhalife@gmail.com
4   Department of Mechanical Engineering, Federal University of Petroleum Resources, Effurun P.M.B. 1221, Delta State, Nigeria; samuel.david@fupre.edu.ng
5   Department of Mechanical Engineering, University of South Africa, Florida 1709, South Africa
6   Faculty of Mechanical Engineering and Ship Technology, Institute of Energy, Gdańsk University of Technology, Gdańsk, Narutowicza 11/12, 80-233 Gdańsk, Poland; milad.amiri@pg.edu.pl
7   Department of Strategy and Business Planning, Faculty of Management, Lublin University of Technology, ul. Nadbystrzycka 38D, 20-618 Lublin, Poland; j.dziwulski@pollub.pl
*   Correspondence: abbaspour@uma.ac.ir (Y.A.-G.); mariusz.szymanek@up.lublin.pl (M.S.)

Citation: Zadhossein, S.; Abbaspour-Gilandeh, Y.; Kaveh, M.; Szymanek, M.; Khalife, E.; D. Samuel, O.; Amiri, M.; Dziwulski, J. Exergy and Energy Analyses of Microwave Dryer for Cantaloupe Slice and Prediction of Thermodynamic Parameters Using ANN and ANFIS Algorithms. *Energies* **2021**, *14*, 4838. https://doi.org/10.3390/en14164838

Academic Editors: Muhammad Sultan, Md Shamim Ahamed, Redmond R. Shamshiri and Muhammad Wakil Shahzad

Received: 5 July 2021
Accepted: 6 August 2021
Published: 9 August 2021

Publisher's Note: MDPI stays neutral with regard to jurisdictional claims in published maps and institutional affiliations.

Abstract: The study targeted towards drying of cantaloupe slices with various thicknesses in a microwave dryer. The experiments were carried out at three microwave powers of 180, 360, and 540 W and three thicknesses of 2, 4, and 6 mm for cantaloupe drying, and the weight variations were determined. Artificial neural networks (ANN) and adaptive neuro-fuzzy inference systems (ANFIS) were exploited to investigate energy and exergy indices of cantaloupe drying using various afore-mentioned input parameters. The results indicated that a rise in microwave power and a decline in sample thickness can significantly decrease the specific energy consumption (SEC), energy loss, exergy loss, and improvement potential (probability level of 5%). The mean SEC, energy efficiency, energy loss, thermal efficiency, dryer efficiency, exergy efficiency, exergy loss, improvement potential, and sustainability index ranged in 10.48–25.92 MJ/kg water, 16.11–47.24%, 2.65–11.24 MJ/kg water, 7.02–36.46%, 12.36–42.70%, 11.25–38.89%, 3–12.2 MJ/kg water, 1.88–10.83 MJ/kg water, and 1.12–1.63, respectively. Based on the results, the use of higher microwave powers for drying thinner samples can improve the thermodynamic performance of the process. The ANFIS model offers a more accurate forecast of energy and exergy indices of cantaloupe drying compare to ANN model.

Keywords: cantaloupe; improvement potential; energy efficiency; exergy; ANN; ANFIS

1. Introduction

Cantaloupe (Cucumis melo) belongs to the family of Cucurbitaceous. Cantaloupe is one of the important agricultural crops of Iran, having the fifth rank after tomato, cucumber, watermelon, and Persian melon. It is more cultivated in the Khorasan Razavi, Khuzestan, and Semnan provinces. Based on FAO reports in 2018, Iran was rated third in the production of various types of cantaloupes [1,2]. Moreover, the agricultural product of cantaloupe possesses medicinal value. In this regard, there is a need to minimize crop loss after harvesting [3].

However, the structure and moisture content of the agricultural products play a decisive role in their life length. In this context, researchers have tried to use various methods to increase the durability of these products while maintaining their quality [4]. Drying with the sun is one of the most primitive methods of keep agricultural products. This method, however, suffers from several drawbacks such as the need for large spaces, environmental pollution, sudden climate change, long drying times, and so on [5]. Various industries have emerged to facilitate the production and processing of crops. Drying is one of these industries which can prolong the life of products, hence enhancing their use in a better and simpler way [6]. The high latent heat of water evaporation and the low efficiency of industrial dryers have led to high energy consumptions. Therefore, attempts have been focused on declining the energy consumption and drying time while enhancing the efficiency of industrial dryers.

To include the mentioned points in the design of industrial dryers, thermodynamic science should be exploited. The first and second laws of thermodynamics analyze energy efficiency [7]. The first law of thermodynamics states that energy is not lost but rather converts from one form to another. The second law of thermodynamics indicates the quality and image of this energy conversion. As this energy conversion is accompanied by a decline in quality, a parameter called exergy is introduced which is defined as the maximum useful work obtained from the energy flow from one system at equilibrium with the environment [7,8].

Drying by microwave (MW) power is one of the drying methods with optimal energy consumption which helps in saving the longevity and quality of products [9]. In this method, products are exposed to electromagnetic waves focused on the products. These waves have a high frequency and can penetrate into the product texture and vibrate the polar molecules such as water and salts. The vibrations of these molecules can lead to heat which will result in the transfer of humidity to the surface and finally its evaporation [10]. Owing to the energy concentration on the product, moisture elimination occurs at higher paces. The use of MW can decline the drying time up to 50% depending on the product type and drying conditions [11]. The drying time and MW power are two important factors in the drying of products by MW method which can influence the drying parameters such as drying time, drying efficiency, and quality of the final product.

Exergy and energy analyses of an assorted dryer for different agricultural produce appears in the literature. For instance, Jafari et al. [11] investigated exergy analyses and mathematical modelling of a rice barn in a semi-industrial MW dryer. Their results showed that the energy and exergy efficiency increased by enhancing the thickness of the seeds. The researchers further highlighted that at constant power, the energy and exergy efficiency increased by enhancing the thickness of the seeds. For the same layers, the rise in MW power declined the energy and exergy efficiencies.

Surendhar et al. [5] examined the kinetics of drying, energy, and exergy parameters for curcumin drying in a microwave dryer. Their results indicated that a rise in the MW power can accelerate the drying process and decline drying time. Energy and exergy values were reported to be enhanced by increasing the MW power.

In another study, Azadbakht et al. [12] studied the energy and exergy of drying in orange slices using an MW dryer with ohmic pretreatment. Their results indicated that the amount of the absorbed energy exceeded the lost energy at higher powers. The exergy efficiency was reported to improve with an increase in MW power and ohmic time.

Darvishi et al. [6] and Al-Harahsheh at al. [13] hinted that modelling of drying equipment enables designers to select suitable operating conditions and ensures effective drying operation. Among high-level optimization methods, a hybrid of an Artificial Neural Network (ANN) and Adaptive Neuro-Fuzzy Inference System (ANFIS) has been the chosen option. The choice of the hybrid is motivated by the amalgamation having mathematical recompenses, emphasized elsewhere [14,15]. The ANFIS is a governing data-driven and adaptive computational means having the fitness of plotting non-linear and multifaceted data [16]. Conversely, the constraint of ANN is its black box which flops to relation input parameters with the response. Jang and Sun [17] related the fiasco of the black box method of the ANN model to the incapacity of the model to accommodate linguistic information.

unswervingly. On the other hand, Yaghoobi et al. [18] ascribed the preeminence of the ANFIS model to its capacity to handle lapses in the ANN model.

Presently, the use of ANN and ANFIS techniques have boosted modelling and simulating food processing. These nonlinear modelling methods have been extensively employed for the evaluation of energy, exergy, and quality of the food industry due to their accuracy, robustness, and high speed [19]. ANNs are powerful computational methods to predict the responses of complex systems. The main idea of these types of networks originates from the biological nervous system performance for processing the data and information to learn and create knowledge [8].

ANFIS has recently drawn a considerable deal of attention. This method is a combination of the fuzzy structures with ANN to identify the systems and predict time series. This model has several advantages, among which the ability to simulate nonlinear systems, high precision, and shorter time of model development can be mentioned [20].

Several researchers have presented various models using ANN and ANFIS to predict the energy and exergy parameters of different dryers for drying various products. Abbaspour-Gilandeh et al. [21] used ANFIS and ANN to predict the energy and exergy of the fruits dried by a convective dryer. Kariman et al. [22] applied ANN to predict the energy and exergy of dried kiwi using MW dryer. Azadbakht et al. [4] optimized and predicted the energy and exergy of drying potato slices by fluid substrate dryer, and Nikbakht et al. [8] modelled the drying of pomegranate in the convective drier with MW pretreatment using ANN and a surface response method at the industrial scale. Kaveh et al. [23] used the ANFIS system for the prediction of the energy and exergy of drying green peas using a convective-rotary dryer.

Taghinezhad et al. [24] investigated the application of ANN and ANFIS in energy and exergy analysis of an infrared-convective dryer with ultrasonic pretreatment for drying blackberry samples. To predict the energy and exergy parameters of the blackberry drying process, the ANN (with one or two hidden layers and two Lonberg-Marquardt algorithms and Bayesian regulation) and the ANFIS model (membership function for each input: trimf and Gaussian, membership function for each output: linear and hybrid algorithm) were explored. Drying time, inlet air temperature, and ultrasonic time in the dryer were considered as inputs, while exergy efficiency, exergy loss, energy consumption, and energy consumption ratio were selected as outputs. The statistical parameters showed that the ANFIS network was more successful than ANN in predicting the energy and exergy of the drying blackberry. The prediction of energy efficiency, exergy efficiency, energy consumption ratio and energy consumption at any time were successfully accomplished with the aid of ANFIS approach. The high speed of obtaining the answer makes this method suitable for modelling and controlling the processes.

Azadbakht et al. [10] employed ANN method to predict osmotic pretreatment based on energy and exergy analysis in drying orange slices using a microwave dryer. An increase in MW power enhanced energy and exergy efficiency and reduced drying time. Moreover, a multilayer perceptron (MLP) neural network model was utilized to predict energy efficiency, specific energy loss, exergy efficiency, and specific exergy loss. MW power and osmotic time were considered as inputs, while energy efficiency, specific energy loss, exergy efficiency, and specific exergy loss were regarded as outputs. The studied artificial neural network in osmotic times and microwave power with 6 neurons in the hidden layer was employed to predict the regression coefficient (R^2) for energy efficiency and specific exergy loss as 0.999 and 0.871, respectively.

Liu et al. [7] adopted a multilayer feed-forward neural network to predict the energy and exergy of a convective dryer to dry mushroom slices. Their study entailed four input variables (drying time, air temperature, air velocity, and thickness of the samples) and four responses (energy consumption, energy consumption ratio, exergy loss, and exergy efficiency). The researchers further adopted the sigmoid tangent activator function as a transfer function and the Levenberg–Marquardt algorithm for network training. The researchers attributed the capability of the ANN model in predicting energy and exergy parameters of convective dryers due to that maximum R^2 (0.966) and the lowest value of MSE (0.001261) and MAE (0.02208).

The drying process is an important operation and observing the technical and scientific principles in the cantaloupe drying process will increase its quality and efficiency. The evaluation of energy and exergy parameters of cantaloupe drying and its modelling in different modes leads to further understanding of how the product dries. Such information can be used in designing and optimizing the drying process. In this study, the effect of MW power and sample thickness on the drying kinetics of cantaloupe slices, effective moisture diffusion coefficient, and energy and exergy parameters are investigated. Then, to model the drying behavior of cantaloupe pieces under different powers and thicknesses in an MW dryer, the ANN and ANFIS models were are used. Finally, the performances of these two models in predicting the energy and exergy parameters of cantaloupe drying in a microwave dryer were evaluated. Likewise, integrating ANN and ANFIS models has been very interesting among researchers since it reinforces the performance of the model and aids robust modelling for actual productivity and sustainability [24–27]. Inopportunely the scrutiny of the survey disclosed that there are (1) no recognized ANN models for the prediction of thermodynamic parameters of a microwave dryer (MD) for cantaloupe slices and (2) comparison capacity of hybrid models such as ANN and ANFIS models for the exergetic parameters of a microwave dryer for cantaloupe slices in the literature is scarce. Henceforth, there is a need to trim the lapses in the knowledge of such reports and launch robust models capable of improving thermodynamic performance and decreasing the environmental penalties of the drying process.

Based on the above mentioned descriptions and the targets of the study, the hypotheses of the study are as follows: (1) higher microwave power increases energy and exergy efficiencies, and (2) higher microwave power and the lower slice thickness reduces drying time, exergy improvement potential, and specific energy consumption.

2. Materials and Methods

2.1. Sample Preparation

Cantaloupe was purchased from a local market in Sardasht (West Azerbaijan, Sardasht Iran). To prevent initial moisture loss, the product was stored at 4 ± 1 in the refrigerator. To perform the experiments, the product was removed from the refrigerator 2 h before cutting to reach ambient temperature. Cantaloupes were cut to 2, 4, and 6 mm thickness using a cutter. To determine the initial moisture content of the samples, the product was placed in an oven (Memmert, UFB 500, Schwabach, Germany) at 70 °C for 24 h [2]. Finally, the initial moisture content of cantaloupe pieces was obtained at 17.94% on a wet basis.

2.2. Dryer Conditions

In the present research, a programmable domestic microwave oven (Sharp R-I96T, Bangkok, Thailand) was used to perform the experiments that were capable of generating microwave waves in the range of 100 to 900 W. The oven has an internal compartment with dimensions of $350 \times 350 \times 220$ mm^3 and a rotating plate with a diameter of 180 mm. For experiments, sliced cantaloupe samples with similar thicknesses were weighed and placed on the rotating plate of the machine. The proposed method was performed in such a way that the samples with three thicknesses of 2, 4, and 6 mm were subjected to MW powers of 360, 180, and 540 W. The drying of 60 g cantaloupe slices (ca. 14 samples) continued until the relative humidity of the samples approached about 0.2 on a wet basis. The temperature of the samples was measured by IR temperature sensor (accuracy of ± 1.5 °C). The reference dead state conditions were considered as $T_0 = 22$ °C and $P_0 = 101.325$ kPa. Each experiment was performed in three replications.

2.3. Drying Kinetics

The moisture ratio of cantaloupe was determined using Equation (1) [10].

$$MR = \frac{M_t - M_e}{M_0 - M_e} \tag{1}$$

The moisture propagation coefficient was assumed to the same at all directions (isotropic material) with negligible shrinkage. Under such conditions, the moisture transfer from the solid phase in the descending period of the rate can be estimated by Equation (2) as described by Fick law [28].

$$\frac{\partial M}{\partial t} = -D_{eff}\frac{\partial^2 M}{\partial r^2} \tag{2}$$

Assuming constant effective moisture diffusion coefficient and by the analytical solution of Fick's second law, the effective moisture diffusion coefficient can be determined using Equation (3) [21].

$$MR = \frac{M_t - M_e}{M_0 - M_e} = \frac{8}{\pi^2}\sum_{n=1}^{\infty}\frac{1}{(2n+1)}\exp\left(\frac{-D_{eff}(2n+1)^2\pi^2 t}{4L^2}\right) \tag{3}$$

By increasing t, all the terms will tend to zero except the first one. The effective moisture diffusion coefficient (D_{eff}) can be obtained from the slope (k) of Ln (MR) vs. t using Equation (4) [20].

$$k = \left(\frac{D_{eff}\pi^2}{4L^2}\right) \tag{4}$$

2.4. Energy Analysis

2.4.1. Specific Energy Consumption, Dryer Efficiency, and Thermal Efficiency

SEC refers to the ratio of the total energy consumption during the drying of cantaloupe slices to the water loss during the drying process. The SEC of cantaloupe slices by microwave method can be determined by Equation (5) [29].

$$SEC = \left(\frac{P.t}{M_w}\right) \tag{5}$$

Dryer and thermal efficiencies, as well as the vaporization latent heat, can be determined by Equations (6)–(9), respectively [23,30].

$$D_e = \left(\frac{E_{evap} + E_{heating}}{SEC}\right) \tag{6}$$

$$TE = \frac{D.A.h_{fg}.(M_i - M_0)}{3600.Z.t.(100 - M_0)} \tag{7}$$

$$E_{eva} = h_{fg}.M_w \tag{8}$$

$$h_{f.g} = 2.503 \times 10^3 - 2.386(T_{abs} - 273.16)$$
$$h_{f.g} = (7.33 \times 10^6 - 16T_{abs}^2)^{0.5} \tag{9}$$

2.4.2. Energy Efficiency and Energy Loss

A thermodynamic analysis is essential for the optimization and design of thermal systems. Based on the first law of thermodynamics, the general mass conservation equation can be expressed by Equation (10) [31].

$$\sum m_{in} = \sum m_{out} \tag{10}$$

Energy equilibrium can be expressed by Equation (11), which states that the input energy is equal to the output energy [32].

$$\sum E_{in} = \sum E_{out} \tag{11}$$

The dryer chamber is considered as the control volume, and the mass conservation energy can be determined by Equation (12) [33].

$$m_{wp} = m_{dp} + m_{wt} \tag{12}$$

The mass of the evaporated water can be calculated by Equation (13) [10].

$$m_{wt} = m_{dp}(M_o - M_t) \tag{13}$$

The energy conservation for the tangible heat, latent heat, and heat source of the microwave can be determined by Equation (14) for the cantaloupe slices [34].

$$P_{in} = P_{abs} + P_{ref} + P_{tra} \tag{14}$$

In the above equation, $E_{loss} = P_{ref} + P_{tra}$ is the lost energy.
The input energy to the microwave can be determined by Equation (15) [35].

$$P_{in} \times t = \left((mC_pT)_{out} - (mC_pT)_{in}\right) + \lambda_k m_w + (E_{ref} + E_{tra}) \tag{15}$$

Equation (15) contains three terms including the absorbed energy, reflected energy and transmitted energy. Equation (16) shows the energy absorbed by the product [33].

$$\left((mC_pT)_{out} - (mC_pT)_{in}\right) + \lambda_k m_w \tag{16}$$

The latent heat of the cantaloupe slices can be also determined by Equation (17) [6].

$$\lambda_k = \lambda_{wf}(1 + 23\exp(-0.4M_t)) \tag{17}$$

The latent heat of evaporation was determined by Darvishi et al. [35] based on Equation (18) [10]:

$$\lambda_{wf} = 2503 - 2.386(T - 273) \tag{18}$$

According to Brooker et al. [36], the heat capacity is a function of the moisture content and can be described by Equation (19).

$$C_p = 840 + 3350 \times \left(\frac{M_t}{1 + M_t}\right) \tag{19}$$

Based on Darvishi et al. [6] and Jafari et al. [33], the energy efficiency in the MW dryer can be calculated by Equation (20).

$$\eta_{en} = \frac{(m.C_p.T)_{out} + m_{ew} + \gamma_{wp}}{(m.C_p.T)_{in} + P.t} \tag{20}$$

The specific energy loss for drying cantaloupe slices can be determined by either Equations (21) or (22) [9].

$$E_{loss} = \left(1 - \eta_{en}.\frac{P.t}{m_w}\right) \tag{21}$$

$$E_{loss} = \frac{E_{in} - E_{abs}}{m_w} \tag{22}$$

The total input and output energy and exergy loss were determined according to the second law of thermodynamics. The main method to analyze the exergy of the dryer chamber relied on the calculation of the exergy values in stable points and the determination of the reason for the changes in the exergy of the process. Generally, Equation (23) shows the exergy balance for an MW dryer [12].

$$EX_{in} = EX_{abs} + \overbrace{EX_{ref} + EX_{tra}}^{\cong EX_{loss}}$$ (23)

The input exergy of the MW dryer can be determined by Equation (24) [35].

$$\underbrace{P_{in} \times t}_{exergy\ input} = \left(\underbrace{(m \times ex)_{dp}}_{exergy\ of\ dry\ poduct} - \underbrace{(m \times ex)_{wp}}_{exergy\ of\ wetproduct} \right) + \underbrace{ex'_{exap} \times t}_{exergy\ evaporation} + \overbrace{EX_{ref} + EX_{tra}}^{\cong EX_{loss}}$$ (24)

The exergy rate (J/s) of the evaporation in the dryer chamber can be determined by [5]:

$$ex'_{exap} = \left(1 - \frac{T_O}{T_p} \right) \times \dot{m}_{wv} \lambda_{wp}$$ (25)

$$\dot{m}_{wv} = \frac{m_{t+\Delta t} - m_t}{\Delta t}$$ (26)

The specific exergy (J/s) can be calculated by Equation (27) [19].

$$Ex = C_p \left[(T - T_\infty) - T_\infty \ln \left(\frac{T}{T_\infty} \right) \right]$$ (27)

The exergy efficiency of the MW dryer can be estimated by Equation (28) [5].

$$\eta_{ex} = \frac{EX_{abs}}{P_{in}}$$ (28)

The specific exergy loss can be also expressed by Equation (29) as reported by Kariman et al. [22].

$$EX_{loss} = \frac{EX_{in} - EX_{abs}}{m_{wv}}$$ (29)

The exergetic sustainability index is defined to express the performance of the exergy. The use of the improvement potential can also be helpful in the evaluation of economic activities [37]. The exergy improvement potential and the exergetic sustainability index can be expressed by Equations (30) and (31), respectively [32].

$$IP = (1 - \eta_{ex}) \times EX_{in} - EX_{out}$$ (30)

$$SI = \frac{1}{(1 - \eta_{ex})}$$ (31)

2.5. Statistical Analysis

SPSS (V.19) and the Duncan test (at the probability level of 5%) were used to investigate the effect of Mw power and thickness of cantaloupe slices on the studied indices.

2.6. Experimental Uncertainty Analysis

Experimental uncertainty was calculated by Equation (32) [38]:

$$U_R = \left[\left(\frac{\partial R}{\partial x_1} U_1 \right)^2 + \left(\frac{\partial R}{\partial x_2} U_2 \right)^2 + \ldots + \left(\frac{\partial R}{\partial x_n} U_n \right)^2 \right]^{\frac{1}{2}}$$ (32)

All uncertainties are displayed in Table 1.

Table 1. Uncertainties in measurement of parameters during onion drying.

Parameter	Unit	Value
Inlet microwave power	W	±1.5
Slice thickness	mm	±0.02
Uncertainty in the measurement of moisture quantity	G	±0.018
Uncertainty in the measurement of relative humidity of air	RH	±0.65
Drying Rate (DR)	g water/g dry matter min	±0.17
Uncertainty in Moisture Ratio (MR)	Dimensionless	±0.14
Uncertainty in Specific Energy Consumption (SEC)	MJ/kg	±1.01
Uncertainty in energy efficiency	Dimensionless	±1.4
Uncertainty in specific energy loss	MJ/kg	±0.004
Uncertainty in exergy loss	MJ/kg	±0.005
Uncertainty in exergy efficiency	Dimensionless	±1.55

2.7. Artificial Neural Network (ANN)

An artificial neural network is composed of countless artificial neurons operating as interconnected, parallel networks. Each neuron acts as a processor in the network and receives and processes neural signals (input) from other neurons or their surroundings. Similar to the human brain, an artificial neural network can learn everything on its own [39]. Neurons are trained by applying a training algorithm to the network. An artificial neural network consists of 3 layers: the input layer that receives the primary data, the hidden layer that processes the received data, and the output layer. Each layer contains a group of neurons, each of which is connected to all the neurons in the other layers, but the neurons in each layer are not in contact with other neurons in the same layer. In this way, neurons act independently, and a superposition of the neurons' behavior reflects the network behavior [40]. The latent layer may be monolayer (perceptron neural networks) or multilayer (multilayer perceptron (MLP) networks).

In this study, a multilayer perceptron artificial neural network was selected for modelling energy parameters (SEC, energy loss, energy efficiency, dryer efficiency, and thermal efficiency), exergy parameters (exergy drop, exergy efficiency, and exergy recovery potential) and duration of heating and drying cantaloupe by a microwave dryer considering different thicknesses. The perceptron multilayer neural network is a feed-forward network with three inputs, one or two hidden layers, and one output layer. This network was selected by one or two hidden layers for the experiment, in which 2–15 neurons were placed in each layer by trial and error. Moreover, Tansig, Logsig, and Purelin activation functions were used in the hidden input and output layer. In this research, the Levenberg−Marquardt optimization was used to train the network. Three iterations were considered the average of the learning cycle for simulation of artificial neural network data to minimize the error rate and maximize network stability. The error estimation algorithm in the formed networks was performed using the error propagation algorithm.

2.8. Adaptive Neuro-Fuzzy Inference System (ANFIS)

The adaptive neuro-fuzzy inference system formulates the behavior of a process using descriptive if-then rules. This system includes 4 main parts: rule base, fuzzification, inference engine, and defuzzification. Each ANFIS model consists of 5 layers which include inputs, membership functions corresponding to inputs, rules, membership functions related to outputs, and outputs [19]. In this study, a hybrid method was used to train ANFIS, which is a combination of the least-squares method and the post-diffusion method. The error limit used to create a training stop criterion was set to zero. To optimize the model, different types and numbers of membership functions were used to determine the optimal number and type. A Sugeno-type fuzzy inference system was employed to find the optimal model; triangular, trapezoidal, and Gaussian membership functions were examined. Regarding the two-variable nature of the model input, 2-2 and 3-3 membership functions were investigated.

30% of the data were used for testing while 70% of them were applied for training. Microwave power, the thickness of samples, and drying time were regarded as inputs of both models (ANN and ANFIS) while SEC, energy loss, energy efficiency, thermal efficiency, dryer efficiency, exergy drop, exergy efficiency, exergy improvement potential, and exergetic sustainability index during microwave drying were the outputs. In this study, Matlab software (Matlab R2019a) was used to model ANN and ANFIS.

To evaluate the network, two criteria of coefficient of determination (R^2) and root mean square error (RMSE) were taken into account. The coefficient of determination determines the degree of correlation between the output data of ANN and ANFIS and the observed data which can be calculated from Equation (34); its ideal value is one. The root mean square of the error determines the difference between the predicted and the actual data and can be calculated by Equation (33) [7]. The goal of a good network is to minimize the amount of this error, and its ideal value is zero.

$$RMSE = \sqrt{\frac{1}{N}\sum_{i=1}^{N}(X_i - X_p)^2} \tag{33}$$

$$R^2 = \frac{\left[\sum_{i=1}^{N}(X_i - X_{mean})^2\right] - \left[\sum_{i=1}^{N}(X_i - X_p)\right]}{\left[\sum_{i=1}^{N}(X_i - X_{mean})^2\right]} \tag{34}$$

$$MRE = \frac{1}{N}\sum_{p=1}^{N}\left|\frac{X_i - X_p}{X_i}\right| \tag{35}$$

3. Results and Discussions

3.1. Kinetics of Drying

The diagram of moisture reduction in cantaloupe pieces with different thicknesses during drying at various MV powers is depicted in Figure 1. MW drying reduced the moisture content of cantaloupe pieces from 17.99 to 0.20 w.b., and the drying time varied from 55 to 180 min (Figure 1). According to Figure 1, a rise in the microwave power and a decrement in the sample's thickness enhanced the slope of the moisture reduction. The time required to reach the moisture content of 0.2 w.b. showed a significant decline by increasing the microwave power and decreasing the sample thickness (Table 2). This time was decremented from 180 min for the power of 180 W and thickness of 6 mm to 55 min for the case of the power of 540 W and thickness of 2 mm.

The mass and heat transfer are faster in higher MW powers and thinner thicknesses. Higher powers, indeed, enhance the kinetic energy and absorbed energy, giving rise to a higher difference in the vapor pressure between the center and surface of the samples; this will eventually increase the rate of moisture elimination [9]. In a similar study, Darvishi et al. [6] dried white mulberry using an MW dryer and declared that the high humidity of the samples led to elevated friction against the rotation of dipoles, resulting in high heat generation within the white mulberries. This phenomenon accelerated the vapor motion, guiding the water toward the surface of the samples. For constant microwave power, a rise in the sample thickness will increase the path the moisture has to pass to reach the surface which will eventually enhance the drying time [41]. Other researchers also addressed the effect of the MW power and sample thickness on the drying time of the agricultural products, among which the works by Beigi and Torki [29] on onion slices, Çınkır and Süfer [42] on red meat radish, and Darvishi et al. [43] on sweet potato can be mentioned.

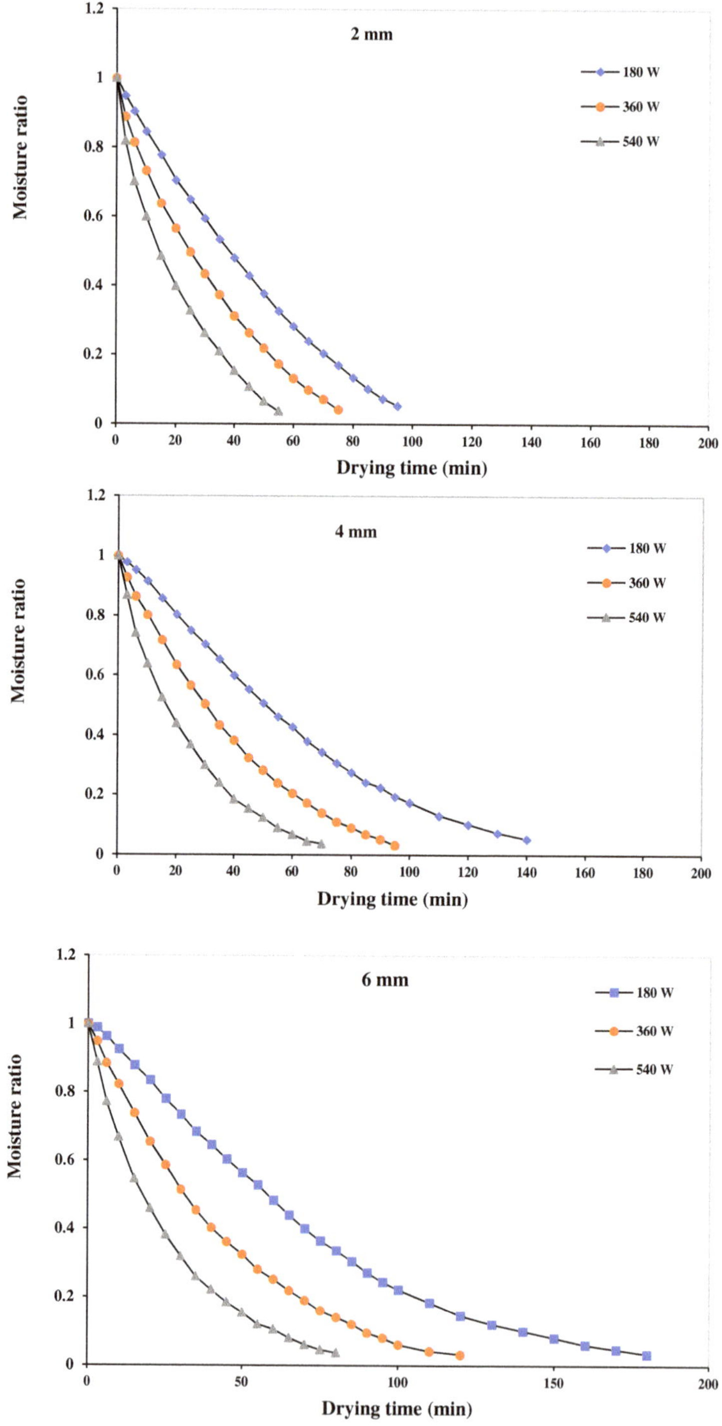

Figure 1. Variation of moisture ratio for different thicknesses and microwave power levels.

Table 2. Variation of drying time, effective moisture diffusivity and specific energy consumption of cantaloupe slices at various slice thickness and microwave power levels.

Microwave Power (W)	Slice Thickness (mm)	Drying Time (min)	D_{eff}	SEC
	2	95 ± 5 [d]	$3.15 \times 10^{-12} \pm 8.04 \times 10^{-13}$ [g]	17.1 ± 1.01 [d]
180	4	140 ± 10 [b]	$9.05 \times 10^{-11} \pm 9.11 \times 10^{-13}$ [e]	23.26 ± 0.92 [b]
	6	180 ± 10 [a]	$1.85 \times 10^{-11} \pm 1.31 \times 10^{-13}$ [c]	25.92 ± 1.14 [a]
	2	75 ± 5 [ef]	$4.22 \times 10^{-12} \pm 7.51 \times 10^{-13}$ [fg]	13.5 ± 0.88 [e]
360	4	95 ± 5 [d]	$1.48 \times 10^{-11} \pm 1.41 \times 10^{-12}$ [d]	17.25 ± 1.21 [d]
	6	120 ± 10 [c]	$2.83 \times 10^{-11} \pm 1.48 \times 10^{-12}$ [b]	21.60 ± 0.80 [c]
	2	55 ± 5 [g]	$6.00 \times 10^{-12} \pm 7.04 \times 10^{-13}$ [f]	10.48 ± 0.57 [f]
540	4	70 ± 5 [f]	$2.05 \times 10^{-11} \pm 1.51 \times 10^{-12}$ [c]	13.34 ± 0.98 [e]
	6	80 ± 5 [e]	$4.04 \times 10^{-11} \pm 1.77 \times 10^{-12}$ [a]	16.20 ± 0.79 [d]

Different letters for the same segment represent statistically significant differences at a confidence level of 95%.

3.2. Effective Moisture Diffusion Coefficient

The D_{eff} of all samples lie within the range reported by Kaveh et al. [44] for the food products (10^{-12}–10^{-7} m^2/s) as indicated in Table 2. The analysis of the D_{eff} values indicated that at a specific thickness, the microwave power has a significant influence ($p < 0.05$) on the moisture diffusion. As observable in Table 2, a rise in the Mw power enhanced the value of D_{eff} which can be assigned to the increment of the heat energy which declined the viscosity of the water present in the samples and hence increased its activity and accelerated the evaporation [29]. Sharabiani et al. [39] and Raj and Dash [45] also reported similar results in drying apple and dragon fruit slices using an MW dryer. Moreover, the statistical analysis indicated that at a given MW power, the sample thickness significantly affected the effective diffusion coefficient ($p < 0.05$); however, the increase in the thickness from 2 to 6 mm elevated the diffusion coefficient, which can be attributed to the surface hardness of the samples as surface hardening occurs more rapidly in thinner samples while the evaporation rate of thinner samples is far higher [46]. The faster surface hardening in the thinner samples can limit the displacement of humidity, hence declining the moisture propagation in the thinner samples [47]. Studied on drying kiwi fruit [19,43] and eggplant slices [48] with various thicknesses also led to similar results.

3.3. Specific Energy Consumption

Table 2 lists the SEC of drying cantaloupe slices using an MW dryer at various thicknesses. The lowest required SEC (10.47 MJ/kg) was achieved at the MW power of 540 W and thickness of 2 mm. The increase in the energy consumption by changing the power from 180 to 540 W was significant at all thicknesses ($p < 0.05$); the results indicated that a rise in the MW power and a decrement in the sample thicknesses can decrement the energy consumption. The reason could be the enhanced destruction of the product thickness and accelerated moisture release [39,41]. Moreover, with increasing MW output power, the thermal gradient inside the sample increased, followed by an increment in evaporation rate and a decrease in drying time. By reducing this time, the energy consumption also decreased [49]. In a study by Azimi-Nejadian and Hoseini [47], a specific energy consumption between 0.68 to 2.59 MJ/kg was obtained for drying potatoes in a microwave dryer at power range of 200–800 W. In another study, Khan et al. [50] reported specific energy consumption for drying fenugreek leaves in various MW powers (30 to 100 W) in the range of 1.86–2.47 MJ/kg.

Furthermore, the energy required to remove one kilogram of water from the cantaloupe pieces during the microwave drying process significantly increased (with a probability level of 5%) by enhancing the thickness of the samples. The rise in energy consumption by increasing the thickness of the sample can be attributed to the processing time. In thicker samples, the water must travel a longer distance inside the product to reach the surface, i.e., it takes longer to remove moisture. Such an increment in drying time led to higher energy consumption [41]. Similar reports were also emphasized by researchers elsewhere [29,43,51] for drying kiwi, apple, and onion by microwave dryer, respectively.

3.4. Dryer Efficiency and Thermal Efficiency

Table 3 depicts the dryer efficiency and the thermal efficiency of the microwave dryer for the cantaloupe slice. As shown, the highest and lowest dryer efficiency and heat efficiency were observed at the power level of 540 and 180 W and thickness of 2 and 6 mm, respectively. Furthermore, by increasing the microwave power and decreasing the sample thickness, both efficiencies (heat and drying) exhibited an increasing trend which can be due to the high drying rate. Such a high drying rate can be assigned to the difference in thermal gradient between the product and the drying temperature. Moreover microwave energy penetrates the cantaloupe samples and generates heat by inducing polarity in the water molecules, hence improving drying and uniformity of heat and water distribution in the sample [28]. Similar results have been reported for drying rice paddy [33], chamomile [30], dragon fruit slicess [45], and apple [52].

Table 3. Variation of drying efficiency, thermal efficiency, energy efficiency, and exergy efficiency of cantaloupe slices at various slice thickness and microwave power levels.

Microwave Power (W)	Slice Thickness (mm)	Drying Efficiency (%)	Thermal Efficiency (%)	Energy Efficiency (%)	Exergy Efficiency (%)
	2	24.64 ± 1.2 [e]	17.40 ± 0.57 [cd]	28.18 ± 1.97 [cd]	24.59 ± 1.77 [c]
180	4	14.64 ± 1.24 [fg]	10.40 ± 0.88 [ef]	19.19 ± 1.36 [ef]	15.22 ± 1.22 [e]
	6	12.36 ± 0.99 [g]	7.02 ± 74 [f]	16.91 ± 1.02 [f]	11.25 ± 1.01 [f]
	2	32.69 ± 1.34 [b]	30.45 ± 0.54 [b]	37.24 ± 2.12 [b]	32.59 ± 2.01 [b]
360	4	21.36 ± 0.88 [e]	17.12 ± 0.77 [d]	25.91 ± 1.15 [d]	22.36 ± 1.54 [c]
	6	17.35 ± 0.91 [a]	12.91 ± 0.88 [e]	21.90 ± 1.22 [e]	18.88 ± 1.08 [d]
	2	42.70 ± 1.44 [c]	36.46 ± 0.90 [a]	47.24 ± 2.31 [a]	38.89 ± 2.04 [a]
540	4	30.14 ± 1.12 [c]	25.90 ± 1.01 [b]	36.69 ± 1.88 [b]	29.99 ± 1.48 [b]
	6	26.31 ± 1.05 [d]	22.40 ± 0.88 [c]	30.86 ± 1.55 [c]	24.66 ± 1.11 [c]

Different letters for the same segment represent statistically significant differences at a confidence level of 95%.

3.5. Energy Efficiency

The energy efficiency of microwave drying of the cantaloupe slices was calculated from the experimental data using Equation (17). The mean energy efficiency obtained in this study (Table 3) varied from 16.91 to 47.24%, which was in line with the reported drying efficiency for various products by microwave as such as chrysanthemum (29.98 to 62.52%), green peas (28.36 to 57.98%), and chamomile (35.83 to 49.99%) reported by Wang et al. [28], Kaveh et al. [31], and Motevali et al. [30], respectively. The experimental results and statistical analysis revealed a significant improvement in the energy efficiency of the process (at a probability level of 5) with increasing microwave power and decreasing the thickness of the samples. Shortening the duration of the drying process at higher MW powers in thinner samples incremented the energy efficiency. This phenomenon may be due to the fact that the moisture content of cantaloupe pieces is generally high, and since the thickness of the samples are low and their surface is higher compared to their mass, they have a high rate of moisture transfer to their surface; thus, a decline in the thickness of the samples will accelerate moisture transfer to the surface. Therefore, reducing the thickness of the samples will increase energy efficiency [53].

Darvishi et al. [43] reported that the energy efficiency of drying kiwifruit slices with different thicknesses (3, 6, and 9 mm) by microwave (at power levels of 200, 300, 400, and 500) were in the range of 15.15 to 32.27%. Their results also established that the increase in the thickness of the samples led to a significant reduction (at a probability level of 5%) in energy efficiency [51].

3.6. Specific Energy Loss

The mean specific energy loss during the MW drying process of cantaloupe slices at different thicknesses and power levels was calculated by Equation (18) as shown in Table 4. Darvishi [35] and Kouhila et al. [9] highlighted that the lost energy for MW-dried soybean and Dosidicus gigas ranged from 8.89–5.04 MJ/kg and 12.9–59.47 MJ/kg, respectively. This implies that the increase in the microwave power of the dryer investigated led to a significant reduction of specific energy loss at the probability level of 5%. Additionally, the lost energy during the drying of thicker samples was significantly higher than the thinner samples at a probability level of 5%. Owing to the governing mechanism of product heating during the MW drying process, the use of higher powers for thinner samples (as much as possible) can decrement the specific energy loss. Since the specific energy loss is inversely related to water discharged from the product, the specific energy loss decreased by enhancing the amount of water discharged from the product [10].

Table 4. Average values of energy and exergy loss of microwave drying process of cantaloupe slices.

Microwave Power (W)	Slice Thickness (mm)	Energy Loss MJ/kg	Exergy Loss MJ/kg
	2	7.44 ± 0.32 [e]	8.14 ± 0.41 [e]
180	4	10.22 ± 0.42 [b]	10.83 ± 0.35 [b]
	6	11.24 ± 0.52 [a]	12.20 ± 0.44 [a]
	2	4.79 ± 0.27 [g]	5.73 ± 0.27 [g]
360	4	8.39 ± 0.39 [d]	8.88 ± 0.39 [d]
	6	9.29 ± 0.44 [c]	10.11 ± 0.29 [c]
	2	2.65 ± 0.31 [h]	3.00 ± 0.24 [h]
540	4	5.24 ± 0.33 [g]	6.26 ± 0.33 [g]
	6	6.67 ± 0.40 [f]	7.41 ± 0.28 [f]

Different letters for the same segment represent statistically significant differences at a confidence level of 95%.

3.7. Specific Exergy Loss

Table 4 summarizes the influence of drying conditions (microwave power and slice thickness) on the specific exergy loss. As noticed, the reduction in the microwave power led to the exergy loss by 3.008 to 12 MJ/kg water. Moreover, the changes in the specific exergy loss with decreasing thickness exhibited a descending trend in various MW powers. This implies that the use of thinner samples reduces the contact time of the product with the microwave power. As the rate of mass and heat transfer increases, the drying time decreases, and less energy is transferred out of the dryer chamber, resulting in reduced specific exergy loss [47]. Additionally, the exergy loss declined with increasing microwave power due to the shorter process time. The lowest exergy loss (3.008 MJ/kg water) was observed at the MW power of 540 W and thickness of 2 mm. Higher microwave powers had lower exergy, and this exergy increased water evaporation or exergy consumption, hence reducing exergy loss [43].

The result is consistent with findings reported elsewhere [10,35]. For instance, Azadbakht et al. [10] reported a reduction in the exergy loss of the MW drying process of orange slices at the power levels of 90 to 900 W, with the osmotic pretreatment ranging from 19.85 to 3.71 MJ/kg water with an increase in microwave power. They stated that the intensity of the process reduction is much greater at higher powers, which reduced the exergy loss.

3.8. Exergy Efficiency

Table 3 shows the mean exergy efficiency. As detected, the exergy efficiency of the cantaloupe slice drying varied from 11.25 to 38.89%. This implies that the adoption of higher powers significantly improved the thickness of the slice samples at the probability level of 5%, implying the lowest exergy efficiency (11.25%) obtained at the power of 180 W and a thickness of 6 mm while the highest exergy efficiency (38.89%) released at the power of 540 W and a thickness of 2 mm. The reduction in the drying process time at higher microwave powers reduced the energy loss and ultimately increased the exergy efficiency of the system [12]. In thinner samples, the drying process occurred faster due to the low internal strength of the specimens, increasing the exergy efficiency [43]. This finding was further corroborated by other researchers [43,51].

3.9. Improvement Potential

Table 5 presents the average exergy improvement rate, as determined from Equation (27). The minimum average value of the exergy improvement ability (1.83 MJ/kg) was obtained with the 2-mm thick samples at the microwave power of 540 W, while the maximum value of the exergy improvement ability (10.83 MJ/kg) was observed with 6-mm-thick samples at the power of 180 W. Furthermore, the increase in the microwave power and decrease in the thickness of the samples enhanced the rate of exergy improvement ability.

Table 5. Average improvement potential and exergetic sustainability index of microwave drying process of cantaloupe slices.

Drying Conditions		IP	SI
	2	6.14 ± 0.44 [e]	1.32 ± 0.05 [c]
180	4	9.18 ± 0.49 [b]	1.17 ± 0.06 [ef]
	6	10.83 ± 0.38 [a]	1.12 ± 0.04 [f]
	2	3.86 ± 0.28 [f]	1.48 ± 0.07 [b]
360	4	6.89 ± 0.33 [d]	1.28 ± 0.05 [cd]
	6	8.20 ± 0.42 [c]	1.23 ± 0.04 [de]
	2	1.83 ± 0.33 [g]	1.63 ± 0.06 [a]
540	4	4.38 ± 0.27 [f]	1.42 ± 0.05 [b]
	6	5.58 ± 0.46 [e]	1.34 ± 0.04 [c]

Different letters for the same segment represent statistically significant differences at a confidence level of 95%.

Microwave power enhancement and the sample's thickness decline, indeed, increased the enthalpy around the dryer chamber. The results are comparable to observations reported elsewhere [35]. For instance, Darvishi et al. [35] reported that the average exergy of soybean ranged from of 1.31 to 5.35 MJ/kg as the microwave power increased from 200 to 600 W. The enhancement is attributed to the increase in the microwave power.

3.10. Exergetic Sustainability Index

Table 5 presents the mean exergetic sustainability index and varied from 1.12 to 1.63. Similar reports were reported by Okunola et al. [54], Arslan and Aktas [37], and Beigi et al. [40] for the exergetic sustainability index range for the drying of okra at 2.14 to 2.77, 1.07 to 1.21, and 1.05 to 1.42, respectively. As the value of the exergetic sustainability index is proportional to the exergy efficiency, the highest exergetic sustainability index values showed a slight impact on the environment, resulting in environmental imbalance and improvement of exergy efficiency.

3.11. Artificial Neural Network

Tables 6 and 7 highlight the data simulated from the predicted values of the ANN and ANFIS models. Table 6 presents the values of R^2, MSE, and MAE as well as network type, network topology, algorithm, and threshold functions for predicting data by ANN to easily understand the consistency between real data and simulation. As noticed, the ANN-obtained values of R^2 for the effective moisture diffusion coefficient, SEC, dryer efficiency, energy efficiency, energy loss, energy efficiency, exergy efficiency, exergy loss, improvement potential, and exergetic sustainability index were 0.8917, 0.9040, 0.9188, 0.9441, 0.8894, 0.9258, 0.9008, 0.9321, 0.8998, and 0.9139, respectively. These results indicated that the points predicted by ANN are less accurate. The statistical indices of ANN obtained from the finding are consistent with those of researchers elsewhere [7,10,22].

Table 6. Performance indices for the ANN models.

Parameters	Network Topology	Training Algorithm	Function	Train			Test			Epoch
				RMSE	MAE	R^2	RMSE	MAE	R^2	
D_{eff}	3-10-10-1	Tansig-Tansig-Tansig	LM	2.77×10^{-24}	3.95×10^{-13}	0.8812	3.57×10^{-24}	5.06×10^{-13}	0.8917	67
SEC	3-10-1	Tansig-Tansig	LM	0.2994	0.1362	0.8892	0.2642	0.1469	0.9040	39
Energy efficiency	3-5-1	Tansig-Logsig	LM	1.0851	0.3311	0.9077	0.7925	0.2839	0.9188	25
Drying efficiency	3.8-8-1	Tansig-Logsig-Tansig	BR	0.7046	0.2669	0.9376	0.7873	0.2827	0.9441	93
Thermal efficiency	3-8-1	Tansig-Tansig	BR	1.2081	0.3605	0.8797	1.1318	0.3464	0.8894	29
Energy loss	3-5-5-1	Tansig-Purelin-Tansig	LM	0.1009	0.0885	0.8978	0.1045	0.0876	0.9258	59
Exergy efficiency	3-10-10-1	Tansig-Logsig-Logsig	LM	1.2561	0.3469	0.8917	0.8423	0.3000	0.9015	55
Exergy loss	3-15-1	Tansig-Logsig-Tansig	LM	0.0995	0.0962	0.9036	0.0749	0.0787	0.9321	38
IP	3-15-10-1	Tansig-Tansig-Tansig	LM	0.1783	0.1236	0.8896	0.0911	0.0909	0.8998	44
SI	3-10-1	Tansig-Tansig	BR	0.0003	0.0060	0.8791	0.0003	0.0056	0.9139	97

Table 7. Performance indices for the ANFIS models.

Parameters	Number of MF		Type of MF		Train			Test		
	Input	Cycle	Input	Output	RMSE	MAE	R^2	RMSE	MAE	R^2
D_{eff}	3-3-3	100	Trimf	Linear	2.62×10^{-24}	3.92×10^{-13}	0.9076	2.08×10^{-24}	3.96×10^{-13}	0.9445
SEC	3-3-3	100	Trimf	Linear	0.1759	0.1296	0.9315	0.1662	0.1230	0.9599
Energy efficiency	3-3-3	100	Trimf	Linear	0.4938	0.2222	0.9702	0.2977	0.1744	0.9763
Drying efficiency	3-3-3	100	Gaussmf	Linear	0.2291	0.1432	0.9775	0.1879	0.1350	0.9811
Thermal efficiency	3-3-3	100	Trimf	Linear	0.4074	0.2098	0.9633	0.4030	0.2106	0.9698
Energy loss	3-3-3	100	Gaussmf	Linear	0.0232	0.0469	0.9804	0.0180	0.0414	0.9830
Exergy efficiency	3-3-3	100	Trimf	Linear	0.0837	0.0938	0.9910	0.0601	0.0777	0.9927
Exergy loss	3-3-3	100	Trimf	Linear	0.0090	0.0283	0.9908	0.0061	0.0225	0.9946
IP	3-3-3	100	Gaussmf	Linear	0.0605	0.0741	0.9390	0.1148	0.0959	0.9507
SI	3-3-3	100	Trimf	Linear	0.00005	0.0019	0.9806	0.00002	0.0016	0.9890

3.12. ANFIS

To obtain the best ANFIS model capable of predicting the kinetics, energy, and exergy indices of cantaloupe drying, different ANFIS structures were tested. Then, the best ANFIS structure with the best results is presented in Table 7. To achieve the best ANFIS structure with the highest precision in predicting kinetics, energy, and exergy indices, changes were applied in various parameters such as number and type of input and output membership functions, optimization methods, and number of epochs. The best ANFIS model had the input and output membership functions, the number of epochs, and the learning algorithm type of Trimmf, linear, 100, and hybrid, respectively. The values of R2, RMSE, and RMSE of the best ANFIS model for predicting kinetic, energy, and exergy indices are shown in Table 7.

As shown in Table 7, R2 values of effective moisture diffusion coefficient, SEC, dryer efficiency, energy efficiency, energy loss, energy efficiency, exergy efficiency, exergy loss, improvement potential, and exergetic sustainability index were 0.9445, 0.9599, 0.9763, 0.9811, 0.9698, 0.9830, 0.9927, 0.9946, 0.9507, and 0.9890, respectively.

3.13. Comparison of ANFIS and ANN Models

The prediction capabilities of the statistical indices for the ANN and ANFIS models of exergy based of microwave dryer for the cantaloupe were compared. Owing to the higher value of R^2 and lower values of other statistical indices of ANN compared to ANFIS, the accuracy of the ANFIS model is better than that of the ANN model. Similar results were observed by [21,24,31] for the ANN-ANFIS based prediction of exergy parameters of onion drying with continuous industrial dryer, quince under hot air dryer, and of blackberry drying by an infrared-convective dryer, respectively. The higher prediction ability of the latter model compared to the former can be attributed to the use of fuzzy inference system.

3.14. Certainty Analysis

For the purpose of certainty, sensitivity tests are often conducted to determine the relative importance of each independent variable in affecting the dependent variables. All independent variables are, in turn, taken into consideration in the analysis of sensitivity. To obtain the sensitivity level of each input variable in the determination of different parameters by ANNs and ANFIS, the omission of each input variable (microwave power, slice thickness, and drying time) was used as a technique [25]. The results of the sensitivity analysis of the input parameters in the drying process of cantaloupe are given in a set of classified data in Table 8. These results indicate that drying time and microwave power had the highest and lowest effects on different parameters of cantaloupe, respectively.

Table 8. Analyzing the effects of indirect independent variables in predicting different parameters in cantaloupe slices.

Parameters	ANN				ANFIS			
	Total	Without Microwave Power	Without Slice Thickness	Without Drying Time	Total	Without Microwave Power	Without Slice Thickness	Without Drying Time
Deff	0.8917	0.8898	0.8823	0.8323	0.9445	0.9325	0.9301	0.9090
SEC	0.9040	0.8952	0.8878	0.8657	0.9599	0.9511	0.9421	0.9237
Energy efficiency	0.9188	0.8995	0.8912	0.8804	0.9763	0.9553	0.9432	0.9025
Drying efficiency	0.9441	0.9425	0.9389	0.9025	0.9811	0.9621	0.9425	0.9166
Thermal efficiency	0.8894	0.8599	0.8468	0.8226	0.9698	0.9258	0.9090	0.8882
Energy loss	0.9258	0.9090	0.9053	0.8888	0.9830	0.9632	0.9489	0.9208
Exergy efficiency	0.9015	0.8680	0.8511	0.8258	0.9927	0.9880	0.9733	0.9494
Exergy loss	0.9321	0.9123	0.9089	0.8974	0.9946	0.9911	0.9722	0.9525
IP	0.8998	0.8931	0.8898	0.8529	0.9507	0.9411	0.9212	0.8859
SI	0.9139	0.9045	0.8969	0.8787	0.9890	0.9833	0.9599	0.9489

4. Conclusions

In the present study, cantaloupe pieces were dried under different conditions using a microwave dryer, and energy and exergy indices were explored. According to thermodynamic analysis, an increase in the microwave power declined the SEC, while the enhancement of sample thickness declined SEC. The minimum and maximum SEC were 10.48 and 25.92 MJ/kg water, respectively. Energy efficiency was in the range of 16.91 to 47.24% and was higher at higher powers and lower thicknesses of the samples. Increasing the microwave power and decreasing the thickness of the samples led to a reduction in specific energy and exergy losses and IP. Dryer and thermal efficiencies were recorded between 12.36 to 42.70% and 7.02 to 36.46%, respectively. The mean exergy efficiency of the cantaloupe drying process ranged from 11.25% for microwave power of 540 W and thickness of 2 mm to 38.98% for microwave power of 180 W and thickness of 6 mm. The minimum and maximum mean exergetic sustainability indices were 1.12 and 1.63, respectively. According to the predicted results, energy and exergy parameters predicted by ANFIS models were much more accurate than ANN as the R^2 values of the ANFIS-predicted variables were closer to one and also exhibited more agreement with the real data. Real and predicted results for energy and exergy parameters could be useful for designing and manufacturing modern dryers with maximum exergy and minimum energy consumption. The use of these results can lead to lower environmental consequences.

Author Contributions: Conceptualization, M.K. and S.Z.; methodology, E.K.; software, O.D.S.; validation, M.A., Y.A.-G., M.S. and J.D.; formal analysis, S.Z.; investigation, E.K.; resources, M.K., E.K. and Y.A.-G.; data curation, Y.A.-G.; writing—original draft preparation, M.K. and O.D.S.; writing—

review and editing, M.S., M.K. and J.D.; visualization, O.D.S. All authors have read and agreed to the published version of the manuscript.

Funding: This research received no external funding.

Institutional Review Board Statement: Not applicable.

Informed Consent Statement: Not applicable.

Data Availability Statement: Data sharing is not applicable to this article.

Conflicts of Interest: The authors declare that they have no known competing financial interests or personal relationships that could have appeared to influence the work reported in this paper.

Abbreviations

A	Tray area (m^2)	m_{wp}	Mass of wet product [kg]
C_p	Heat capacity [J/kg K]	m_{dp}	Mass of dry product [kg]
D	Weight density (kg/m^2)	M_w	Mass of water evaporated [kg]
D_{eff}	Effective diffusivity coefficient [m^2/s]	Mt	Moisture content of the product at any level and at any time [g water/g dry matter]
D_e	Drying efficiency (%)	Mo	Initial moisture content [g water/g dry matter]
$E_{heating}$	Energy for the material heating (kJ)	m_d	Mass of dry sample [kg]
E_{ref}	Energy reflected [J]	m	Mass [kg]
E_{tre}	Energy transmitted [J]	N	Total the data [-]
E_{in}	Energy input	P_{tra}	Microwave power transmitted (W)
E_{out}	Energy output	P_{in}	Microwave power emitted by the magnetron [W]
E_{loss}	Specific energy loss [J/kg water]	P_{abs}	Microwave power absorbed [W]
EX_{in}	Exergy input [J]	P_{ref}	Microwave power reflected [W]
EX_{abs}	Exergy absorbed [J]	P	Microwave output power (kW)
EX_{ref}	Exergy reflected [J]	SEC	Specific energy consumption [J/kg water]
EX_{tra}	Exergy transmitted [J]	SI	Exergetic sustainability index
ex_{exap}	Exergy of evaporation water [J/kg water]	TE	Thermal efficiencies
EX_{loss}	Specific exergy loss [J/kg water]	T_o	Ambient temperature [K]
Ex	Exergy [J]	T	Temperature [K]
ex	Specific exergy [J/kg water]	t	Drying time [min]
E_{evap}	Energy consumed to evaporate moisture from drying samples (kJ)	Xi	Measured values
$h_{f.g}$	Latent heat of vaporization (kJ/kg)	X_p	Predicted values
IP	Improvement potential [MJ/kg water]	X_{mean}	Average predicted values
L	Product thickness [m]	λ_{wf}	Latent heat of free water [J/kg]
m_{in}	Mass input [kg]	λ_K	Latent heat of sample [J/kg]
m_{out}	Mass output [kg]	η_{ex}	Exergy efficiency [%]
Me	Equilibrium moisture content [g water/g dry matter]	η_e	Energy efficiency [%]
MR	Moisture ratio [-]		

References

1. FAOSTAT. FAO Statistics Data Base on the World Wide Web. Available online: http://faostat.fao.org (accessed on 1 May 2018).
2. Chayjan, R.A.; Agha-Alizade, H.H.; Barikloo, H.; Soleymani, B. Modeling some drying characteristics of cantaloupe slices. *Cercet. Agron. Mold.* **2012**, *2*, 5–14. [CrossRef]
3. Li, T.S.; Sulaiman, R.; Rukayadi, Y.; Ramli, S.; Rukayadi, Y. Effect of gum arabic concentrations on foam properties, drying kinetics and physicochemical properties of foam mat drying of cantaloupe. *Food Hydrocoll.* **2021**, *116*, 106492. [CrossRef]
4. Azadbakht, M.; Aghili, H.; Ziaratban, A.; Torshizi, M.V. Application of artificial neural network method to exergy and energy analyses of fluidized bed dryer for potato cubes. *Energy* **2017**, *120*, 947–958. [CrossRef]

5. Surendhar, A.; Sivasubramanian, V.; Vidhyeswari, D.; Deepanraj, B. Energy and exergy analysis, drying kinetics, modeling and quality parameters of microwave-dried turmeric slices. *J. Ther. Anal. Calor.* **2019**, *136*, 185–197. [CrossRef]

6. Darvishi, H.; Zarein, M.; Minaei, S.; Khafajeh, H. Exergy and Energy Analysis, Drying Kinetics and Mathematical Modeling of White Mulberry Drying Process. *Int. J. Food Eng.* **2014**, *10*, 269–280. [CrossRef]

7. Liu, Z.-L.; Bai, J.-W.; Wang, S.-X.; Meng, J.-S.; Wang, H.; Yu, X.-L.; Gao, Z.-J.; Xiao, H.-W. Prediction of energy and exergy of mushroom slices drying in hot air impingement dryer by artificial neural network. *Dry. Technol.* **2020**, *38*, 1959–1970. [CrossRef]

8. Nikbakht, A.M.; Motevali, A.; Minaei, S. Energy and exergy investigation of microwave assisted thin-layer drying of pomegranate arils using artificial neural networks and response surface methodology. *J. Saudi Soc. Agr. Sci.* **2014**, *13*, 81–91. [CrossRef]

9. Kouhila, M.; Moussaoui, H.; Bahammou, Y.; Tagnamas, Z.; Lamsyehe, H.; Lamharrar, A.; Idlimam, A. Exploring drying kinetics and energy exergy performance of Mytilus Chilensis and Dosidicus gigas undergoing microwave treatment. *Heat Mass Transf.* **2020**, *56*, 2985–2999. [CrossRef]

10. Azadbakht, M.; Torshizi, M.V.; Noshad, F.; Rokhbin, A. Application of artificial neural network method for prediction of osmotic pretreatment based on the energy and exergy analyses in microwave drying of orange slices. *Energy* **2018**, *165*, 836e845. [CrossRef]

11. Jafari, H.; Kalantari, D.; Azadbakht, M. Energy consumption and qualitative evaluation of a continuous band microwave dryer for rice paddy drying. *Energy* **2018**, *142*, 647–654. [CrossRef]

12. Azadbakht, M.; Vahedi Torshizi, M.; Noshad, F.; Rokhbin, A. Energy and exergy analyses in microwave drying of orange slices *Iran. Food Sci. Technol. Res. J.* **2020**, *16*, 1–13.

13. Al-Harahsheh, M.; Ala'a, H.; Magee, T.R. Microwave drying kinetics of tomato pomace: Effect of osmotic dehydration. *Chem Eng. Process. Process Intensif.* **2009**, *48*, 524–531. [CrossRef]

14. Okwu, M.O.; Adetunji, O. A comparative study of artificial neural network (ANN) and adaptive neuro-fuzzy inference system (ANFIS) models in distribution system with nondeterministic inputs. *Int. J. Eng. Bus. Manag.* **2018**, *10*, 1–17. [CrossRef]

15. Samuel, O.; Okwu, M.; Tartibu, L.; Giwa, S.; Sharifpur, M.; Jagun, Z. Modelling of Nicotiana Tabacum L. Oil Biodiesel Production: Comparison of ANN and ANFIS. *Front. Energy Res.* **2021**, *8*, 377. [CrossRef]

16. Rao, P.; Srinivas, K.; Mohan, A. A Survey on Stock Market Prediction Using Machine Learning Techniques. In *ICDSMLA 2019*, Springer: Singapore, 2019; pp. 923–931.

17. Jang, J.S.; Sun, C.T. Neuro-fuzzy modeling and control. *Proc. IEEE* **1995**, *83*, 378–406. [CrossRef]

18. Yaghoobi, A.; Bakhshi-Jooybari, M.; Gorji, A.; Baseri, H. Application of adaptive neuro fuzzy inference system and genetic algorithm for pressure path optimization in sheet hydroforming process. *Int. J. Adv. Manuf. Technol.* **2016**, *86*, 2667–2677 [CrossRef]

19. Taghinezhad, E.; Kaveh, M.; Khalife, E.; Chen, G. Drying of organic blackberry in combined hot air- infrared dryer with ultrasound pretreatment. *Dry. Technol.* **2020**, 1–17. [CrossRef]

20. Kaveh, M.; Chayjan, R.A.; Golpour, I.; Poncet, S.; Seirafi, F.; Khezri, B. Evaluation of exergy performance and onion drying properties in a multi-stage semi-industrial continuous dryer: Artificial neural networks (ANNs) and ANFIS models. *Food Bioprod. Process.* **2021**, *127*, 58–76. [CrossRef]

21. Abbaspour-Gilandeh, Y.; Jahanbakhshi, A.; Kaveh, M. Prediction kinetic, energy and exergy of quince under hot air dryer using ANNs and ANFIS. *Food Sci. Nutr.* **2020**, *8*, 594–611. [CrossRef]

22. Kariman, M.; Tabarsa, F.; Zamani, S.; Kashi, P.K.; Torshizi, M.V. Classification of the energy and exergy of microwave dryers in drying kiwi using artificial neural networks. *Carpath. J. Food Sci. Technol.* **2019**, *11*, 29–45.

23. Kaveh, M.; Abbaspour-Gilandeh, Y.; Chen, G. Drying kinetic, quality, energy and exergy performance of hot air-rotary drum drying of green peas using adaptive neuro-fuzzy inference system. *Food Bioprod. Process.* **2020**, *124*, 168–183. [CrossRef]

24. Taghinezhad, E.; Kaveh, M.; Szumny, A. Thermodynamic and Quality Performance Studies for Drying Kiwi in Hybrid Hot Air-Infrared Drying with Ultrasound Pretreatment. *Appl. Sci.* **2021**, *11*, 1297. [CrossRef]

25. Jahanbakhshi, A.; Kaveh, M.; Taghinezhad, E.; Sharabiani, V.R. Assessment of kinetics, effective moisture diffusivity, specific energy consumption, shrinkage, and color in the pistachio kernel drying process in microwave drying with ultrasonic pretreatment. *J. Food Process Preserv.* **2020**, *44*, e14449. [CrossRef]

26. Chasiotis, V.; Nadi, F.; Filios, A. Evaluation of multilayer perceptron neural networks and adaptiveneuro-fuzzy inference systems for the mass transfer modeling of Echium amoenum Fisch. & C. A. Mey. *J. Sci. Food Agric.* **2021**. [CrossRef]

27. Bakhshipour, A.; Zareiforoush, H.; Bagheri, I. Mathematical and intelligent modeling of stevia (Stevia Rebaudiana) leaves drying in an infrared-assisted continuous hybrid solar dryer. *Food Sci. Nutr.* **2021**, *9*, 532–54328. [CrossRef]

28. Wang, Y.; Li, X.; Chen, X.; Li, B.O.; Mao, X.; Miao, J.; Zhao, C.; Huang, L.; Gao, W. Effects of hot air and microwave-assisted drying on drying kinetics, physicochemical properties, and energy consumption of chrysanthemum. *Chem. Eng. Process. Process Intensif.* **2018**, *129*, 84–94. [CrossRef]

29. Beigi, M.; Torki, M. Experimental and ANN modeling study on microwave dried onion slices. *Heat Mass Transfer.* **2020**. [CrossRef]

30. Motevali, A.; Minaei, S.; Banakar, A.; Ghobadian, B.; Khoshtaghaza, M.H. Comparison of energy parameters in various dryers. *Energy Convers. Manag.* **2014**, *87*, 711–725. [CrossRef]

31. Kaveh, M.; Abbaspour-Gilandeh, Y.; Nowacka, M. Comparison of different drying techniques and their carbon emissions in green peas. *Chem. Eng. Process. Process Intensif.* **2021**, *160*, 108274. [CrossRef]

32. Kumar, D.; Mahanta, P.; Kalita, P. Performance analysis of a solar air heater modified with zig-zag shaped copper tubes using energy-exergy methodology. *Sustain. Energy Technol. Assess.* **2021**, *46*, 101222. [CrossRef]

43. Jafari, H.; Kalantari, D.; Azadbakht, M. Semi-industrial continuous band microwave dryer for energy and exergy analyses, mathematical modeling of paddy drying and it's qualitative study. *Energy* **2017**, *138*, 1016–1029. [CrossRef]

44. Jindarat, W.; Rattanadecho, P.; Vongpradubchai, S. Analysis of energy consumption in microwave and convective drying process of multi-layered porous material inside a rectangular wave guide. *Exp. Thermal Fluid Sci.* **2011**, *35*, 728–737. [CrossRef]

45. Darvishi, H. Quality, performance analysis, mass transfer parameters and modeling of drying kinetics of soybean. *Braz. J. Chem. Eng.* **2017**, *34*, 143–158. [CrossRef]

46. Brooker, D.B.; Bakker-Arkema, F.W.; Hall, W. *Drying and Storage of Grains and Oilseeds*; Van Nostrand Reinhold: New York, NY, USA, 1992; Volume 49, p. 450.

47. Arslan, E.; Aktas, M. 4E analysis of infrared-convective dryer powered solar photovoltaic thermal collector. *Solar Energy* **2020**, *208*, 46–57. [CrossRef]

48. Dolgun, E.C.; Karaca, G.; Aktaş, M. Performance analysis of infrared film drying of grape pomace using energy and exergy methodology. *Int. Commun. Heat Mass Transf.* **2020**, *118*, 104827. [CrossRef]

49. Sharabiani, V.R.; Kaveh, M.; Abdi, R.; Szymanek, M.; Tanaś, W. Estimation of moisture ratio for apple drying by convective and microwave methods using artificial neural network modeling. *Sci. Rep.* **2021**, *11*, 9155. [CrossRef]

50. Beigi, M.; Tohidi, M.; Torki-Harchegani, M. Exergetic analysis of deep-bed drying of rough rice in a convective dryer. *Energy* **2017**, *140*, 374–382. [CrossRef]

51. Nagvanshi, S.; Venkata, S.K.; Goswami, T.K. Study of color kinetics of banana (Musa cavendish) under microwave drying by application of image analysis. *Food Sci. Technol. Int.* **2020**. [CrossRef]

52. Çınkır, N.I.; Süfer, O. Microwave drying of TURKISH red meat (watermelon) radish (Raphanus Sativus l.): Effect of osmotic dehydration, pre-treatment and slice thickness. *Heat Mass Transfer.* **2020**, *56*, 3303–3313. [CrossRef]

53. Darvishi, H.; Zarein, M.; Farhudi, Z. Energetic and exergetic performance analysis and modeling of drying kinetics of kiwi slices. *J. Food Sci. Technol.* **2016**, *53*, 2317–2333. [CrossRef]

54. Kaveh, M.; Amiri Chayjan, R.; Taghinezhad, E.; Abbaspour Gilandeh, Y.; Younesi, A.; Sharabiani, V.R. Modeling of thermodynamic properties of carrot product using ALO, GWO, and WOA algorithms under multi-stage semi-industrial continuous belt dryer. *Eng. Comput.* **2019**, *35*, 1045–1058. [CrossRef]

55. Raj, G.V.S.B.; Dash, K.K. Microwave vacuum drying of dragon fruit slice: Artificial neural network modelling, genetic algorithm optimization, and kinetics study. *Comp. Electron. Agric.* **2020**, *178*, 105814. [CrossRef]

56. Nguyen, M.H.; Price, W.E. Air-drying of banana: Influence of experimental parameters, slab thickness, banana maturity and harvesting season. *J. Food Eng.* **2007**, *79*, 200–207. [CrossRef]

57. Azimi-Nejadian, H.; Hoseini, S.S. Study the effect of microwave power and slices thickness on drying characteristics of potato. *Heat Mass Transfer.* **2019**, *55*, 2921–2930. [CrossRef]

58. Doymaz, I.; Göl, E. Convective drying characteristics of eggplant slices. *J. Food Proc. Eng.* **2011**, *34*, 1234–1252. [CrossRef]

59. Motevali, A.; Tabatabaei, S.R. A Comparison between Pollutants and Greenhouse Gas Emissions from Operation of Different Dryers based on Energy Consumption of Power Plants. *J. Clean. Prod.* **2017**, *154*, 445–461. [CrossRef]

60. Khan, M.K.I.; Ghauri, Y.M.; Alvi, T.; Amin, U.; Khan, M.I.; Nazir, A.; Saeed, F.; Aadil, R.M.; Nadeem, M.T.; Babu, I.; et al. Microwave assisted drying and extraction technique; kinetic modelling, energy consumption and influence on antioxidant compounds of fenugreek leaves. *Food Sci. Technol.* **2021**. [CrossRef]

61. Azimi-Nejadian, H.; Houshyar, E. Thermodynamic analysis of potato drying process in a microwave dryer. *Food Sci. Technol.* **2020**, *106*, 1–12. (In Farsi)

62. Motevali, A.; Hashemi, S.J.; Taghinezhad, E. Investigation of Energy Parameters, Environment and Social Costs for Drying Process (Case Study: Apple Slices). *Agric. Mech. Syst. Res.* **2019**, *20*, 36–54. (In Farsi)

63. Nwakuba, N.R.; Chukwuezie, O.C.; Asonye, G.U.; Asoegwu, S.N. Energy analysis and optimization of thin layer drying conditions of okra. *Innov. Technol. Sustain. Agric. Prod. Food Suffic.* **2018**, *14*, 129–148.

64. Okunola, A.; Adekanye, T.; Idahosa, E. Energy and exergy analyses of okra drying process in a forced convection cabinet dryer. *Res. Agric. Eng.* **2021**, *67*, 8–16. [CrossRef]

Article

Study on Adsorption Properties of Modified Corn Cob Activated Carbon for Mercury Ion

Yuyingnan Liu [1], **Xinrui Xu** [1,2], **Bin Qu** [1], **Xiaofeng Liu** [3], **Weiming Yi** [4] and **Hongqiong Zhang** [1,*]

[1] College of Engineering, Northeast Agriculture University, Harbin 150030, China; tomtom12138@outlook.com (Y.L.); 13258587435@163.com (X.X.); qubin@neau.edu.cn (B.Q.)
[2] CAS Key Laboratory of Renewable Energy, Guangzhou Institute of Energy Conversion, Guangzhou 510640, China
[3] Key Laboratory of Environmental and Applied Microbiology, Environmental Microbiology Key Laboratory of Sichuan Province, Chengdu Institute of Biology, Chinese Academy of Sciences, Chengdu 610041, China; lxf3636@163.com
[4] College of Agricultural Engineering and Food Science, Shandong University of Technology, Zibo 255000, China; yiweiming@sdut.edu.cn
* Correspondence: zhhqiong@163.com

Abstract: In this study, corn cob was used as raw material and modified methods employing KOH and $KMnO_4$ were used to prepare activated carbon with high adsorption capacity for mercury ions. Experiments on the effects of different influencing factors on the adsorption of mercury ions were undertaken. The results showed that when modified with KOH, the optimal adsorption time was 120 min, the optimum pH was 4; when modified with $KMnO_4$, the optimal adsorption time was 60 min, the optimal pH was 3, and the optimal amount of adsorbent and the initial concentration were both 0.40 g/L and 100 mg/L under both modified conditions. The adsorption process conforms to the pseudo-second-order kinetic model and Langmuir model. Scanning electron microscopy and energy-dispersive X-ray spectroscopy (SEM-EDS), Fourier transform infrared spectroscopy (FTIR), X-ray photoelectron spectroscopy (XPS) and Zeta potential characterization results showed that the adsorption process is mainly physical adsorption, surface complexation and ion exchange.

Keywords: mercury ion; corn cob; activated carbon; adsorption

Citation: Liu, Y.; Xu, X.; Qu, B.; Liu, X.; Yi, W.; Zhang, H. Study on Adsorption Properties of Modified Corn Cob Activated Carbon for Mercury Ion. *Energies* **2021**, *14*, 4483. https://doi.org/10.3390/en14154483

Academic Editor: Muhammad Sultan

Received: 22 June 2021
Accepted: 19 July 2021
Published: 24 July 2021

Publisher's Note: MDPI stays neutral with regard to jurisdictional claims in published maps and institutional affiliations.

1. Introduction

In recent years, with the development of industrialization and the discharge of various wastes, a large amount of mercury-containing wastewater has been discharged into soil, rivers, lakes, oceans and other water bodies, causing serious soil and water pollution [1–4]. However, because mercury is difficult to degrade in the environment and has high fluidity, it will flow into animals and plants through soil and water, and enter the human body through the food chain, seriously endangering human life and ecological safety [5]. The main symptom of mercury poisoning is damage to the nervous system and this can lead to death in severe cases. Therefore, mercury pollution has become a serious threat to human life and safety [6].

Corn is one of the most productive food crops in China, and its output is increasing year by year [7]. The corn cob is the part of corn after threshing treatment, which accounts for about 0.21 of the weight of the corn. The treatments of corn cob are mainly landfill, incineration and other methods [8]. These treatment methods not only fail to rationally use resources, they cause waste, and may also pollute the environment. Biomass had the advantages of high carbon content, easy activation and low degradation rate [9,10]. Activated carbon products prepared with biomass not only have the characteristics of high purity, high specific surface area, and good adsorption effect, but also realize the recycling of carbon and are green and environmentally friendly [11–14]. The gradual scarcity of non-renewable energy has made the use of low-cost and huge reserves of biomass resources as raw materials for the preparation of activated carbon gradually gain widespread popularity [15,16].

At present, the commonly used methods to remove mercury ions in wastewater include chemical precipitation, flocculation, ion exchange, and adsorption using activated carbon [17,18]. Activated carbon is a kind of porous material processed by a special process with carbon atoms as the main component [19,20]. Activated carbon has the characteristics of well-developed pores, large specific surface area, low price, strong adsorption capacity, extensive raw material sources, safety and environmental protection, not reacting easily with acids or alkalis, insoluble in water and organic solvents, renewable and recyclable [21]. However, due to limited resources for preparing activated carbon, leading to high cost, this limits the large-scale application of activated carbon. Therefore, it is necessary to find suitable and low-cost raw materials for the preparation of activated carbon. Sajjadi et al. used pistachio wood waste to prepare activated carbon, using ammonium nitrate as a new activator. The mercury ion adsorption effect was significantly higher than commercial activated carbon, and the preparation cost was lower [22]; Hsu et al. found that copper/sulfur co-impregnated activated carbon exhibited greater Hg(II) adsorption than both raw activated carbon and sulfur-impregnated activated carbon at an initial Hg(II) concentration of higher than 8000 ng/L [23]. Yagmur et al. used oleaster fruit as a substrate and KOH and $ZnCl_2$ to prepare activated carbon, and found that the two play key roles in producing activated carbon with fairly different properties [24].

Using corn cob as raw material, the effects of activation temperature, activation time, the KOH to carbon ratio on a specific surface area and mercury ion adsorption capacity of activated carbon were explored by single factor experiment and orthogonal experiment, and the best preparation process was selected. The mercury ion adsorption performance of the prepared activated carbon under different factors such as initial concentration of mercury ions, amount of adsorbent added, pH, adsorption time, and interference of metal ions were investigated using a variety of characterization methods such as scanning electron microscopy and energy-dispersive X-ray spectroscopy (SEM-EDS), BET, Fourier transform infrared spectroscopy (FTIR), X-ray photoelectron spectroscopy (XPS), Zeta potential, etc., The analysis of the characterization results of activated carbon before and after raw material modification and mercury ion adsorption was carried out.

2. Materials and Methods

2.1. Materials and Reagents

Corn cobs were collected from the experimental base of Xiangyang farm of Northeast Agricultural University. The corn cobs were crushed to 80 mesh, washed with deionized water and anhydrous ethanol, dried to constant weight at 105 °C, and sterilized for use. Potassium hydroxide, 1000 mg/L mercury ion standard solution, cadmium nitrate, lead nitrate, copper nitrate, nickel nitrate, zinc nitrate were purchased from Shanghai Macklin Biochemical Co., Ltd. (Shanghai, China). Potassium permanganate was purchased from Guangzhou Chemical Reagent Factory. Hydrochloric acid was purchased from Shenzhen Chemical Reagent Technology Co., Ltd. (Shenzhen, China). Sodium hydroxide was purchased from Shanghai Acmec Biochemical Technology Co., Ltd. (Shanghai, China).

2.2. Preparation of Activated Carbon

2.2.1. Single Factor Experimental Design

The dried biomass materials were heated with nitrogen in a tubular furnace. The heating rate was 10 °C/min, and the carbonization temperature was 500 °C for 1 h. After the biochar was cooled to room temperature, it was then taken out, cleaned with ultrapure water, and dried in 105 °C oven to constant weight to obtain corn cob carbon (CCC). Under the conditions of different activation time, activation temperature and alkali–carbon ratio, KOH and CCC were evenly mixed according to the set mass and heated with nitrogen in a tubular furnace. Firstly, KOH was heated to 400 °C at 10 °C/min for 30 min, and then heated for a period of time at the same heating rate to the experimental set temperature.

After the activation experiment, the activated carbon was reduced to room temperature. In order to remove the KOH and other impurities that were not completely reacted and washing with ultrapure water, 1 mol/L hydrochloric acid and absolute ethanol to neutral. KOH modified corn cob activated carbon (CCAC) was prepared by putting the sample into a 105 °C oven until constant weight.

2.2.2. Orthogonal Experimental Design

The orthogonal experiment was designed by orthogonal test table L9(3^3). Taking the mercury ion adsorption capacity and specific surface area of activated carbon as analysis indexes, the range method and variance method were used to analyze the primary and secondary relationship of different influencing factors so as to determine the best experimental combination KOH modified corn cob activated carbon preparation methods (CCAC).

2.3. Experiment on KMnO₄ Modified Activated Carbon

A certain amount of CCAC was mixed with different concentrations of $KMnO_4$ solution (0.02–0.18 mol/L). After soaking at room temperature for 12 h, it was dried in a 105 °C oven. The it was washed with ultrapure water (the pH of the washing liquid is neutral), and placed it in a drying box to constant weight. Finally, the $KMnO_4$ modified corn cob activated carbon (CKAC) was optimized by a mercury ion adsorption experiment.

2.4. Characterization of Activated Carbon

SEM and EDS are often used to analyze the micro morphology of various samples and to analyze the elements qualitatively and quantitatively. In this experiment, the conductive adhesive was glued on the sample table, a small amount of samples was smeared on the conductive adhesive, and the gold spraying treatment was carried out. Then the surface characteristics and different elemental compositions of activated carbon were explored by SEM (SEM SU-70) and EDS. The BET parameters of activated carbon were measured and analyzed by specific surface area and a porosity analyzer (ASAP 2460). Before the experiment, the samples were degassed at 200 °C for 12 h, and the nitrogen adsorption/desorption isotherm was measured. The BET method was used to calculate the parameters of specific surface area, pore volume and pore size. FTIR was used to analyze the chemical composition of organic or inorganic samples qualitatively, as well as the characteristic vibration frequencies of functional groups or chemical bonds. Before the experiment, the completely dried activated carbon sample was mixed with KBr at a certain mass ratio and pressed into thin sheets under a certain pressure. This experiment used Nicolet iS50 to determine the groups contained in the sample. XPS was used to clarify the element composition and molecular structure of the material. The sample preparation process of this experiment was as follows: cut the double-sided tape into small pieces, paste the sample on one side, and paste the aluminum foil on the other side. After the samples were compressed into tablets, ESCALAB 250Xi was finally used for testing. The measurement conditions were Al Kα rays, the scanning range was 1300~0 eV, and the carbon C 1s peak (284.60 eV) was used for correction. The 40 mL ultrapure water was put into a 150 mL beaker, and 10 mg of well ground sample was added. Then, the pH of the sample was adjusted to 2–6 with 0.10 mol/L HCl and 0.10 mol/L NaOH, respectively. The Zeta potential of the sample was measured by a nano ZS Mastersizer 2000E.

2.5. Mercury Ion Adsorption Experiment

(1) Effect of adsorption time on mercury ion adsorption

We put 50 mL of 100 mg/L mercury ion solution into 10 different conical flasks, then added 0.40 g/L activated carbon, respectively. The adsorption experiment was carried out in a constant temperature oscillation water bath. The experiment temperature was 30 °C and the oscillation rate was 150 r/min. The adsorption time was set to 15–1440 min. After adsorption, solid–liquid separation was carried out, and the liquid phase was collected to test the concentration of mercury ion to determine the best adsorption time.

(2) Effect of adsorbent dosage on mercury ion adsorption

We took 50 mL solution with initial concentration of mercury ion of 100 mg/L into five different 100 mL conical flasks, then added 0.20–1.00 g/L activated carbon respectively. The adsorption experiment was carried out in a constant temperature oscillation water bath. The experiment temperature was 30 °C and the oscillation rate was 150 r/min. After adsorption, solid–liquid separation was carried out, and the liquid phase was collected to test the concentration of mercury ion to determine the dosage of adsorbent.

(3) Effect of initial concentration on mercury ion adsorption

We took 50 mL solution with initial concentration of mercury ion of 40, 60, 80, 100, 120 mg/L into five different 100 mL conical flasks, then added 0.40 g/L activated carbon respectively. The adsorption experiment was carried out in a constant temperature oscillation water bath. The experimental temperature was 30 °C and the oscillation rate was 150 r/min. After adsorption, the solid-liquid separation was carried out, and the liquid phase was collected to test the concentration of mercury ion to determine the optimal initial concentration of mercury ion.

(4) Effect of pH on mercury ion adsorption

We put 50 mL of 100 mg/L mercury ion solution into five different conical flasks, and then added 0.40 g/L activated carbon. The experiment was carried out in a constant temperature oscillating water bath. The experimental temperature was 30 °C, and the oscillation rate was 150 r/min. We used 0.10 mol/L hydrochloric acid and 0.10 mol/L NaOH to adjust the pH to 2, 3, 4, 5 and 6, respectively. After adsorption, the solid-liquid separation was carried out, and the liquid phase was collected to test the mercury ion concentration to determine the optimal pH value.

(5) Effect of interfering ions on mercury adsorption

In order to study the interference of single cation on mercury ion adsorption, 5 mL of Cd^{2+}, Pb^{2+}, Cu^{2+}, Zn^{2+}, Ni^{2+} with initial concentration of 1.00 g/L were added into five different 50 mL volumetric flasks respectively, and then 5 mL of mercury ion solution with initial concentration of 1.00 g/L was added respectively. The solution after constant volume was added into five different conical flasks, and then 0.40 g/L activated biochar was added respectively. The solution was placed in a constant temperature oscillation water bath for adsorption experiment. The temperature was 30 °C, and the oscillation rate was 150 r/min. After adsorption, the solid–liquid separation was carried out, and the liquid phase was collected to test the concentration of metal cations.

In order to simulate the actual wastewater situation better, the interference of various cations on mercury ion adsorption was explored. We added 5 mL Hg^{2+}, Cd^{2+}, Pb^{2+}, Cu^{2+}, Zn^{2+}, Ni^{2+} with initial concentration of 1.00 g/L to a 50 mL constant volume bottle for constant volume. We added the constant volume solution to a 100 mL conical flask, and added 0.40 g/L activated biochar. Then we put the solution in a constant temperature oscillating water bath for adsorption experiment. The temperature was 30 °C, and the oscillation rate was 150 r/min. After adsorption, the solid–liquid separation was carried out, and the liquid phase was collected to test the concentration of metal cations.

The adsorption capacity and removal rate of mercury ion were calculated by the following formulas, respectively [25].

$$q_e = \frac{(C_0 - C_e)V}{m} \tag{1}$$

$$R = \frac{(C_0 - C_e)}{C_0} * 100\% \tag{2}$$

q_e—adsorption capacity of activated carbon for metal ions, mg/g
R—removal rate of metal ions by biochar, %
C_0—initial concentration of metal ions, mg/L
C_e—residual concentration of metal ions, mg/L
V—volume of solution, L
m—mass of activated carbon, g

2.6. Adsorption Kinetics

The study of adsorption kinetics was based on the measurement and description of adsorption rate and adsorption process, and then qualitative analysis of adsorption type, quantitative calculation of adsorption equilibrium time, etc., and finally we explored the adsorption mechanism of adsorbent for pollutants. We took 50 mL of a solution with an initial concentration of 100 mg/L of mercury ions, set the time range of the adsorption experiment to 15–360 min, and added 0.02 g of activated carbon to different experimental groups. The experiment temperature was set to 30 °C. After experiment was completed, the sample was taken out to filter the solid, the concentration of mercury ions in the liquid was measured, and q_e and q_t were calculated to complete the fitting of the adsorption kinetic equation. In this experiment, the pseudo-first- and –second-order kinetic models, Elovich model and intra particle diffusion model were used.

(1) Pseudo-first-order kinetic model

The pseudo-first-order kinetics is understood as that the mass transfer resistance in the particle had a great influence on the adsorption process at the beginning of the adsorption reaction, which was dominated by the monolayer adsorption by boundary diffusion. The equation can be expressed as [26]:

$$\log(q_e - q_t) = \log q_e - \frac{k_1 t}{2.303} \tag{3}$$

q_t—the adsorption capacity of activated carbon to mercury ions at t, mg/g
k_1—the reaction rate constant of the pseudo-first-order kinetics, 1/min

(2) Pseudo-second-order adsorption kinetic model

The pseudo-second-order adsorption kinetic model assumed that there is a chemical reaction between the adsorbent and the adsorbate, and the adsorption rate is related to the chemical adsorption, which might include the gain, loss or sharing of electrons. The equation can be expressed as [27]:

$$\frac{t}{q_t} = \frac{1}{k_2 q_e^2} + \frac{t}{q_e} \tag{4}$$

k_2—the reaction rate constant of the pseudo-second-order kinetics, g/(mg·min)

(3) Elovich model

The Elovich model generally describes the adsorption behavior of heterogeneous solid surface, and the rate of the adsorption process will decrease exponentially with the increase of adsorbent adsorption capacity. The equation can be expressed as:

$$q_t = \alpha + \beta \ln t \tag{5}$$

α/β—parameters of Elovich model

(4) Intra-particle diffusion model

There are three assumptions in the intra-particle diffusion model, that the liquid film diffusion resistance can be ignored or only affected in a very short time; the diffusion direction of adsorbate is random and the concentration does not change with the position of the particles; the internal diffusion coefficient is constant and does not change with time and position. The equation is as follows [26]:

$$q_t = k_{id} t^{\frac{1}{2}} + C \tag{6}$$

k_{id}—Intra particle diffusion rate constant, (mg/(g·min$^{-1/2}$))
C—Constants related to the thickness of the boundary layer

2.7. Adsorption Isotherm

The adsorption characteristics of activated carbon for heavy metal ions could be expressed through adsorption isotherms, which played an important role in exploring adsorption mechanisms and practical applications. This experiment used the Langmuir isotherm model and Freundlich isotherm model to explore the adsorption mechanism. We prepared 50 mL mercury ion solutions with different concentrations (20, 40, 60, 80, 100, 120 mg/L) and placed them in 100 mL conical flasks. The addition of activated carbon was 0.40 g/L, the adsorption time was 24 h, the temperature was 30 °C, and the oscillation rate was 150 r/min. After the experiment, we took out the sample, filtered the solid, measured the concentration of mercury ion in the liquid, and calculated the value of q_e.

(1) Langmuir adsorption isotherm model

The derivation and establishment of the Langmuir adsorption isotherm model are based on the following four assumptions: adsorption is a single molecular layer, and no other molecular cover exists; the possibility of adsorption sites being used is the same; the surface of the adsorbent is the same; the possibility of molecules attached to a certain location is not related to whether the adjacent space has been occupied by other molecules. The equation can be expressed as [28]:

$$\frac{C_e}{q_e} = \frac{1}{q_m K_L} + \frac{C_e}{q_m} \tag{7}$$

q_m—the maximum theoretical adsorption amount of mercury ions adsorbed by activated carbon, mg/g

K_L—Langmuir isotherm adsorption rate constant, L/mg

(2) Freundlich adsorption isotherm model

The Freundlich adsorption isotherm equation is derived from the Langmuir adsorption isotherm equation, which is an empirical equation. It is assumed that the adsorption sites on the adsorbent surface are unevenly distributed and have different energies, and the adsorption process is mainly multi-layer adsorption, which is mainly suitable for physical adsorption. The equation can be expressed as follows [29]:

$$\ln q_e = \ln K_F + (1/n_F) \ln C_e \tag{8}$$

K_F—Freundlich isotherm adsorption equation constant
n_F—Freundlich adsorption intensity

3. Results and Discussion

3.1. Single Factor Experiment

3.1.1. Effect of Activation Temperature

Figure 1a shows the effect of activation temperature on the adsorption of mercury ions by corn cob activated carbon. It can be seen from the figure that with the increase of activation temperature, the adsorption capacity of corn cob activated carbon gradually increased. When the activation temperature was 800 °C, the adsorption capacity of mercury ions reached 180.02 mg/g. This was because in the process of heating, the carbon materials gradually decomposed and the reaction rate increased. The corrosion and pore-forming ability of KOH on the activated carbon was enhanced, which promoted the development of the pore structure, produced a large number of micropores and mesopores, and gradually increased the adsorption capacity.

When the activation temperature exceeds 800 °C, the pore structure of the activated carbon will be destroyed and the specific surface area of the activated carbon will be reduced, resulting in the decrease of mercury ion adsorption [30].

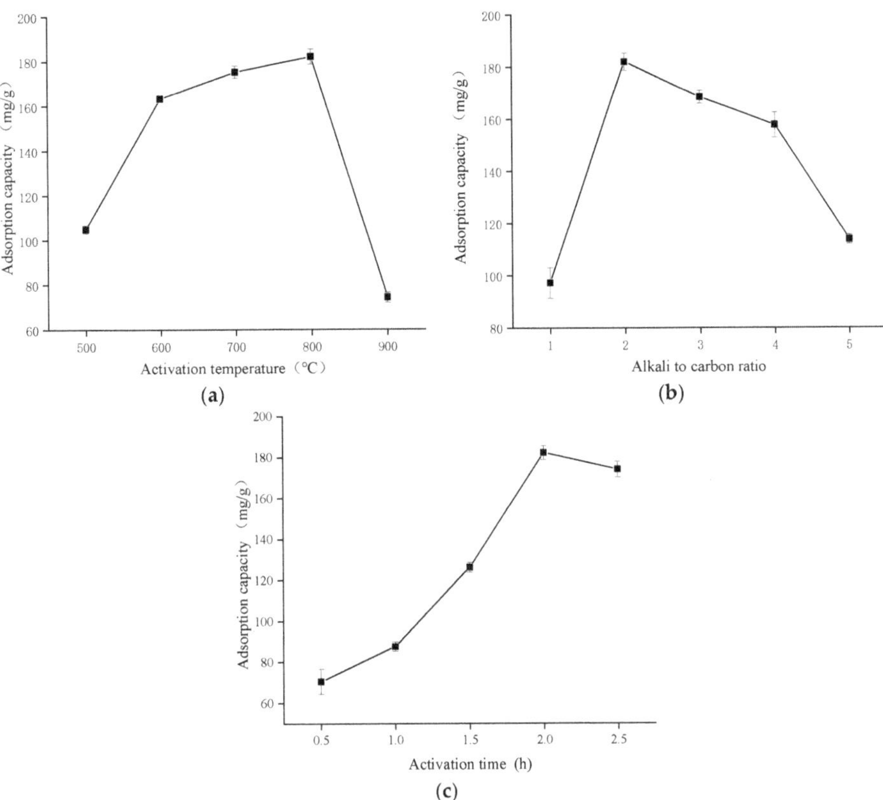

Figure 1. Influence of activation temperature (**a**), alkali-carbon ratio (**b**), activation time (**c**) on mercury ion adsorption.

3.1.2. Effect of Alkali–Carbon Ratio

It can be seen from Figure 1b that when the KOH carbon ratio of corn cob activated carbon was less than 2:1, the mercury ion adsorption capacity increases continuously. This is because with the increase of KOH ratio, the pore structure and adsorption sites of activated carbon increased continuously, and the specific surface area increased continuously, which promoted the adsorption process. When the KOH to carbon ratio is 2:1, it reached the optimal alkali to carbon ratio, and the adsorption capacity of mercury ion was 180.02 mg/g. When the KOH carbon ratio was higher than the optimal ratio, the adsorption capacity of mercury ions will be significantly reduced. The reason was that the excess activator will destroy the pore structure which has been formed, and have a negative impact on the adsorption process, resulting in the reduction of mercury ion adsorption.

3.1.3. Effect of Activation Time

It can be seen from Figure 1c that with the increase of activation time, the adsorption capacity of corn cob activated carbon and mercury ion first increased and then decreased. When the activation time of CCAC was 2 h, the adsorption capacity of mercury ions reached the maximum value of 182.02 mg/g. This was because with the increase of activation time, KOH gradually eroded the carbon skeleton of raw materials, and more oxygen-containing functional groups and new pore structures were formed in the activated carbon, which further improved the activation degree and increased the adsorption capacity. When the activation time exceeded the optimal activation time, the activated carbon was eroded excessively, and the number of pore structures destroyed was greater than the number of newly formed pores, resulting in the decrease of adsorption capacity.

3.2. Orthogonal Experiment

The orthogonal experiment of KOH modified corn cob activated carbon was used to clarify the changes of mercury ion adsorption capacity, specific surface area of activated carbon under the influence of different activation time, activation temperature and alkali carbon ratio. The results of the orthogonal experiment and extreme value analysis are shown in Tables 1 and 2.

Table 1. Orthogonal experimental results of corn cob activated carbon (CCAC) modified by potassium hydroxide.

	Activation Time (h)	Activation Temperature (°C)	Alkali-Carbon Ratio	Adsorption Capacity (mg/g)	Specific Surface Area (m^2/g)
	A	B	C		
1	1	750	1.5:1	104.61	1539.76
2	1	800	2.0:1	168.21	1927.48
3	1	850	2.5:1	54.48	1567.00
4	1.5	850	1.5:1	140.22	1753.69
5	1.5	750	2.0:1	60.69	1910.83
6	1.5	800	2.5:1	126.22	2551.00
7	2	800	1.5:1	61.96	1831.84
8	2	850	2.0:1	89.17	1993.16
9	2	750	2.5:1	183.36	2509.61

Table 2. Extreme value analysis of CCAC modified by potassium hydroxide.

	Adsorption Capacity (mg/g)			Specific Surface Area (m^2/g)		
	A	B	C	A	B	C
K_1	327.30	348.66	306.79	5034.24	5960.20	5125.29
K_2	327.13	356.39	318.07	6215.52	6310.32	5831.47
K_3	334.49	283.87	364.06	6334.61	5313.85	6627.61
k'_1	109.10	116.22	102.26	1678.08	1986.73	1708.43
k'_2	109.04	118.80	106.02	2071.84	2103.44	1943.82
k'_3	111.50	94.62	121.35	2111.54	1771.28	2209.20
R	2.45	24.17	19.09	433.46	332.16	500.77

According to the range analysis results of the mercury ion adsorption capacity, the order of influence degree from large to small was activation temperature, KOH carbon ratio and activation time, which was B > C > A. According to the comparison of K values, the best preparation conditions were as follows: activation time was 2 h, activation temperature was 800 °C, KOH carbon ratio was 2.5:1, which was $B_2C_3A_3$. According to the range analysis results of specific surface area of activated carbon, the order of influence degree from large to small was KOH carbon ratio, activation time and activation temperature, which was C > A > B. According to the size of K value, the best preparation conditions were as follows: activation time was 2 h, activation temperature was 800 °C, KOH carbon ratio was 2.5:1, which was $C_3A_3B_2$, and the best preparation conditions were the same. Since there was no such combination in the orthogonal experiment, the mercury ion adsorption capacity and specific surface area of the combination were 184.76 mg/g and 2797.75 m^2/g, respectively, which were higher than those of the orthogonal experiment group. The results showed that the scheme was the best preparation scheme, and the activated carbon prepared by this process was named CCAC.

3.3. Adsorption Experiment of Activated Carbon Modified by KMnO$_4$

It can be seen from Figure 2 that the adsorption capacity of activated carbon modified by KOH and KMnO$_4$ was higher than that of CCAC. This was because KMnO$_4$ increased the number of oxygen-containing functional groups on the surface of activated carbon, and mercury ion reacted with oxygen-containing functional groups. The optimal $KMnO_4$ concentration of CKAC was 0.14 mol/L, the maximum adsorption capacity was 222.22 mg/g, and the removal rate of mercury ion was 88.89%, which was 14.99% higher than that of activated carbon without $KMnO_4$ modification. When the concentration of $KMnO_4$ was higher than the optimal concentration, the adsorption capacity decreased. This may be because of the strong oxidation of $KMnO_4$ made some original adsorption sites disappear. It may also be because excessive $KMnO_4$ covered the original adsorption sites, resulting in the reduction of the adsorption capacity of mercury ions.

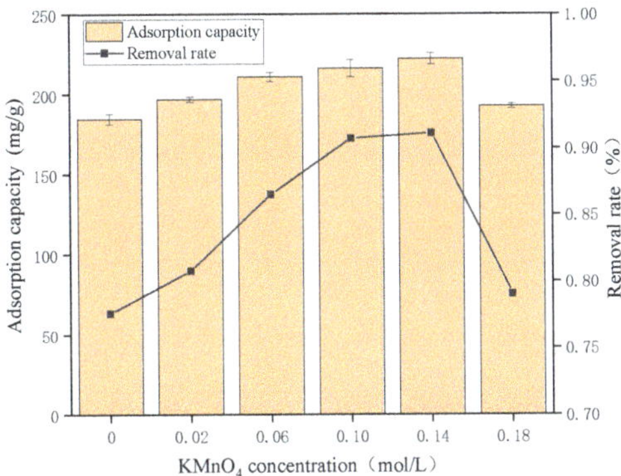

Figure 2. Influence of $KMnO_4$ concentration on mercury ion adsorption.

3.4. Adsorption Experiments

3.4.1. The Effect of Adsorption Time

It can be seen from Figure 3a that the initial stage of adsorption (0–120 min) was the rapid adsorption stage, and the adsorption capacity and removal rate of activated carbon for mercury ions increased rapidly. At this time, the mass transfer effect in the solution was the largest, and the adsorption rate of mercury ions was also faster. When the adsorption reaction reached 120–1440 min, the adsorption rate decreased, and finally reached the dynamic adsorption equilibrium. At equilibrium, the maximum adsorption capacity of CCAC, CKAC reached 184.76 mg/g, 222.22 mg/g, respectively, and the maximum removal rates were 73.91%, 88.89%, respectively. It can be seen that the two activated carbon had strong adsorption capacity, and the adsorption capacity of CKAC was higher than that of CCAC. At this stage, the adsorption sites on the surface of biochar were occupied by mercury ions, and the repulsion between adsorbates hindered the adsorption process. According to the experimental data, in order to take into account the adsorption rate and economic benefits, the adsorption times of CCAC, CKAC were 120 min and 60 min, respectively.

3.4.2. The Effect of pH

It can be seen from Figure 3b that the adsorption capacity of the two activated carbons was weak at pH < 2. The reason was that the adsorption environment was more acidic, and hydrogen ions occupied the adsorption sites, which made the surface of biochar positively charged. Mercury ions need to compete with hydrogen ions in the adsorption process, resulting in low adsorption capacity of mercury ions. When pH reached 3–4, the adsorption capacity of activated carbon for mercury increased gradually. The results showed that the pHPZC of CKAC was less than 2, and the pHPZC of CCAC was equal to 3.49. When pH of CKAC and CCAC was 3 and 4, respectively, the adsorption of mercury ions reached the maximum, which was consistent with the characterization of Zeta potential of activated carbon. The maximum adsorption capacity of CKAC was 214.58 mg/g, and the maximum adsorption capacity of CCAC was 171.69 mg/g. When the pH was too high, mercury ions will precipitate, which will reduce the adsorption effect of activated carbon. Therefore, the subsequent adsorption experiments of CKAC and CCAC were carried out with pH of 3 and 4

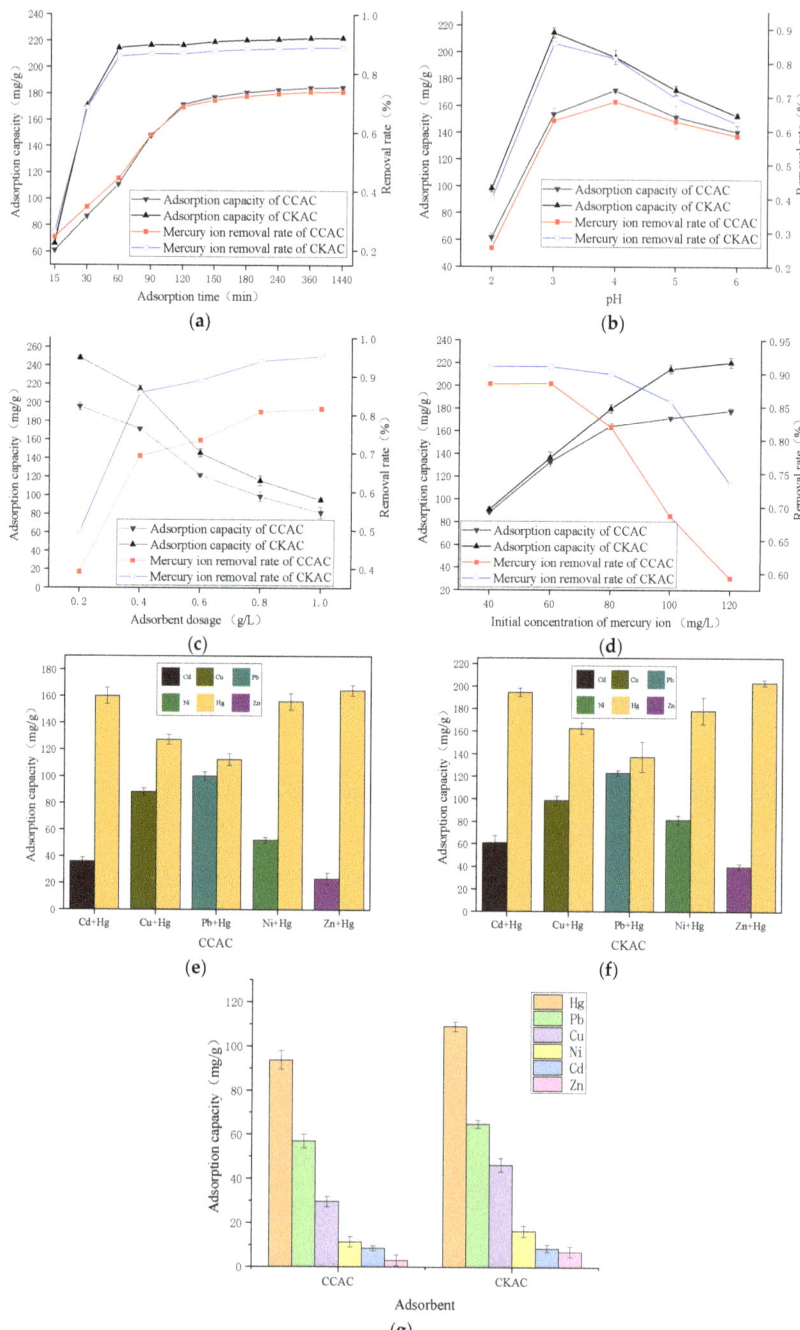

Figure 3. Effect of adsorption time (**a**), pH (**b**), adsorbent dosage (**c**), initial concentration (**d**), single ion interference (**e**,**f**), multiple ion interference (**g**) on the adsorption capacity and removal rate.

3.4.3. The Effect of Adsorbent Dosage

It can be seen from Figure 3c that when the adsorbent dosage of CCAC and CKAC was 0.20 g/L, the maximum adsorption capacity was 195.77 mg/g and 248.22 mg/g, respectively. Subsequently, the adsorption capacity decreased with the increase of adsorbent dosage, but the removal rate of mercury ions increased gradually. When the adsorbent dosage was 1.00 g/L, the removal rates of CCAC were 81.63% and 95.30%, respectively. This was because the addition of activated carbon provided more adsorption sites for mercury ion. When the dosage of adsorbent was more than 0.40 g/L, the rising rate of mercury removal became slow, but with the addition of adsorbent, a large number of adsorption sites on the surface of adsorbent were not utilized, resulting in the decrease of adsorption capacity. Considering the experimental results and economic efficiency, 0.40 g/L activated carbon was selected in the experiment.

3.4.4. The Effect of Initial Concentration

From Figure 3d, when the initial concentration of mercury ion was small, the removal rates of mercury ions by CCAC and CKAC were 88.45% and 91.04%. This was due to the low initial concentration of mercury ions, and the number of adsorption sites on activated carbon was enough to adsorb these mercury ions. When the initial concentration of mercury ions increased from 40 mg/L to 100 mg/L, the adsorption capacity of activated carbon for mercury ions increased rapidly. The reason was that mercury ions would occupy the adsorption sites on activated carbon rapidly during the increase of concentration. When the initial concentration was 120 mg/L, the number of adsorption sites was limited, which was not enough to adsorb them completely. Therefore, the change of adsorption capacity was very small at higher concentration, but the removal rate of mercury ion will continue to decrease, which was below 59.32% and 73.43%, respectively. For the effectiveness of mercury ion treatment, 100 mg/L was selected as the initial concentration of mercury ion.

3.4.5. The Effect of Heavy Metal Ion Interference
The Effect of Single Heavy Metal Ion Interference

It can be seen from Figure 3e,f, that the adsorption effect of mercury ion in each group of experiments was better than that of other metal ions, indicating that the activated carbon had good selectivity for mercury adsorption. However, each metal ion had different effects on the adsorption of mercury ions, and the degree of influence from large to small was as follows: $Pb^{2+} > Cu^{2+} > Ni^{2+} > Cd^{2+} > Zn^{2+}$. Since the water and ion radius of Pb^{2+} is smaller than that of other ions, the ability of Pb^{2+} to compete for adsorption sites is higher than other ions, and the stronger the repulsive force to other metal ions after adsorption, so Pb^{2+} have the greatest impact on Hg^{2+} adsorption. The results showed that the different metal ions in the solution have different competitive adsorption effects. The smaller the radius of hydrated ion, the stronger the affinity and the better the adsorption effect.

The Effect of Various Heavy Metal Ion Interference

It can be seen from Figure 3g that compared with the binary adsorption system, the adsorption capacity of activated carbon for each metal ion was reduced, and the competitiveness of metal ions was the same as that of the binary adsorption system. This may be because the number of adsorption sites was not enough to completely adsorb it, and there was fierce competition among a variety of metal ions. The adsorption capacity of activated carbon for mercury ions was higher than that of other metal ions, indicating that they have good selectivity for mercury ions.

3.5. Adsorption Kinetics

In this paper, the data of CCAC and CKAC at different adsorption times were used for kinetic analysis by using pseudo-first-order and pseudo-second-order kinetic models, the Elovich model and the intra particle diffusion model, respectively. The linear fitting results and parameters of the four adsorption kinetic equations are shown in Figure 4 and Table 3.

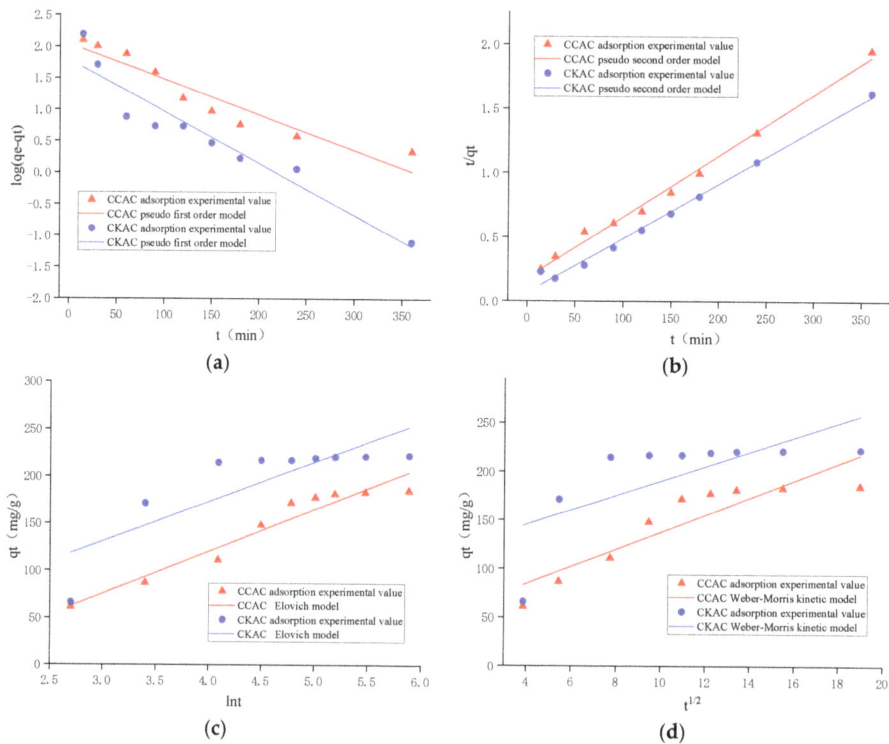

Figure 4. Pseudo-first-order model (**a**), pseudo-second-order model (**b**), Elovich model (**c**), and intraparticle diffusion model (**d**) for mercury ion adsorption on CCAC, CKAC.

Table 3. The fitting parameters of adsorption kinetics.

Model	Parameter	Value	
		CCAC	CKAC
pseudo-first-order	k_1 (min^{-1})	0.013	0.019
	q_e (mg/g)	110.03	61.94
	R^2	0.8838	0.8956
pseudo-second-order	k_2 (g/(mg·min))	0.00013	0.00029
	q_e (mg/g)	208.77	235.77
	R^2	0.9917	0.9908
Elovich	α (g/(mg·min))	−59.39	4.05
	β (mg/g)	44.79	42.18
	R^2	0.9269	0.6494
intra-particle diffusion	k_{id} (mg/(g·min$^{-1/2}$))	8.85	7.46
	R^2	0.7946	0.4130

According to Table 3, the fitting results of the pseudo-second-order kinetic model were better than those of the other three models. R^2 was 0.9917 and 0.9908, and the actual adsorption capacity obtained through the experiment was compared with the fitting result. The theoretical adsorption capacity was close, which could more truly reflect the adsorption process of activated carbon to mercury ions. This showed that mercury ions have good stability during the adsorption process. The reaction process was related to the adsorption rate and the reaction was irreversible. The adsorption of mercury ions by activated carbon was mainly chemical adsorption.

3.6. Adsorption Isotherm

The Langmuir and Freundlich isotherm models were used to fit the adsorption process of mercury ions by CCAC and CKAC. The curve fitting results and parameters are shown in Figure 5 and Table 4.

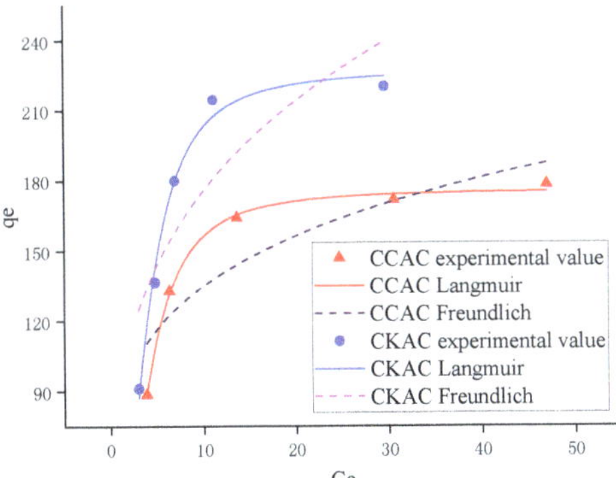

Figure 5. Langmuir, Freundlich isotherm model for mercury ion adsorption on CCAC, CKAC.

Table 4. The fitting parameters of the adsorption isotherm.

Model	Parameter	Value	
		CCAC	**CKAC**
Langmuir	K_L (L/mg)	0.05	0.06
	q_m (mg/g)	175.88	227.32
	R^2	0.9939	0.9781
Freundlich	K_F (mg/g)	82.67	90.54
	$1/n$	0.21	0.29
	R^2	0.7403	0.6502

It can be seen from the above table that the correlation coefficients R^2 fitted by the Langmuir adsorption isotherms of the two kinds of activated carbon were 0.9939 and 0.9781, respectively, which were all larger than the coefficients fitted by Freundlich. The Langmuir model was used to fit the maximum adsorption of mercury ions by activated carbon to 175.88 mg/g and 227.32 mg/g, which were equivalent to the actual adsorption value. Therefore, it can be explained that the adsorption of mercury ions by the activated carbon was mainly the chemical adsorption on the monolayer.

3.7. Characterization of Activated Carbon

3.7.1. Scanning Electron Microscopy (SEM) Analysis

In the SEM of unmodified corn cob biochar shown in Figure 6a the surface of the biochar is smooth and there is no obvious pore structure. From the EDS diagram Figure 7a, it can be seen that C and O were the main element components of biochar, and there were still small amounts of Mg, Si, K elements. It can be seen from the SEM pictures in Figure 6b that there were a lot of pore structures on the surface of activated carbon modified by KOH. This was because during the activation process the carbon material reacted with KOH, and the carbon surface was eroded, which increased the specific surface area of activated carbon and provided a large number of adsorption sites for mercury ion adsorption [31]. It can be seen from the SEM images of Figure 6c that CKAC reacted with $KMnO_4$ solution in the modification process, and the manganese oxides formed adhered to the surface of activated carbon. Although these manganese oxides will cover the original pores, a large number of oxygen-containing functional groups will be formed on the surface of modified activated carbon, which will play an important role in the adsorption process and enhanced the adsorption capacity of activated carbon. It can be seen from the EDS diagram in Figure 7b that the main elements of KOH modified activated carbon are C, O and a small amount of K, while in Figure 7c it can be seen that the main elements of KOH+KMnO$_4$ modified activated carbon were C, O, K and Mn. Among them, the proportion of O increased significantly, which further indicated that KMnO$_4$ could increase the number of oxygen-containing functional groups on the surface of activated carbon. It can be seen from Figure 6d,e that after adsorption of mercury ions by activated carbon, white substances will appear on the activated carbon and adhere to the carbon surface. It can be found from the EDS spectrum in Figure 7d,e that the main components of CCAC were C, O and Hg, while the main components of CKAC were C, O, Hg and Mn, which further indicated that mercury ion was adsorbed by activated carbon. The content of C increased, while the content of O and Mn decreased. This might be due to the reaction of functional groups containing O and Mn during the mercury ion adsorption process, which involved the release of O and Mn. Moreover, the amount of other elements on the activated carbon decreased, which might be due to the ion exchange between mercury and other elements in the adsorption process. In addition, there were Cl on the surface of CCAC, which may be due to the formation of compounds with Hg and attachment to the activated carbon.

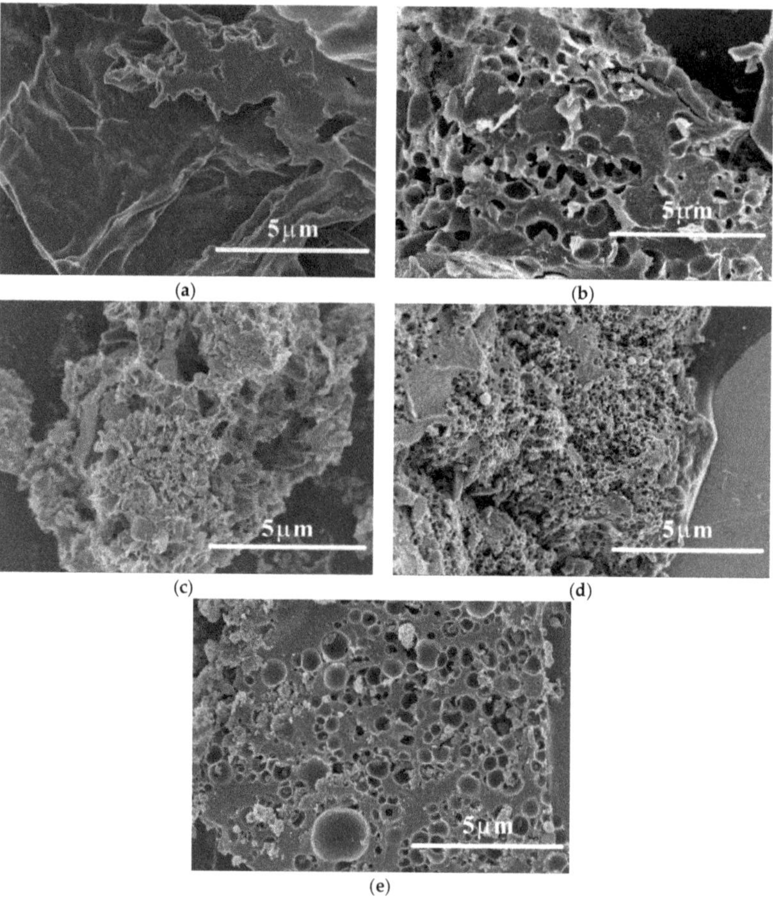

Figure 6. Scanning electron microscope (SEM) image of corn cob carbon (CCC) (**a**), CCAC (**b**), CKAC (**c**), CCAC after adsorption of mercury ions (**d**), CKAC after adsorption of mercury ions (**e**).

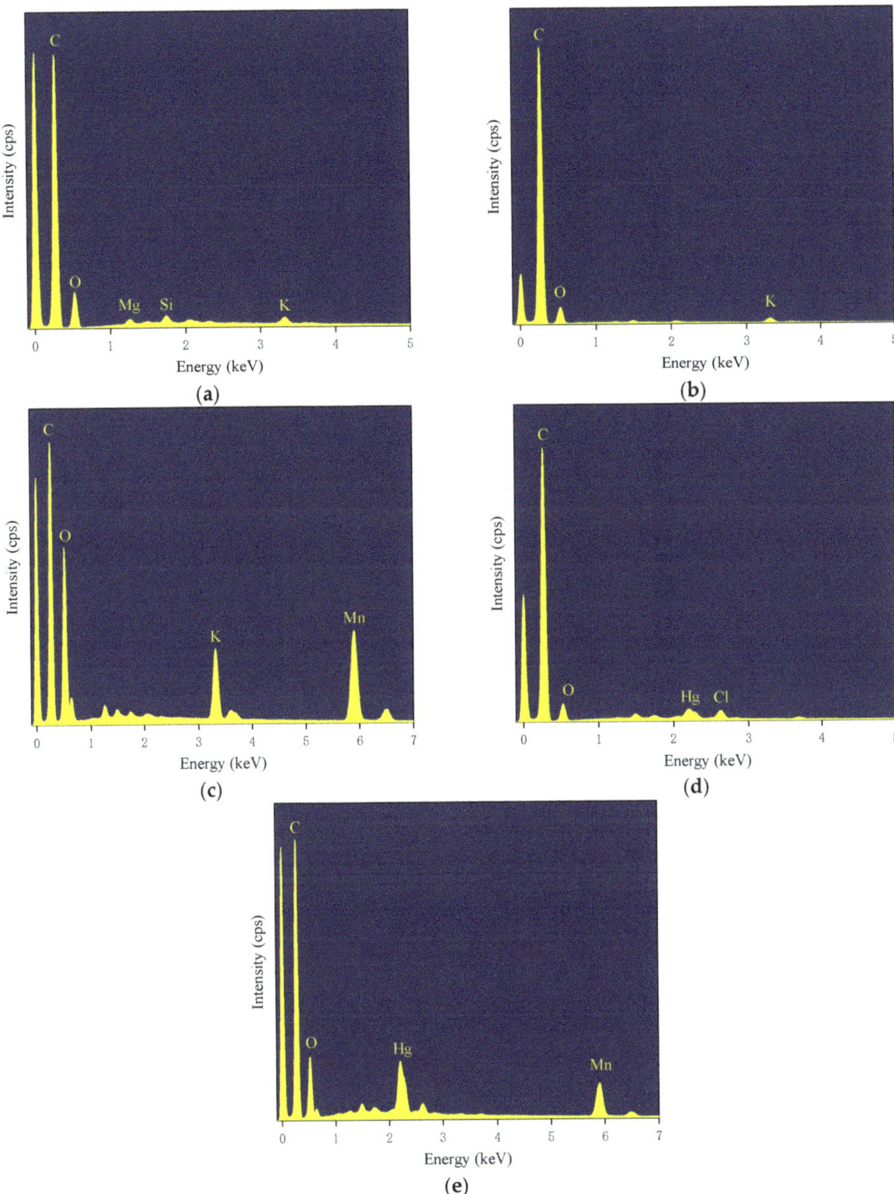

Figure 7. Energy-dispersive X-ray spectroscopy (EDS (image of CCC (**a**), CCAC (**b**), CKAC (**c**), CCAC after adsorption of mercury ions (**d**), CKAC after adsorption of mercury ions (**e**).

3.7.2. BET Analysis

It could be seen from Table 5 that the specific surface area of the biochar obtained after the corn cob was carbonized at 500 °C is relatively small, which was only 11.16 m^2/g, and the total pore volume was 0.003 cm^3/g. The specific surface area value of activated carbon after KOH activation was obvious, reaching 2797.75 m^2/g, and the proportion of micropores was extremely large, which played an important role in the adsorption [32]. At the same time, the pore volume of CCAC also increased to 1.36 cm^3/g. The increase in pore volume could provide more adsorption sites and promote the adsorption of mercury ions. The specific surface area, pore volume and micropores of the activated carbon modified by potassium permanganate were reduced compared to CCAC. This was because the manganese oxide formed during the modification process was attached to the surface of the activated carbon, causing pore blockages. As seen in Figure 8a, both activated carbons belonged to the type I adsorption isotherm. The N_2 adsorption and desorption curves of CCAC increased rapidly in the low pressure region (P/P_0 = 0–0.05), during which monolayer adsorption was mainly carried out in micropores, indicating that there were a large number of microporous structures in both activated carbons. However, the rate of increase of CKAC was slower than that of CCAC, indicating that the number of micropores was relatively small. As was seen from Table 5, the specific surface area of CCAC and CKAC decreased by 418.60 m^2/g and 319.90 m^2/g, respectively, and the micropore area decreased by 2216.28 m^2/g and 156.96 m^2/g, respectively, through comparison before and after adsorption. It can be seen that the specific surface area and micropore area of CCAC decreased more obviously, indicating that the specific surface area and micropore played an important role in the adsorption of mercury ions, and the total pore volume and micropore volume also decreased to different degrees. Through the above data analysis, it can be inferred that mercury ions were adsorbed to the outer surface and the pore surface of KOH modified activated carbon. However, the specific surface area and micropore area of CKAC decreased slightly. The modification of $KMnO_4$ has covered some original pore structures, and the adsorption of mercury ions may be mainly caused by the oxygen-containing functional groups on the surface.

Table 5. The BET parameters of activated carbon.

Sample	S_{BET} (m^2/g)	S_{mic} (m^2/g)	V_{tot} (cm^3/g)	V_{mic} (cm^3/g)	V_{mic}/V_{tot} (%)
CCC	11.16	10.08	0.003	0.0002	6.67%
CCAC	2797.75	2551.15	1.36	1.16	85.29%
CKAC	1212.93	360.71	0.59	0.16	27.12%
CCAC after adsorption	2379.15	334.87	1.20	0.14	11.67%
CKAC after adsorption	893.03	203.75	0.49	0.13	26.53%

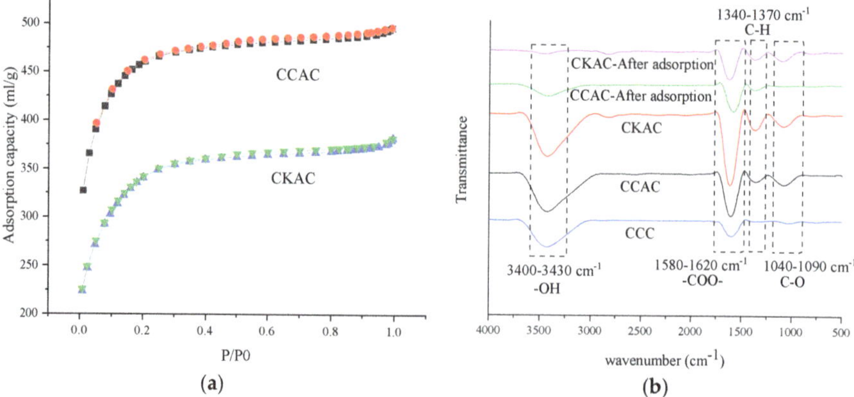

Figure 8. *Cont.*

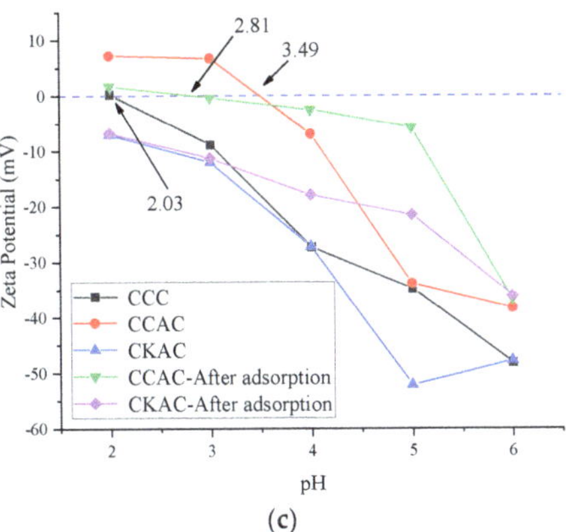

(c)

Figure 8. Nitrogen adsorption–desorption isotherm (**a**), Fourier transform infrared (FTIR) spectra (**b**), Zeta potential (**c**) of corn cob activated carbon before and after modification and before and after adsorption.

3.7.3. Fourier Transform Infrared (FTIR) Analysis

From Figure 8b, the characteristic peak at 3420 cm^{-1} was caused by the stretching vibration of the hydroxyl (-OH) on the surface of the activated carbon [33]. The characteristic peaks at 1600 cm^{-1} were the stretching vibration peaks of the carbonyl group (-C=O) and the carboxyl group (-COO-). The activated carbon modified by KOH and KOH + KMnO$_4$ showed new characteristic peaks at 1340 cm^{-1} and 1030 cm^{-1}, respectively. Among them, the characteristic peaks at 1340 cm^{-1} and 1390 cm^{-1} corresponded to the bending vibration of the C–H bond. The characteristic peaks at 1030 cm^{-1} and 1040 cm^{-1} were caused by the stretching vibration of the C–O bond [34]. The analysis results showed that the modification methods of KOH and KOH+KMnO$_4$ enriched the types of oxygen-containing functional groups on the surface of activated carbon, and the existence of these functional groups could improve the adsorption capacity of activated carbon for mercury ions. After adsorption, the positions of the characteristic peaks of -OH shifted to 3400 cm^{-1} and 3430 cm^{-1}, and the intensity of the peaks became weaker, indicating that -OH played an important role in the adsorption of mercury ions. The -C=O peak position shifted to 1580 cm^{-1} and 1620 cm^{-1}, which might be caused by the reaction of -C=O with mercury ions. The position of the C–H peak shifted from 1340 cm^{-1} and 1390 cm^{-1} to 1360 cm^{-1} and 1370 cm^{-1}, which meant that C–H promoted the adsorption of mercury ions to a certain extent. The position of the characteristic C–O peak was shifted to 1090 cm^{-1}. It is speculated that the C–O bond was involved in the adsorption process of mercury ions.

3.7.4. Zeta Potential Analysis

As was shown in Figure 8c, the pHPZC of CCC and CCAC were 2.03 and 3.49, respectively. When the pH of the solution was less than pHPZC, ions on the surface of the carbon material will protonate and become positively charged. At this time, ions on the surface of the carbon material will repel each other with the positively charged mercury ions in the solution, resulting in a low adsorption capacity of activated carbon. When the pH of the solution was more than pHPZC, the surface of the carbon material was negatively charged. At this time, the activated carbon and mercury ions in the solution attracted each other to promote the adsorption. However, in the pH range of 2 to 6, Zeta potential value of CKAC was always less than 0, indicating that the pHPZC of the activated carbons was less than 2, indicating that CKAC could adapt to a wider pH range, and the adsorption of mercury ions was easier.

3.7.5. X-ray Photoelectron Spectroscopy (XPS) Analysis

Figure 9a shows the XPS full spectrum of biochar before and after modification. It can be seen that the main elements of biochar before and after modification by KOH were C and O. The main elements modified by KOH + $KMnO_4$ were C, O and Mn. Figure 9b shows the C 1s spectra of corn cob carbon before and after modification, respectively. It can be seen from the figure that C mainly exists in the form of C–C, C–O and C=O. The C–C ratio of unmodified biochar was 80.71% and 83.73%, respectively. However, the proportion of C–C in the modified activated carbon decreased, and the proportion of C–O and C=O increased, indicating that during the modification process, the original C–C was transformed into C–O or C=O and other oxygen-containing functional groups through chemical reactions. Figure 9c shows the O 1s spectra of corn cob carbon before and after modification, respectively. It can be seen that the O element in the modified activated carbon mainly existed in the form of O–H and C–OH, and a few existed in the form of C=O. In the XPS spectra of CKAC which is shown in Figure 9c there are characteristic peaks (Mn-O) formed by metal and O and oxygen at 529.59 eV, indicating that manganese oxide was attached to the surface of activated carbon, which was consistent with the results of EDS spectra. Compared with the full spectrum before adsorption, in Figure 10, the contents of O 1s and Mn 2p in XPS decreased after adsorption of mercury ions, which might be due to the reaction of mercury ions with the chemical bonds of O and Mn during the adsorption process, and the emergence of a new Hg 4f peak in Figure 10d, which meant that mercury ions were successfully adsorbed on activated carbon. The intensity of the O–H peak decreased from 97.77% to 18.95%, which indicated that the adsorption of mercury ions consumed a lot of O–H in modified activated carbon. At the same time, the position of the characteristic peak formed by metal oxide and O before adsorption shifted from 529.59 eV to 529.63 eV, and the peak area increased from 52.62% to 54.60%, indicating that Hg–O was formed on the surface of activated carbon after the adsorption of mercury ion. The new peak of Hg 4f was fitted by peak software. The peak of Hg $4f_{7/2}$ was in the range of 98–103 eV, and the peak of Hg $4f_{5/2}$ was in the range of 103–108 eV. Four groups of characteristic peaks were obtained by fitting, which were located at 100.78–101.18 eV, 102.45–102.79 eV, 105.02–105.22 eV and 106.38–106.77 eV, respectively. The two peaks in the range of 100.78–101.18 eV and 105.02–105.22 eV indicated that mercury existed on the surface of the material in the form of monovalent or divalent mercury ions, which may be in the form of HgCl or $HgCl_2$, which was consistent with the characterization results of EDS. Two peaks in the range of 102.45–102.79 eV and 106.38–106.77 eV indicate that mercury was complexed by functional groups on the material in the form of divalent ions to form Hg–O bonds, which was consistent with the results of O 1s.

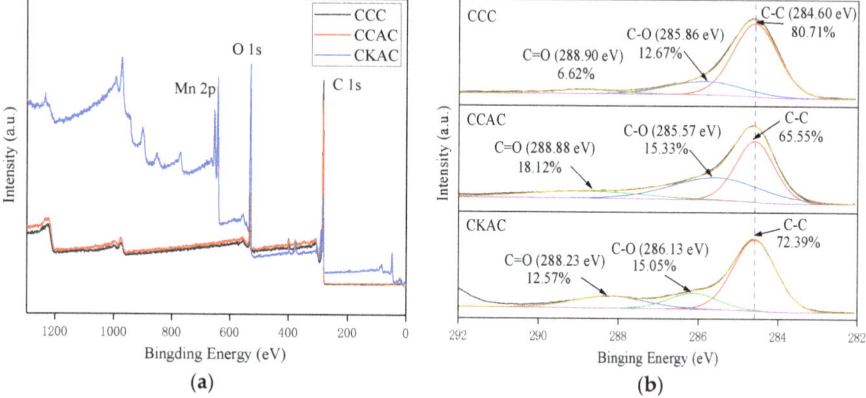

Figure 9. *Cont.*

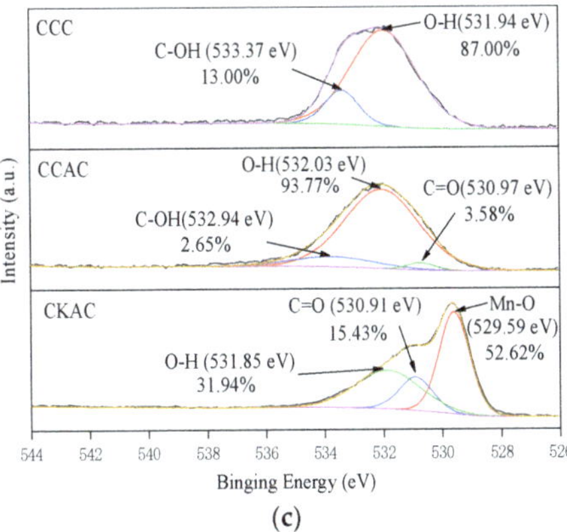

Figure 9. X-ray photoelectron spectroscopy (XPS) spectra of survey (**a**), C 1s (**b**), O 1s (**c**) of CCC, CCAC, CKAC.

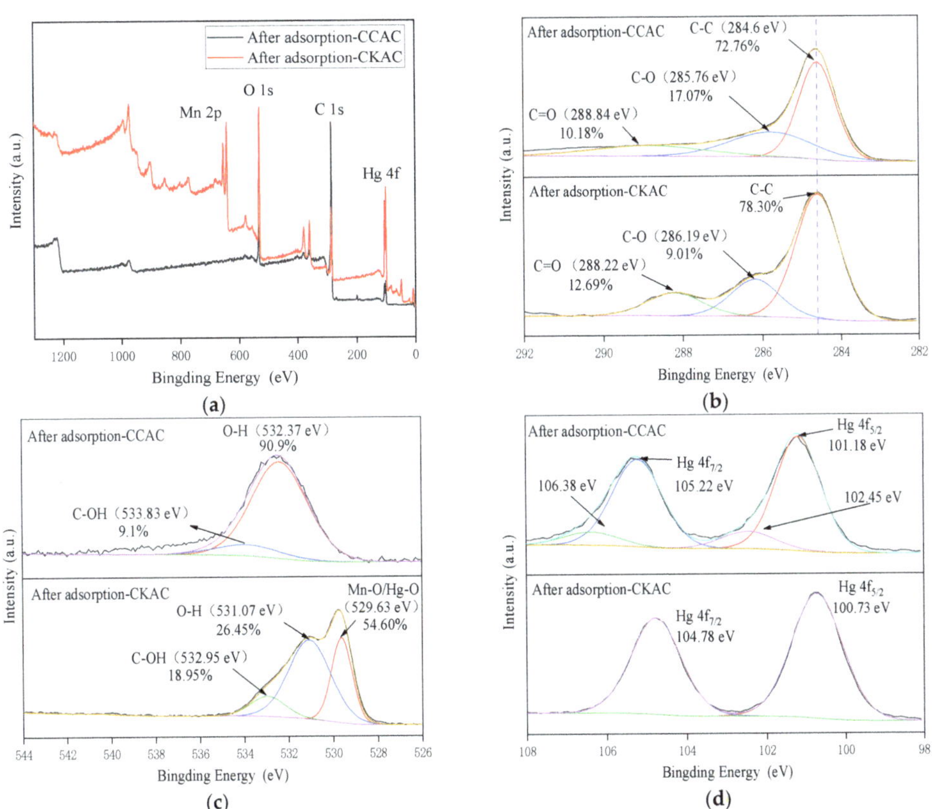

Figure 10. XPS spectra of survey (**a**), C 1s (**b**), O 1s (**c**), Hg 4f (**d**) of CCAC, CKAC after adsorption of mercury ions.

3.8. Comparison with Other Adsorbents

Table 6 compares the adsorption capacity of various adsorbents for mercury ions. By comparison, the maximum adsorption capacity of activated carbon prepared from corncobs is relatively better. The structural characteristics of various adsorbents, as well as the dosage and pH of the adsorbents under the optimal adsorption conditions, are shown in Table 6.

Table 6. Adsorption capacity of different adsorbents.

Adsorbent	Surface Area (m²/g)	m/V$_{adsorbant}$ (g/L)	pH	q_m (mg/g)	Reference
Modified multiwall carbon nanotube	-	0.4	6	84.66	[35]
Thiol-incorporated activated carbon-based fir wood	1162	0.5	6	129	[36]
Mix-ZC activated carbon	1492.4	2	8	147.1	[37]
pistachio wood wastes activated carbon	1448	2.5	7	201.09	[22]
KOH modified corn cob activated carbon	2797.75	0.4	4	184.76	this study
KMnO₄ modified corn cob activated carbon	1212.93	0.4	3	222.22	this study

4. Conclusions

In this study, corn cob was used as raw material to prepare an activated carbon adsorbent with low cost and strong adsorption capacity by means of carbonization and chemical modification. SEM-EDS, FTIR, XPS and other methods were used to characterize the activated carbon before and after modification and adsorption. The results show that the modification method of KOH and KMnO₄ can improve the specific surface area and the number of oxygen-containing functional groups, and increase the number of adsorption sites, so as to improve the adsorption capacity of activated carbon for mercury ions. The maximum adsorption capacity of CCAC and CKAC are 184.76 mg/g and 222.22 mg/g, respectively. The adsorption process is in accordance with the Langmuir isotherm and pseudo-second-order kinetic model, which indicates that the adsorption process is dominated by chemical adsorption.

Author Contributions: Conceptualization, B.Q.; Investigation, X.X.; Project administration, X.L and W.Y.; Supervision, H.Z.; Writing—original draft, Y.L. All authors have read and agreed to the published version of the manuscript.

Funding: This research received no external funding.

Acknowledgments: The authors are grateful for the support of National Key R&D Program of China (2019YFD1100603) and Heilongjiang Province Science and Technology Plan, Provincial Academy Science and Technology Cooperation Project (YS20B01).

References

1. Liu, Z.; Sun, Y.; Xu, X.; Meng, X.; Qu, J.; Wang, Z.; Liu, C.; Qu, B. Preparation, characterization and application of activated carbon from corn cob by KOH activation for removal of Hg(II) from aqueous solution. *Bioresour. Technol.* **2020**, *306*, 123154. [CrossRef]
2. Li, G.; Bai, X.; Li, H.; Lu, Z.; Zhou, Y.; Wang, Y.; Cao, J.; Huang, Z. Nutrients removal and biomass production from anaerobic digested effluent by microalgae: A review. *Int. J. Agric. Biol. Eng.* **2019**, *12*, 8–13. [CrossRef]
3. Zhao, J.; Wu, E.; Zhang, B.; Bai, X.; Lei, P.; Qiao, X.; Li, Y.-F.; Li, B.; Wu, G.; Gao, Y. Pollution characteristics and ecological risks associated with heavy metals in the Fuyang river system in North China. *Environ. Pollut.* **2021**, *281*, 116994. [CrossRef]
4. Li, G.; Zhang, J.; Li, H.; Hu, R.; Yao, X.; Liu, Y.; Zhou, Y.; Lyu, T. Towards high-quality biodiesel production from microalgae using original and anaerobically-digested livestock wastewater. *Chemosphere* **2021**, *273*, 128578. [CrossRef]
5. Melnyk, L.J.; Lin, J.; Kusnierz, D.H.; Pugh, K.; Durant, J.T.; Suarez-Soto, R.J.; Venkatapathy, R.; Sundaravadivelu, D.; Morris, A.; Lazorchak, J.M.; et al. Risks from mercury in anadromous fish collected from Penobscot River, Maine. *Sci. Total Environ.* **2021**, *781*, 146691. [CrossRef]
6. Kadam, A.R.; Nair, G.B.; Dhoble, S.J. Insights into the extraction of mercury from fluorescent lamps: A review. *J. Environ. Chem. Eng.* **2019**, *7*, 103279. [CrossRef]
7. Wang, S.; Huang, X.; Zhang, Y.; Yin, C.; Richel, A. The effect of corn straw return on corn production in Northeast China: An integrated regional evaluation with meta-analysis and system dynamics. *Resour. Conserv. Recycl.* **2021**, *167*, 105402. [CrossRef]
8. Yang, X.; Cheng, L.; Huang, X.; Zhang, Y.; Yin, C.; Lebailly, P. Incentive mechanism to promote corn stalk return sustainably in Henan, China. *Sci. Total Environ.* **2020**, *738*, 139775. [CrossRef]
9. Vassilev, S.V.; Vassileva, C.G.; Vassilev, V.S. Advantages and disadvantages of composition and properties of biomass in comparison with coal: An overview. *Fuel* **2015**, *158*, 330–350. [CrossRef]
10. Li, G.; Ji, F.; Bai, X.; Zhou, Y.; Dong, R.; Huang, Z. Comparative study on thermal cracking characteristics and bio-oil production from different microalgae using Py-GC/MS. *Int. J. Agric. Biol. Eng.* **2019**, *12*, 208–213. [CrossRef]

11. Ajay, K.M.; Dinesh, M.N.; Byatarayappa, G.; Radhika, M.G.; Kathyayini, N.; Vijeth, H. Electrochemical investigations on low cost KOH activated carbon derived from orange-peel and polyaniline for hybrid supercapacitors. *Inorg. Chem. Commun.* **2021**, *127*, 108523. [CrossRef]

12. Han, Z.; Yu, H.; Li, C.; Zhou, S. Mulch-assisted ambient-air synthesis of oxygen-rich activated carbon for hydrogen storage: A combined experimental and theoretical case study. *Appl. Surf. Sci.* **2021**, *544*, 148963. [CrossRef]

13. Juan, Z.; Kaixuan, F.; Pingping, W.; Yue, Z.; Yongke, Z. Enhancement of the adsorption of bilirubin on activated carbon via modification. *Results Mater.* **2021**, *9*, 100172. [CrossRef]

14. Narvekar, A.A.; Fernandes, J.B.; Naik, S.P.; Tilve, S.G. Development of glycerol based carbon having enhanced surface area and capacitance obtained by KOH induced thermochemical activation. *Mater. Chem. Phys.* **2021**, *261*, 124238. [CrossRef]

15. Li, G.; Bai, X.; Huo, S.; Huang, Z. Fast pyrolysis of LERDADEs for renewable biofuels. *IET Renew. Power Gener.* **2020**, *14*, 959–967. [CrossRef]

16. Li, G.; Lu, Z.; Zhang, J.; Li, H.; Zhou, Y.; Mohammed Ibrahim Zayan, A.; Huang, Z. Life cycle assessment of biofuel production from microalgae cultivated in anaerobic digested wastewater. *Int. J. Agric. Biol. Eng.* **2020**, *13*, 241–246. [CrossRef]

17. Shrestha, R.; Ban, S.; Devkota, S.; Sharma, S.; Joshi, R.; Tiwari, A.P.; Kim, H.Y.; Joshi, M.K. Technological trends in heavy metals removal from industrial wastewater: A review. *J. Environ. Chem. Eng.* **2021**, *9*, 105688. [CrossRef]

18. Yumak, T. Surface characteristics and electrochemical properties of activated carbon obtained from different parts of Pinus pinaster. *Colloids Surf. A Physicochem. Eng. Asp.* **2021**, *625*, 126982. [CrossRef]

19. Feng, C.; Chen, Y.-A.; Yu, C.-P.; Hou, C.-H. Highly porous activated carbon with multi-channeled structure derived from loofa sponge as a capacitive electrode material for the deionization of brackish water. *Chemosphere* **2018**, *208*, 285–293. [CrossRef] [PubMed]

20. Sun, Y.; Zhang, Z.Z.; Sun, Y.M.; Yang, G.X. One-pot pyrolysis route to Fe-N-Doped carbon nanosheets with outstanding electrochemical performance as cathode materials for microbial fuel cell. *Int. J. Agric. Biol. Eng.* **2020**, *13*, 207–214. [CrossRef]

21. Oginni, O.; Singh, K.; Oporto, G.; Dawson-Andoh, B.; McDonald, L.; Sabolsky, E. Influence of one-step and two-step KOH activation on activated carbon characteristics. *Bioresour. Technol. Rep.* **2019**, *7*, 100266. [CrossRef]

22. Sajjadi, S.-A.; Mohammadzadeh, A.; Tran, H.N.; Anastopoulos, I.; Dotto, G.L.; Lopičić, Z.R.; Sivamani, S.; Rahmani-Sani, A.; Ivanets, A.; Hosseini-Bandegharaei, A. Efficient mercury removal from wastewater by pistachio wood wastes-derived activated carbon prepared by chemical activation using a novel activating agent. *J. Environ. Manag.* **2018**, *223*, 1001–1009. [CrossRef] [PubMed]

23. Hsu, C.-J.; Xiao, Y.-Z.; Hsi, H.-C. Simultaneous aqueous Hg(II) adsorption and gaseous Hg0 re-emission inhibition from SFGD wastewater by using Cu and S co-impregnated activated carbon. *Chemosphere* **2021**, *263*, 127966. [CrossRef] [PubMed]

24. Yagmur, E.; Gokce, Y.; Tekin, S.; Semerci, N.I.; Aktas, Z. Characteristics and comparison of activated carbons prepared from oleaster (Elaeagnus angustifolia L.) fruit using KOH and ZnCl2. *Fuel* **2020**, *267*, 117232. [CrossRef]

25. Zakaria, R.; Jamalluddin, N.A.; Abu Bakar, M.Z. Effect of impregnation ratio and activation temperature on the yield and adsorption performance of mangrove based activated carbon for methylene blue removal. *Results Mater.* **2021**, *10*, 100183. [CrossRef]

26. Medhat, A.; El-Maghrabi, H.H.; Abdelghany, A.; Abdel Menem, N.M.; Raynaud, P.; Moustafa, Y.M.; Elsayed, M.A.; Nada, A.A. Efficiently activated carbons from corn cob for methylene blue adsorption. *Appl. Surf. Sci. Adv.* **2021**, *3*, 100037. [CrossRef]

27. Kharrazi, S.M.; Soleimani, M.; Jokar, M.; Richards, T.; Pettersson, A.; Mirghaffari, N. Pretreatment of lignocellulosic waste as a precursor for synthesis of high porous activated carbon and its application for Pb (II) and Cr (VI) adsorption from aqueous solutions. *Int. J. Biol. Macromol.* **2021**, *180*, 299–310. [CrossRef]

28. Tran, T.H.; Le, H.H.; Pham, T.H.; Nguyen, D.T.; La, D.D.; Chang, S.W.; Lee, S.M.; Chung, W.J.; Nguyen, D.D. Comparative study on methylene blue adsorption behavior of coffee husk-derived activated carbon materials prepared using hydrothermal and soaking methods. *J. Environ. Chem. Eng.* **2021**, *9*, 105362. [CrossRef]

29. Kharrazi, S.M.; Mirghaffari, N.; Dastgerdi, M.M.; Soleimani, M. A novel post-modification of powdered activated carbon prepared from lignocellulosic waste through thermal tension treatment to enhance the porosity and heavy metals adsorption. *Powder Technol.* **2020**, *366*, 358–368. [CrossRef]

30. Wang, J.; Lei, S.; Liang, L. Preparation of porous activated carbon from semi-coke by high temperature activation with KOH for the high-efficiency adsorption of aqueous tetracycline. *Appl. Surf. Sci.* **2020**, *530*, 147187. [CrossRef]

31. Cheng, H.; Ye, G.; Wang, X.; Su, C.; Zhang, W.; Yao, F.; Wang, Y.; Jiao, Y.; Huang, H.; Ye, D. Micro-mesoporous carbon fabricated by Phanerochaete chrysosporium pretreatment coupling with chemical activation: Promoting effect and toluene adsorption performance. *J. Environ. Chem. Eng.* **2021**, *9*, 105054. [CrossRef]

32. Su, C.; Guo, Y.; Yu, L.; Zou, J.; Zeng, Z.; Li, L. Insight into specific surface area, microporosity and N, P co-doping of porous carbon materials in the acetone adsorption. *Mater. Chem. Phys.* **2021**, *258*, 123930. [CrossRef]

33. Altalhi, A.A.; Mohammed, E.A.; Morsy, S.S.M.; Negm, N.A.; Farag, A.A. Catalyzed production of different grade biofuels using metal ions modified activated carbon of cellulosic wastes. *Fuel* **2021**, *295*, 120646. [CrossRef]

34. Sellaoui, L.; Silva, L.F.O.; Badawi, M.; Ali, J.; Favarin, N.; Dotto, G.L.; Erto, A.; Chen, Z. Adsorption of ketoprofen and 2-nitrophenol on activated carbon prepared from winery wastes: A combined experimental and theoretical study. *J. Mol. Liq.* **2021**, *333*, 115906. [CrossRef]

35. Hadavifar, M.; Bahramifar, N.; Younesi, H.; Li, Q. Adsorption of mercury ions from synthetic and real wastewater aqueous solution by functionalized multi-walled carbon nanotube with both amino and thiolated groups. *Chem. Eng. J.* **2014**, *237*, 217–228. [CrossRef]
36. Kazemi, F.; Younesi, H.; Ghoreyshi, A.A.; Bahramifar, N.; Heidari, A. Thiol-incorporated activated carbon derived from fir wood sawdust as an efficient adsorbent for the removal of mercury ion: Batch and fixed-bed column studies. *Process Saf. Environ. Prot.* **2016**, *100*, 22–35. [CrossRef]
37. Asasian, N.; Kaghazchi, T.; Soleimani, M. Elimination of mercury by adsorption onto activated carbon prepared from the biomass material. *J. Ind. Eng. Chem.* **2012**, *18*, 283–289. [CrossRef]

Article

Effect of Operating Parameters and Energy Expenditure on the Biological Performance of Rotating Biological Contactor for Wastewater Treatment

Muhammad Irfan [1], Sharjeel Waqas [2,3,*], Javed Akbar Khan [4,*], Saifur Rahman [1], Izabela Kruszelnicka [5], Dobrochna Ginter-Kramarczyk [5], Stanislaw Legutko [6], Marek Ochowiak [7], Sylwia Włodarczak [7] and Krystian Czernek [8]

1 Electrical Engineering Department, College of Engineering, Najran University Saudi Arabia, Najran 11001, Saudi Arabia; miditta@nu.edu.sa (M.I.); srrahman@nu.edu.sa (S.R.)
2 Chemical Engineering Department, Universiti Teknologi PETRONAS, Bandar Seri Iskandar 32610, Perak, Malaysia
3 School of Chemical Engineering, The University of Faisalabad, Faisalabad 37610, Pakistan
4 Mechanical Engineering Department, Universiti Teknologi PETRONAS, Bandar Seri Iskandar 32610, Perak, Malaysia
5 Department of Water Supply and Bioeconomy, Faculty of Environmental Engineering and Energy, Poznan University of Technology, 60-965 Poznan, Poland; izabela.kruszelnicka@put.poznan.pl (I.K.); dobrochna.ginter-kramarczyk@put.poznan.pl (D.G.-K.)
6 Faculty of Mechanical Engineering, Poznan University of Technology, 60-965 Poznan, Poland; stanislaw.legutko@put.poznan.pl
7 Department of Chemical Engineering and Equipment, Poznan University of Technology, 60-965 Poznan, Poland; marek.ochowiak@put.poznan.pl (M.O.); sylwia.wlodarczak@put.poznan.pl (S.W.)
8 Department of Process and Environmental Engineering, Opole University of Technology, 45-271 Opole, Poland; k.czernek@po.edu.pl
* Correspondence: sharjeel_17000606@utp.edu.my (S.W.); javedakbar.khan@utp.edu.my (J.A.K.)

Citation: Irfan, M.; Waqas, S.; Khan, J.A.; Rahman, S.; Kruszelnicka, I.; Ginter-Kramarczyk, D.; Legutko, S.; Ochowiak, M.; Włodarczak, S.; Czernek, K. Effect of Operating Parameters and Energy Expenditure on the Biological Performance of Rotating Biological Contactor for Wastewater Treatment. *Energies* 2022, 5, 3523. https://doi.org/10.3390/en15103523

Academic Editors: Redmond R. Shamshiri, Muhammad Wakil Shahzad, Muhammad Sultan and Md Shamim Ahamed

Received: 28 March 2022
Accepted: 8 May 2022
Published: 11 May 2022

Publisher's Note: MDPI stays neutral with regard to jurisdictional claims in published maps and institutional affiliations.

Abstract: The rotating biological contactor (RBC) is resistant to toxic chemical and shock loadings, and this results in significant organic and nutrient removal efficiencies. The RBC system offers a low-energy footprint and saves up to 90% in energy costs. Due to the system's low-energy demand, it is easily operable with renewable energy sources, either solar or wind power. An RBC was employed to degrade pollutants in domestic wastewater through biodegradation mechanisms in this study. The high microbial population in the RBC bioreactor produced excellent biological treatment capacity and higher effluent quality. The results showed that the RBC bioreactor achieved an average removal efficiency of 73.9% of chemical oxygen demand (COD), 38.3% of total nitrogen (TN), 95.6% of ammonium, and 78.9% of turbidity. Investigation of operational parameters, disk rotational speed, HRT, and SRT, showed the biological performance impact. Disk rotational speed showed uniform effluent quality at 30–40 rpm, while higher values of disk rotational speed (>40 rpm) resulted in lower effluent quality in COD, TN, and turbidity. The longer hydraulic retention time and sludge retention time (SRT) facilitated higher biological performance efficiency. The longer SRTs enabled the higher TN removal efficiency because of the higher quantity of microbial biomass retention. The longer SRT also resulted in efficient sludge-settling properties and reduced volume of sludge production. The energy evaluation of the RBC bioreactor showed that it consumed only 0.14 kWh/m^3, which is significantly lower than the conventional treatment methods; therefore, it is easily operable with renewable energy sources. The RBC is promising substitute for traditional suspended growth processes as higher microbial activity, lower operational and maintenance costs, and lower carbon foot print enhanced the biological performance, which aligns with the stipulations of ecological evolution and environment-friendly treatment.

Keywords: wastewater treatment; biological process; aerobic process; attached growth; biofilm; rotating biological contactor

1. Introduction

Nowadays, the world is facing a major issue, which is freshwater availability. There is no availability of potable water for about one-sixth of the total population of the world. An enormous quantity (30–70 mm^3) of wastewater per person per year is generated by industries and cities of developing countries [1,2]. Due to these problems, industries have to solve serious issues to meet the rigorous wastewater removal requirements [1,3]. Therefore, attention has been paid to the rotating biological contactor (RBC). The RBC has emerged as a low-energy-consuming technology used in wastewater treatment which compares favorably with other treatment methods [4].

The RBC is a fixed-film bioreactor employing rotating discs to provide a support medium for microbial growth and supply dissolved oxygen. The advantages of RBCs over other attached growth systems include, but are not limited to, being able to handle high hydraulic and organic loading rates, being resistant to shock loadings, having low area and maintenance requirements and low-energy demand and operational costs, being able to handle specific contaminants, e.g., hydrocarbons, textile dyes, colored wastewater, heavy metals, and xenobiotics, and being able to produce energy through the anaerobic process along with wastewater treatment [5–9]. The RBC bioreactor characterizes a practical source for nitrification and denitrification processes as the microbial biomass is abundant and stable [10–12]. Nitrifying bacteria responsible for nitrification are one of the major microorganisms in the biofilm. Previous literature has shown that the RBC has successfully treated domestic, municipal, and industrial wastewater [13,14].

The operational parameters influencing the functioning of the RBC bioreactor, such as disk rotational speed, hydraulic retention time (HRT), sludge retention time (SRT), and carrier media type have been extensively reported [15,16]. The selection and optimization of parameters strictly depend on the influent wastewater and effluent quality requirements. Disk rotational speed is an important parameter for acclimatizing the microorganisms and developing a full-grown biofilm to digest the organics and nutrients at the carrier surface. It is also responsible for maintaining sufficient dissolved oxygen (DO) levels inside the bioreactor to facilitate the degradation process [17,18]. The selection of SRT and loading rates relies on the wastewater strength and effluent requirements. A short SRT only facilitates carbon deduction, whereas a longer SRT results in increased sludge concentration hindering the oxygen transfer [19].

Microbial community characteristics are highly dependent on the selected disk rotational speed, SRT, DO levels, and microorganisms present in the bioreactor [20,21]. Carbonaceous and ammonia-oxidizing bacteria (AOB) are aerobic and cultivate in the outer layers of the biofilm, while nitrite-oxidizing bacteria (NOB) dominate the inner parts of the biofilm where anoxic or anaerobic conditions prevail [22]. The competition between the carbonaceous bacteria, AOB, and NOB depends on the influent wastewater and other operating conditions [23]. The results of previous studies have shown that both disk rotational speed and SRT influence biological performance.

Disk rotational speed, HRT, and SRT are significant parameters influencing the performance of the RBC bioreactor. In the present study, to determine the biological performance of domestic wastewater, the effect of disk rotational speed, HRT, and SRT on the chemical oxygen demand (COD), total nitrogen (TN), and turbidity were investigated. The objective of the current study was to explore the biological performance of the RBC bioreactor treating domestic wastewater, focusing on the effect of disk rotational speed, HRT, and SRT.

2. Materials and Methods

2.1. Wastewater Preparation and Bioreactor Acclimatization

The domestic wastewater was prepared by blending food left over from the campus canteen, as described in our previous research [24]. The blended mixture was left for 2 h to settle the suspended particles. The supernatant was then filtered through Whatman filter paper (Grade 1 Qualitative Filter Paper Standard Grade, GE Whatman, Kent, UK). The filtered solution was then diluted to meet the domestic wastewater concentration. The prepared wastewater was analyzed to determine the COD, TN, total Kjeldahl nitrogen (TKN), ammonium, turbidity, pH, and nitrate, as shown in Table 1 [25].

Table 1. Properties of influent wastewater.

Sr #	Contaminant	Unit	Concentration
1	COD	mg/L	281 ± 8.5
2	Ammonium	mg/L	0.66 ± 0.03
3	TN	mg/L	2.5 ± 0.19
4	TKN	mg/L	0.91 ± 0.09
5	Nitrate	mg/L	0.49 ± 0.04
6	Turbidity	NTU	14.6 ± 0.55
7	pH	–	6.28 ± 0.21

2.2. Bioreactor Set-Up

The RBC bioreactor was operated in accordance with the layout depicted in Figure 1. The bioreactor consisted of a 45 L storage tank, a 6.5 L bioreactor, and 6 L settling tank. The feed wastewater from the storage tank was constantly pumped with a peristaltic pump to the RBC bioreactor. The RBC bioreactor with $25 \times 25 \times 30$ cm dimensions was fabricated in-house with methyl methacrylate sheets. In the RBC bioreactor, five disks fabricated from methyl methacrylate sheets with 18 cm diameter were attached to a stainless-steel shaft. The disks were attached to polyurethane sheets (1.22–1.27 g/cm^3 density) to colonize the microbial population. The polyurethane-coated disks, 3 cm apart from each other (corresponding to a net surface area of 2034 cm^2), were placed inside the RBC bioreactor at 40% submergence. The wastewater from the storage tank was fed continuously to the RBC bioreactor, and treated effluent was transferred to the settling tank. A mechanical stirrer continuously stirred the feed wastewater at 100 rpm to keep the concentration uniform. In the settling tank, sludge settled at the bottom of the tank, and was removed frequently while clear effluent flowed out from the top.

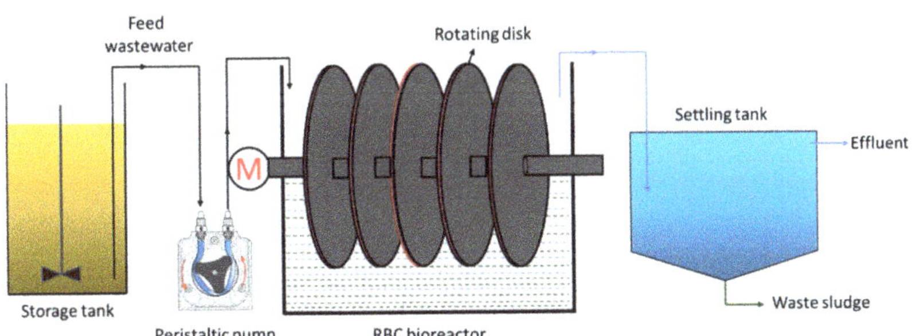

Figure 1. Schematic diagram of RBC configuration.

2.3. Bioreactor Operation

During the acclimatization phase, the operational parameters were constant at 30 rpm disk rotational speed, 9 h HRT, and 15 d SRT. The biofilm layer was observed daily while effluent wastewater concentration was checked every three days. After completing the first phase of the experiments, the bioreactor achieved a steady-state effluent concentration. After biofilm acclimatization, the system was investigated for the effect of disk rotational speed, HRT, and SRT on effluent wastewater. The organic loading rates were kept constant throughout the experiments. The disk rotational speed was increased from 30 to 50 rpm with an interval increase of 5 rpm, while SRT was increased from 5 to 15 d with an interval increase of 2.5 d. The SRT was set by wasting a calculated amount of sludge from the bioreactor each day.

2.4. Analytical Methods

COD, TN, TKN, ammonium, and nitrate were measured using each compound's specific Hach digestion solution (HACH, Loveland, CO, USA). The solution was diluted to fall into the range of the digestion vials being used for the study. The values were determined through Hach DR3900 Spectrophotometer (HACH, Loveland, CO, USA). A Hach 2100Q portable turbidimeter (HACH, Loveland, CO, USA) and Hach HQ411D benchtop PH/MV meter (HACH, Loveland, CO, USA) were used to determine turbidity and pH, respectively [25].

2.5. Scanning Electron Microscope

After the acclimatization stage, a 1 cm^2 piece of biofilm was analyzed using SEM analysis. The biofilm sample was carefully cut from the rotating disk. The foam was then treated with formaldehyde for biofilm impregnation according to the method detailed earlier [26] to maintain the biofilm structure. The biofilm sample was then dehydrated by consecutive immersions in 20, 40, 60, 80, and 100% ethanol solution, each step for 5 min, to avoid shrinkage, followed by a drying process. The dried non-conductive sample was sprayed with conductive gold nanoparticles using an ion-sputter instrument to create a conductive layer on the sample that reduced thermal damage, inhibited charging, and improved the secondary electron signal required for topographic examination in the SEM. The conductive biofilm sample was loaded onto the SEM sample stage under vacuum conditions and an electron gun shot out a beam of high-energy electrons.

2.6. Energy Estimation

The energy consumption of the RBC bioreactor was estimated using an estimation method proposed elsewhere [27]. The RBC treatment capacity is supposed to be the same as that of the referenced bioreactor to evade the effect of plant capability on the energy estimation. Some energy values, such as wastewater pre-treatment, influent pumping, and sludge disposal were assumed constant to depict a fair comparison [28]. The activated-sludge process and MBR utilize coarse-bubble aeration to maintain DO concentration in the bioreactor, accounting for 30–50% of total energy consumption [29]. On the other hand, the RBC bioreactor utilizes disk rotation to maintain DO levels. The energy for shaft rotation was projected from the RBC bioreactor data. The energy contribution of the RBC bioreactor was compared with an optimized referenced MBR with an overall energy consumption of 0.64 kWh/m^3.

The energy consumption for pre-treatment of wastewater (0.02 kWh/m^3) was assumed to be similar to the referenced treatment system. The influent (0.03 kWh/m^3) and effluent (0.02 kWh/m^3) pumping energy demand were also assumed similar because the influent and effluent pumping has no relation to the biological treatment. However, due to system differences and mode of operation, some energy factors, for instance, sludge recycling, coarse-bubble aeration, mixers, bioreactor aeration, and air compression, were not present in the RBC bioreactor.

The major energy requirement of the RBC is the mechanical energy for rotation. The energy for mechanical rotation of the disk was estimated based on the full-scale date of the RBC plant reported elsewhere [27]. The energy consumption was estimated through the energy consumed by the shaft. The shaft energy is a function of flowrate and media surface area. The mechanical energy for shaft rotation is a function of plant capacity and increases exponentially with capacity. The shaft rotation energy was projected using Equations (1) and (2):

$$E_{MR} = \frac{N E_S}{V_O} \tag{1}$$

$$N = \frac{V_O}{H_O A_S} \tag{2}$$

where, E_{MR} is the energy for mechanical rotation (kWh/m^3), N is the total number of shafts, E_S is the energy consumption/shaft (kWh), V_O is the influent flow rate (1774 m^3/h), H_O is the hydraulic loading (gpd/ft^2), and A_S is the media surface area/shaft (m^2).

3. Results and Discussion

3.1. Biofilm Analysis

Figure 2 shows the biofilm developed at the surface of the disk visualized using SEM. The SEM images were obtained at 40× and 5000× magnification levels. Figure 2a shows a birds-eye view of the biofilm established on the carrier media at 40× magnification. The SEM images show the well-established biofilm of microorganisms at the media surface. It can be identified as a mature biofilm that occupied all the sponge media surface. Its excellent biological performance in removing organics from the wastewater detailed in Section 3.3. A mature biofilm with characteristic mushroom formed of polysaccharides can be seen in Figure 2b. At this stage, cells were starting to detach, reverting to planktonic cells that stuck to the new surface to develop another biofilm layer.

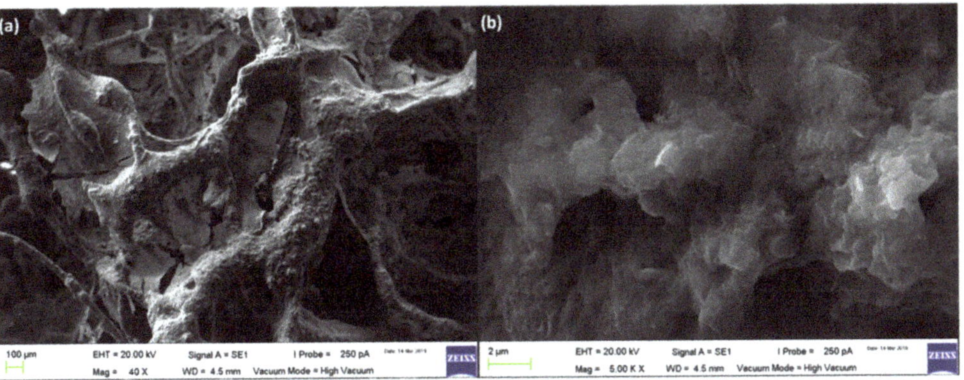

Figure 2. SEM results of biofilm developed at the surface of the rotating disk; (**a**) at 40× and (**b**) at 5000× resolution.

3.2. Biological Performance

Figure 3 and Table 2 show the biological performance in terms of COD, TN, TKN ammonium, and turbidity removal for the 30 d operation of the bioreactor. After biofilm acclimatization, the bioreactor was operated for 30 d to witness the organics and nutrient removal. The RBC bioreactor maintained an average COD removal efficiency of 73.93% with 73.4 mg/L effluent concentration at 9 h HRT and 30 rpm disk rotational speed. The removal efficiency of 38.3% with 1.54 mg/L effluent was obtained for TN, while 95.6% with 0.03 mg/L effluent was obtained for ammonium. The RBC bioreactor significantly reduced the turbidity values, and an average effluent value of 3.1 NTU was obtained with 78.9% removal efficiency. Significant removal efficiencies show the effectiveness of the RBC bioreactor for the treatment of domestic wastewater.

Carbonaceous bacteria are responsible for COD removal, while nitrifying bacteria undergo nitrification. The RBC bioreactor is an aerobic biofilm process with a high DO concentration (8–10 mg/L). Carbonaceous bacteria are heterotrophic and acclimatize in 3–5 d under aerobic conditions. The outer layer of the rotating disks was abundant in carbonaceous bacteria. The substrate from the influent wastewater encountered the microorganisms, and degradation occurred at the disk surface.

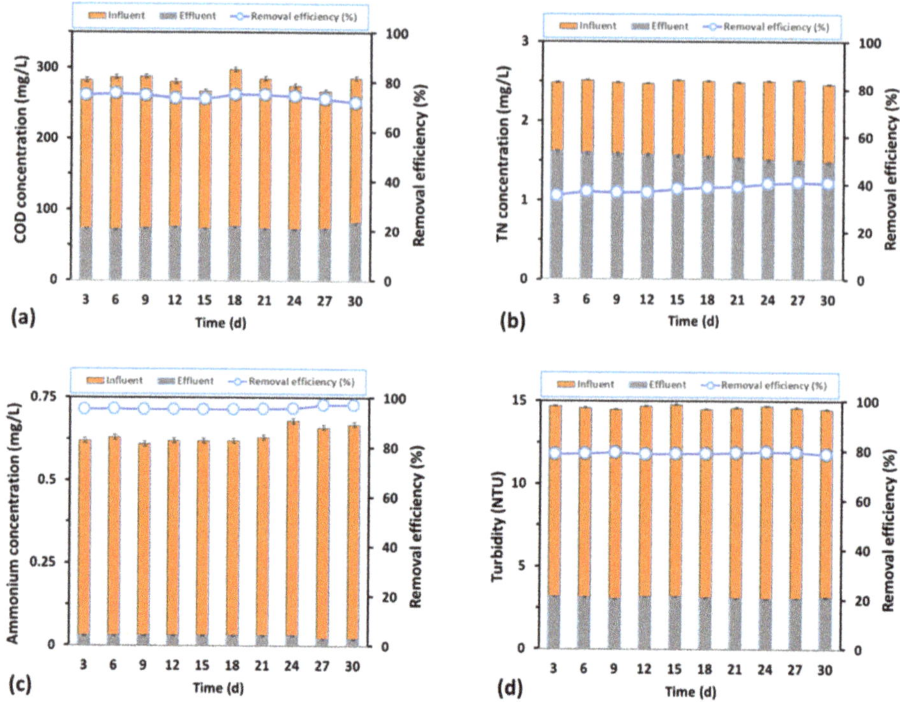

Figure 3. Influent, effluent, and percentage removal efficiency over time of (**a**) COD, (**b**) TN, (**c**) ammonium, and (**d**) turbidity.

Table 2. Properties of effluent wastewater.

Sr #	Contaminant	Unit	Concentration
1	COD	mg/L	73.4 ± 3.0
2	Ammonium	mg/L	0.03 ± 0.0
3	TN	mg/L	1.54 ± 0.05
4	TKN	Mg/L	0.05 ± 0.01
5	Nitrate	mg/L	1.36 ± 0.07
6	Turbidity	NTU	3.09 ± 0.06
7	pH	–	7.35 ± 0.11

The results showed a high COD removal efficiency during the experiments, with average effluent values higher than 73%. Despite the minor alternations of the feeding COD, the bioreactor exhibited an acceptable activity of the biological consortium. The above COD removal efficiency agreed with what was achieved by [28], who found a COD removal efficiency close to 75% in a similar bioreactor. A slight variation could be explained by the fact that both bioreactors were operated at different SRT. The attained results underline the robustness of the RBC bioreactor regarding COD removal.

The attained results emphasize the ammonium removal efficiency of the bioreactor, with values higher than 95% for most of the experiments. The presence of nitrifying bacteria and favorable aerobic conditions enhanced ammonium oxidation conversion to nitrite and nitrate. The bioreactor maintained a constant effluent concentration of 0.03 mg/L depicting an abundance of nitrifying bacteria activity.

Nitrification, an aerobic process, is a two-step process—oxidation of ammonium to nitrite through ammonia-oxidizing bacteria (AOB) and then conversion of nitrite to nitrate through nitrite-oxidizing bacteria (NOB) [30]. The RBC developed abundant AOB and NOB throughout the biofilm along with carbonaceous bacteria. The RBC exhibited excellent ammonium-removal efficiency throughout the experimentation period. In wastewater with a high organics concentration, heterotrophic bacteria significantly diminish the growth of nitrifiers. Therefore, nitrification occurs after organics removal during the last stage of the RBC bioreactor [31].

The system obtained TN removal efficiency values higher than 38%. The RBC bioreactor maintained a high DO concentration due to the disk-rotation mechanism. The conversion of nitrate into the quasi-inert gases, nitrogen and nitrous oxide, by microbial oxidation of organic matter is called denitrification. High nitrate concentration indicates the absence of anoxic and anaerobic conditions in the bioreactor [32]. However, it is understood that deep inside the biofilm, anoxic conditions prevail due to the DO-concentration-gradient profile. A low C/N ratio also hinders the growth of nitrifying bacteria as carbonaceous bacteria outranks nitrifying bacteria in terms of feed availability [33]. The obtained results showed an average turbidity removal efficiency of more than 78%, indicating high microbial activity. The resultant higher removal efficiency of turbidity indicates excellent sludge-settling properties.

3.3. Effect of Disk Rotational Speed

Figure 4 shows the effect of disk rotational speed on the COD, TN, and turbidity removal from the wastewater. Disk rotational speed, being an important RBC parameter, controls the DO concentration. Higher DO concentration favors biological activity; however, excessive disk rotation will increase energy demand. Higher disk rotational speed also generates shear that shreds off the attached biofilm. The detached biofilm remains suspended and does not easily settle. Excessive shear generation can result in loss of complete microbial activity and washout of microbial biomass from the bioreactor. The bioreactor was operated at constant 9 h HRT and 15 d SRT to study the effect of disk rotational speed.

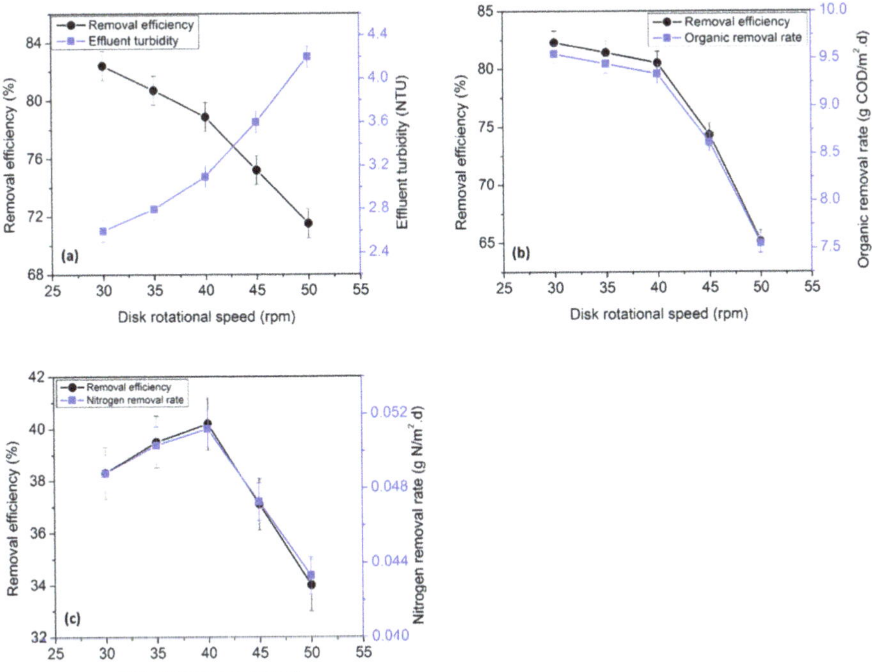

Figure 4. Effect of disk rotational speed on (**a**) turbidity, (**b**) COD, and (**c**) TN removal.

Figure 4a depicts the relationship of effluent turbidity and removal efficiency with the disk rotational speed. The detached floc floats at a higher disk rotational speed and does not settle easily, increasing the effluent turbidity value [22,34]. Lower COD removal efficiency and organic removal rate were obtained at higher disk rotational speed. The bioreactor maintained a constant COD removal efficiency and removal rate as disk rotational speed increased for 30 to 40 rpm. However, a significant change was observed in COD removal efficiency as disk rotational speed increased from 40 to 50 rpm, validating the part of boosted shear rate by the disk rotation. Similar results were obtained for the TN removal efficiency. The bioreactor maintained a constant TN removal efficiency and nitrogen removal rate as rotational speed boosted from 30 to 40 rpm. Further rise in disk rotational speed generates an enhanced shear rate that deteriorates the biofilm and lowers TN removal efficiency.

3.4. Effect of Sludge Retention Time

Figure 5 demonstrates the influence of SRT on the COD, TN, and turbidity removal efficiency. Figure 5a depicts the impact of SRT on turbidity removal efficiency and effluent turbidity. The bioreactor effluent turbidity value increased from 71.5 to 82.4% as SRT increased from 5 to 15 d. The effluent turbidity value also decreased from 4.2 to 2.6 NTU with an increase in SRT. A higher value of turbidity indicates good sludge-settling properties. The results show that the higher the SRT, the higher the turbidity removal efficiency.

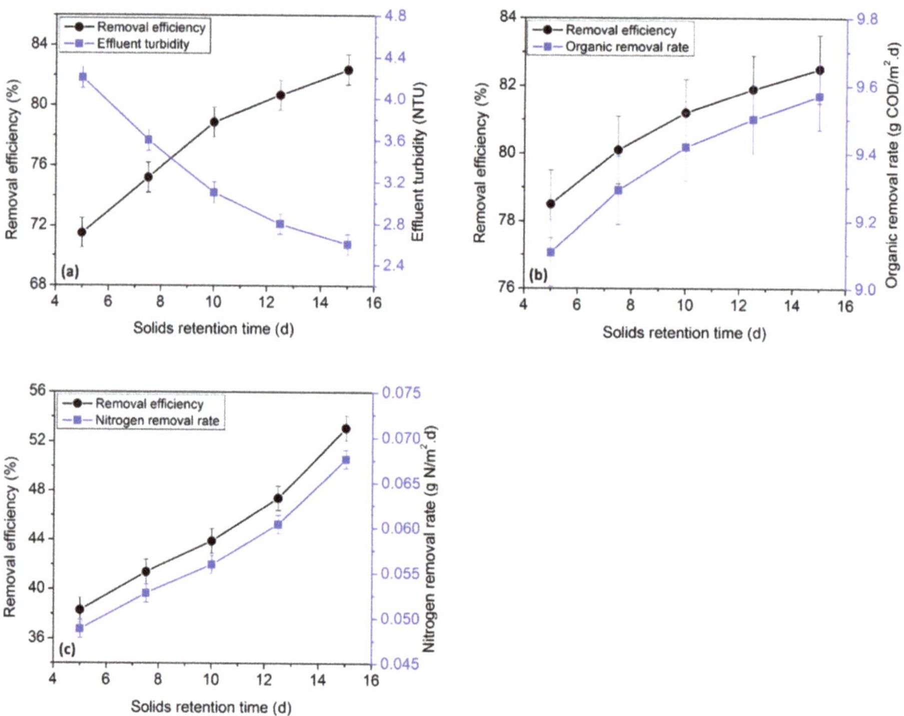

Figure 5. Effect of sludge retention time on (**a**) turbidity, (**b**) COD, and (**c**) TN removal.

Figure 5b shows that excellent treatment efficiency was obtained in COD removal throughout the experiments with little variation with SRT. The COD removal efficiency boosted from 78.5 to 82.5% as SRT increased from 5 to 15 d. This result could be related to the fact that carbonaceous bacteria take 3–5 d to fully acclimatize to the bioreactor. The increase in SRT from 5 to 15 d does not affect the carbonaceous bacterial growth as they are already fully acclimatized. Therefore, the system maintained a high COD removal efficiency and organic removal rate throughout the experiments. Slight changes in COD removal efficiency are due to an increase in sludge age which gives bacteria more time to stay in the bioreactor. The high COD removal efficiency attained in the current study is in good agreement with prior studies on the RBC bioreactor.

The RBC bioreactor showed enhanced nitrification performance throughout the investigation with TN removal efficiency increasing from 38.3 to 53.1% with a rise in SRT from 5 to 15 d as shown in Figure 5c. Meanwhile, the nitrogen removal rate increased from 0.049 to 0.068 g TN/m^2 d. Despite the inherent disadvantage of the absence of anaerobic conditions and low influent TN concentration, the bioreactor showed good nitrification performance. The biofilm growth is significantly affected by the selection of SRT. Higher SRT means microbial biomass is retained in the bioreactor for more time. Nitrogenous bacteria are slow-growing bacteria and take about 14–17 days to fully acclimatize. Higher SRT favors the TN removal as a higher quantity of nitrifying biomass can be retained in the bioreactor and undergo biodegradation. Previous studies highlighted the influence of SRT on microbial activity and enhanced TN removal efficiency along with the higher nitrogen removal rate [35].

3.5. Effect of Hydraulic Rentention Time

Figure 6 describes the influence of HRT on the COD, TN, and turbidity removal. The HRT varied from 6 to 18 h with an equal interval of 3 h at constant 40 rpm disk rotational speed and 15 d SRT. After changing the HRT values, the system was allowed to attain steady-state conditions (48 h). The result indicates that HRT has a significant effect on effluent quality. Figure 5a shows that with the increase in HRT, turbidity removal efficiency increased, and effluent turbidity decreased. The turbidity removal increased from 71.3% to 82.6% as HRT increased from 6 to 18 h. A maximum of 82.6% turbidity removal efficiency with effluent values at 2.62 NTU was achieved at 18 h HRT. The high turbidity removal efficiency at higher HRT results from higher microbial activity and the production of compact sludge that can settle easily.

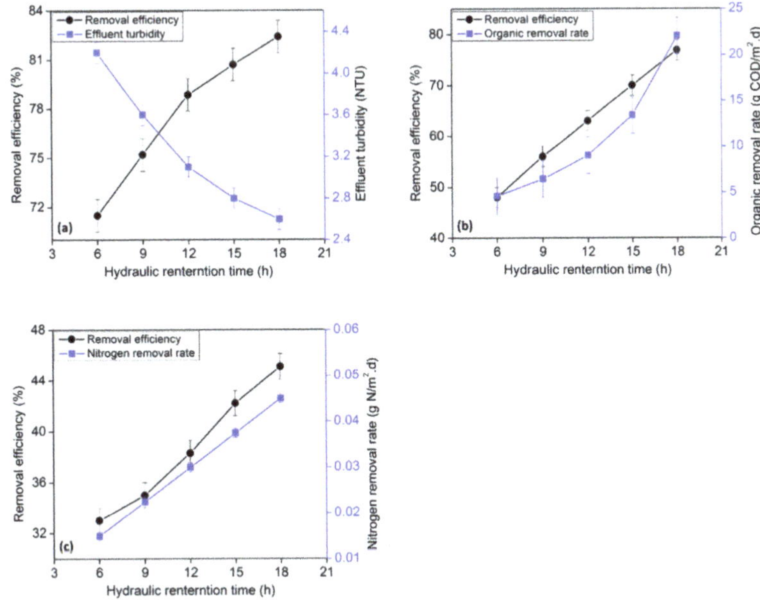

Figure 6. Effect of hydraulic retention time on (**a**) turbidity, (**b**) COD, and (**c**) TN removal.

Figure 6b shows the impact of HRT on the organic removal rate and COD biodegradation in the bioreactor. The COD removal efficiency improved with HRT while the organic removal rate also increased. The COD removal efficiency increased from 48% to 76.9% as HRT increased from 6 to 18 h, and the organic removal rate increases from 4.6 to 22.1 g COD/m^2.d. The organic removal rate improves with HRT until another parameter becomes limiting. The COD removal limitations include insufficient surface area, limited oxygen transfer rate, and hindered carbonaceous bacteria activity.

Figure 6a shows that with the increase in HRT, turbidity removal efficiency increased, and effluent turbidity decreases. Domestic wastewater contains limited TN, which hinders the activity of nitrifying bacteria. The TN removal efficiency and nitrogen removal rate increased with HRT increase. The TN removal increased from 33% to 45.1% as HRT increased from 6 to 18 h. A maximum of 45.1% turbidity removal efficiency with a nitrogen removal rate of 0.084 g N/m^2.d was achieved at 18 h HRT.

3.6. Biomass Characterization

Biofilm systems are rich in microbial biomass containing a variety of microorganisms depending on the operating conditions and influent wastewater. Biofilm acclimatization and the diversity of microorganisms influence the performance of the bioreactor [12]. Physical observation validates the development of a mature biofilm at the carrier medium At this stage, the microbial biomass can completely undergo the biodegradation process and efficiently undertake the removal of the organics and nutrients.

The biofilm consists of two parts—an aerobic biofilm layer at the outer part of the rotating disk and deep inside the biofilm, anaerobic conditions help the dominance of autotrophic bacteria. Carbonaceous bacteria responsible for carbon removal dominate the biofilm due to the abundance of the substrate in the influent wastewater. The nitrifying bacteria develop slowly and dominate the inner layers of the biofilm. Both the carbonaceous and nitrifying bacteria compete to dominate the biofilm. Domestic wastewater is abundant in carbon sources, which facilitates the growth of carbonaceous bacteria, and the substrate in the influent wastewater comes in contact with the microorganisms [29].

Oxidation of ammonium to nitrate is conducted by AOB under aerobic conditions, while NOB reduce nitrate to nitrogen gas in anaerobic conditions, which only prevail deep inside the biofilm. The abundant carbonaceous bacteria in an RBC biofilm are *Bacteroidetes* and *Proteobacteria*, as indicated by various researchers [22]. The *Bacteroidetes* and *Proteobacteria* dominate the biofilm and form around two-thirds of carbonaceous bacteria. They are abundantly found in the outer layer, where aerobic conditions and the availability of high substrate concentrations are readily available [36].

The influent ammonium concentration and aerobic conditions support AOB proliferation, mainly *Nitrosomonadaceae*, in the biofilm [37]. The *Nitrosomonadaceae* present in the biofilm oxidize the ammonium and convert it to nitrite, which is further oxidized to nitrate. The biofilm analysis suggests that Nitrospira is the main NOB microorganism present in denitrification [38].

3.7. Energy Audit

The energy consumption of the projected RBC was calculated by comparison with a well-optimized full-scale MBR. By adopting the same assumptions and estimation method proposed earlier, the estimated energy consumption optimum conditions in this study was 0.14 kWh/m^3. This value is much lower than the referenced full-scale MBR of 0.64 kWh/m^3. This energy consumption obtained was even lower than the one reported in our earlier study (0.18 kWh/m^3).

The RBC showed a substantially lower energy demand than the referenced bioreactor (Figure 7). The primary energy contributor in the RBC bioreactor is mechanical energy for disk rotation. Disk rotation accounted for only 0.07 kWh/m^3 of energy dissipation compared to coarse-bubble aeration (0.23 kWh/m^3). The overall energy demand of the lab-scale RBC was 0.14 kWh/m^3, which is very low compared to conventional activated-sludge treatment process energy consumption of 0.3–0.4 kWh/m^3. However, the mechanical energy for disk rotation increased exponentially as the number of shafts and the treatment capacity of the bioreactor increased.

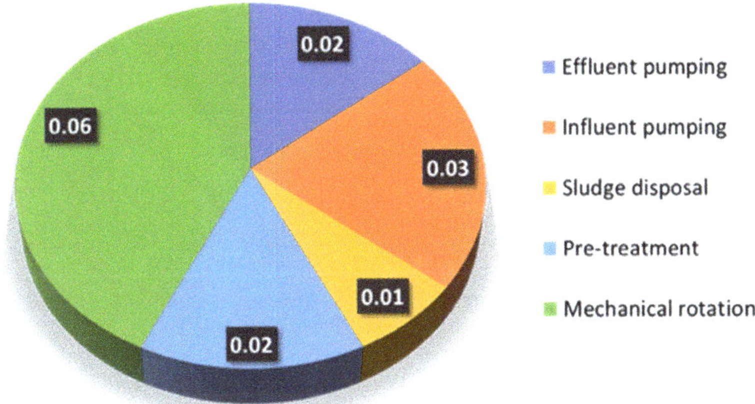

Figure 7. Specific energy demands of the RBC bioreactor.

Submerged MBR adds additional aeration to control membrane fouling, accounting for higher than 50% of energy utilization [39]. Most industrial membrane processes focus on diminishing the total energy requirement by reducing power for membrane aeration while keeping the membrane permeability. A conventional MBR spends around 0.6–0.8 kWh/m^3 of energy; however, it produces less sludge and maintains strict effluent quality [40]. The energy demand in the RBC can be further reduced by employing gravity-driven flow to eliminate the influent and effluent pumping energy demand.

The RBC bioreactor has the capability for energy-efficient treatment of municipal and industrial wastewaters. This outcome is certainly incredibly enthralling and provides the opportunity for considerable improvements to the conventional RBC. The reported energy consumption of the RBC is exceedingly appealing, associated with the CAS for wastewater treatments that utilize the specific energy of 0.3–0.4 kWh/m^3 [40]. An interesting alternative, which may have an additional impact on reducing energy use, will be using new materials, such as polyethylene foam as disc material for rotating biological contactors (RBC). It is important to remember that due to the system's low-energy demand, it is easily operable with renewable energy sources, either solar or wind power [41].

4. Conclusions

The RBC bioreactor achieved superior effluent quality and removal efficiency for domestic wastewater treatment. The average removal efficiency of 73.9% COD, 38.3% TN, 95.6% ammonium, and 78.9% turbidity was attained at a fixed 19 g COD/m^2 d organic loading rate, 9 h HRT, and 30 rpm disk rotational speed. COD, TN, and turbidity removal efficiencies decreased with increased disk rotational, while HRT and SRT positively affected biological performance. At disk rotational speeds higher than 40 rpm, biofilm detachment occurred due to increased shear rate resulting in decreased biological performance and loss of microbial biomass. Higher SRTs increased nitrifying bacteria activity and decreased sludge volumes. The energy audit of the RBC bioreactor shows that it consumed only 0.14 kWh/m^3. Using the disk rotation mechanism to supply oxygen to the microbial activity results in an energy-efficient wastewater treatment process. Therefore, the RBC provides an attractive alternative for treating various types of wastewaters in a decentralized system where issues of a large footprint are less significant.

Author Contributions: M.I., S.R., S.L., I.K., D.G.-K., M.O., S.W. (Sylwia Włodarczak) and K.C.: visualization, analysis, conceptualization, technical editing, re-writing, project management, and resource management; S.W. (Sharjeel Waqas) and J.A.K.: methodology, results, analysis, and manuscript writing. All authors have read and agreed to the published version of the manuscript.

Funding: The APC of the journal was paid by the Opole University of Technology and Poznan University of Technology (from the funds of the Ministry of Education and Science).

Institutional Review Board Statement: Not applicable.

Informed Consent Statement: Not applicable.

Data Availability Statement: Not applicable.

Acknowledgments: The authors acknowledge the support from the Deanship of Scientific Research, Najran University, Saudi Arabia, for funding this work under the Research Groups funding program grant code number (NU/RG/SERC/11/3). The authors also acknowledge Universiti Teknologi PETRONAS, Malaysia for providing the research facilities.

Conflicts of Interest: The authors declare no conflict of interest.

References

1. Meena, M.; Yadav, G.; Sonigra, P.; Shah, M.P. A comprehensive review on application of bioreactor for industrial wastewater treatment. *Lett. Appl. Microbiol.* **2022**, *74*, 131–158. [CrossRef] [PubMed]
2. Edokpayi, J.N.; Odiyo, J.O.; Durowoju, O.S. Impact of wastewater on surface water quality in developing countries: A case study of South Africa. *Water Qual.* **2017**, *10*, 66561.
3. Leonard, P.; Clifford, E.; Finnegan, W.; Siggins, A.; Zhan, X. Deployment and Optimisation of a Pilot-Scale IASBR System for Treatment of Dairy Processing Wastewater. *Energies* **2021**, *14*, 7365. [CrossRef]
4. Waqas, S.; Bilad, M.R. A review on rotating biological contactors. *Indones. J. Sci. Technol.* **2019**, *4*, 241–256. [CrossRef]
5. Enayathali, S.S. Mathematical Modelling For Rotating Biological Contactor for Treatment of Grey wastewater. *SSRG Int. J. Civ. Eng.* **2019**, *6*, 6–10. [CrossRef]
6. Hendrasarie, N.; Trilta, M. Removal of nitrogen-phosphorus in food wastewater treatment by the Anaerobic Baffled Reactor (ABR) and Rotating Biological Contactor (RBC). In Proceedings of the IOP Conference Series: Earth and Environmental Science, Moscow, Russia, 27 May–6 June 2019; p. 012017.
7. Vairavel, P.; Murty, V.R. Decolorization of Congo red dye in a continuously operated rotating biological contactor reactor. *Desalination Water Treat.* **2020**, *196*, 299–314. [CrossRef]
8. Irfan, M.; Waqas, S.; Arshad, U.; Khan, J.A.; Legutko, S.; Kruszelnicka, I.; Ginter-Kramarczyk, D.; Rahman, S.; Skrzypczak, A. Response Surface Methodology and Artificial Neural Network Modelling of Membrane Rotating Biological Contactors for Wastewater Treatment. *Materials* **2022**, *15*, 1932. [CrossRef]
9. Vergara-Araya, M.; Hilgenfeldt, V.; Steinmetz, H.; Wiese, J. Combining Shift to Biogas Production in a Large WWTP in China with Optimisation of Nitrogen Removal. *Energies* **2022**, *15*, 2710. [CrossRef]
10. Ghalehkhondabi, V.; Fazlali, A.; Fallah, B. Performance analysis of four-stage rotating biological contactor in nitrification and COD removal from petroleum refinery wastewater. *Chem. Eng. Processing-Process Intensif.* **2021**, *159*, 108214. [CrossRef]
11. Ito, T.; Aoi, T.; Miyazato, N.; Hatamoto, M.; Fuchigami, S.; Yamaguchi, T.; Watanabe, Y. Diversity and abundance of denitrifying bacteria in a simultaneously nitrifying and denitrifying rotating biological contactor treating real wastewater at low temperatures. *H2Open J.* **2019**, *2*, 58–70. [CrossRef]
12. Waqas, S.; Bilad, M.R.; Aqsha, A.; Harun, N.Y.; Ayoub, M.; Wirzal, M.D.H.; Jaafar, J.; Mulyati, S.; Elma, M. Effect of membrane properties in a membrane rotating biological contactor for wastewater treatment. *J. Environ. Chem. Eng.* **2021**, *9*, 104869. [CrossRef]
13. Waqas, S.; Bilad, M.R.; Huda, N.; Harun, N.Y.; Md Nordin, N.A.H.; Shamsuddin, N.; Wibisono, Y.; Khan, A.L.; Roslan, J. Membrane Filtration as Post-Treatment of Rotating Biological Contactor for Wastewater Treatment. *Sustainability* **2021**, *13*, 7287. [CrossRef]
14. Waqas, S.; Harun, N.Y.; Bilad, M.R.; Samsuri, T.; Nordin, N.A.H.M.; Shamsuddin, N.; Nandiyanto, A.B.D.; Huda, N.; Roslan, J. Response Surface Methodology for Optimization of Rotating Biological Contactor Combined with External Membrane Filtration for Wastewater Treatment. *Membranes* **2022**, *12*, 271. [CrossRef] [PubMed]
15. Mohammadi, M.; Mohammadi, P.; Karami, N.; Barzegar, A.; Annuar, M.S.M. Efficient hydrogen gas production from molasses in hybrid anaerobic-activated sludge-rotating biological contactor. *Int. J. Hydrog. Energy* **2019**, *44*, 2592–2602. [CrossRef]
16. Hamedi, S.; Babaeipour, V.; Rouhi, M. Design, construction and optimization a flexible bench-scale rotating biological contactor (RBC) for enhanced production of bacterial cellulose by Acetobacter Xylinium. *Bioprocess Biosyst. Eng.* **2021**, *44*, 1071–1080.
17. Li, N.; Zeng, W.; Yang, Y.; Wang, B.; Li, Z.; Peng, Y. Oxygen mass transfer and post-denitrification in a modified rotating drum biological contactor. *Biochem. Eng. J.* **2019**, *144*, 48–56. [CrossRef]
18. Subhadarsini, L.; Dash, R.R. Treatment of Industrial Waste Water using Single-Stage Rotating Biological Contactor. In Proceedings of the IOP Conference Series: Earth and Environmental Science, Moscow, Russia, 27 May–6 June 2019; p. 012025.

19. Delgado, N.; Navarro, A.; Marino, D.; Peñuela, G.; Ronco, A. Removal of pharmaceuticals and personal care products from domestic wastewater using rotating biological contactors. *Int. J. Environ. Sci. Technol.* **2019**, *16*, 1–10. [CrossRef]

20. Spasov, E.; Tsuji, J.M.; Hug, L.A.; Doxey, A.C.; Sauder, L.A.; Parker, W.J.; Neufeld, J.D. High functional diversity among Nitrospira populations that dominate rotating biological contactor microbial communities in a municipal wastewater treatment plant. *ISME J.* **2020**, *14*, 1857–1872. [CrossRef]

21. Tauber, J.; Ramsbacher, A.; Svardal, K.; Krampe, J. Energetic Potential for Biological Methanation in Anaerobic Sewage Sludge Digesters in Austria. *Energies* **2021**, *14*, 6618. [CrossRef]

22. Waqas, S.; Bilad, M.R.; Man, Z.; Wibisono, Y.; Jaafar, J.; Mahlia, T.M.I.; Khan, A.L.; Aslam, M. Recent progress in integrated fixed-film activated sludge process for wastewater treatment: A review. *J. Environ. Manag.* **2020**, *268*, 110718. [CrossRef]

23. Ziembińska-Buczyńska, A.; Ciesielski, S.; Żabczyński, S.; Cema, G. Bacterial community structure in rotating biological contactor treating coke wastewater in relation to medium composition. *Environ. Sci. Pollut. Res.* **2019**, *26*, 19171–19179. [CrossRef] [PubMed]

24. Waqas, S.; Bilad, M.R.; Man, Z.B.; Suleman, H.; Nordin, N.A.H.; Jaafar, J.; Othman, M.H.D.; Elma, M. An energy-efficient membrane rotating biological contactor for wastewater treatment. *J. Clean. Prod.* **2021**, *282*, 124544. [CrossRef]

25. APHA; WPCF. *Standard Methods for the Examination of Water and Wastewater*; American Public Health Association: Washington, DC, USA, 2012.

26. Bilad, M.R.; Declerck, P.; Piasecka, A.; Vanysacker, L.; Yan, X.; Vankelecom, I.F. Treatment of molasses wastewater in a membrane bioreactor: Influence of membrane pore size. *Sep. Purif. Technol.* **2011**, *78*, 105–112. [CrossRef]

27. Fenu, A.; Roels, J.; Wambecq, T.; De Gussem, K.; Thoeye, C.; De Gueldre, G.; Van De Steene, B. Energy audit of a full scale MBR system. *Desalination* **2010**, *262*, 121–128. [CrossRef]

28. Waqas, S.; Bilad, M.R.; Man, Z.B.; Klaysom, C.; Jaafar, J.; Khan, A.L. An integrated rotating biological contactor and membrane separation process for domestic wastewater treatment. *Alex. Eng. J.* **2020**, *59*, 4257–4265. [CrossRef]

29. Waqas, S.; Bilad, M.R.; Man, Z.B. Performance and Energy Consumption Evaluation of Rotating Biological Contactor for Domestic Wastewater Treatment. *Indones. J. Sci. Technol.* **2021**, *6*, 101–112. [CrossRef]

30. Vázquez-Padín, J.R.; Mosquera-Corral, A.; Campos, J.L.; Méndez, R.; Carrera, J.; Pérez, J. Modelling aerobic granular SBR at variable COD/N ratios including accurate description of total solids concentration. *Biochem. Eng. J.* **2010**, *49*, 173–184. [CrossRef]

31. Brazil, B.L. Performance and operation of a rotating biological contactor in a tilapia recirculating aquaculture system. *Aquac. Eng.* **2006**, *34*, 261–274. [CrossRef]

32. Annavajhala, M.K.; Kapoor, V.; Santo-Domingo, J.; Chandran, K. Comammox functionality identified in diverse engineered biological wastewater treatment systems. *Environ. Sci. Technol. Lett.* **2018**, *5*, 110–116. [CrossRef]

33. Waqas, S.; Bilad, M.R.; Man, Z.B. Effect of organic and nitrogen loading rate in a rotating biological contactor for wastewater treatment. In Proceedings of the Journal of Physics: Conference Series, Nanchang, China, 26–28 October 2021; p. 012063.

34. Ebrahimi, M.; Kazemi, H.; Mirbagheri, S.; Rockaway, T.D. Integrated approach to treatment of high-strength organic wastewater by using anaerobic rotating biological contactor. *J. Environ. Eng.* **2018**, *144*, 04017102. [CrossRef]

35. Choi, W.-H.; Shin, C.-H.; Son, S.-M.; Ghorpade, P.A.; Kim, J.-J.; Park, J.-Y. Anaerobic treatment of palm oil mill effluent using combined high-rate anaerobic reactors. *Bioresour. Technol.* **2013**, *141*, 138–144. [CrossRef] [PubMed]

36. Choi, J.; Kim, E.-S.; Ahn, Y. Microbial community analysis of bulk sludge/cake layers and biofouling-causing microbial consortia in a full-scale aerobic membrane bioreactor. *Bioresour. Technol.* **2017**, *227*, 133–141. [CrossRef] [PubMed]

37. Prinčič, A.; Mahne, I.; Paul, E.A.; Tiedje, J.M. Effects of pH and oxygen and ammonium concentrations on the community structure of nitrifying bacteria from wastewater. *Appl. Environ. Microbiol.* **1998**, *64*, 3584–3590. [CrossRef] [PubMed]

38. Kogler, A.; Farmer, M.; Simon, J.A.; Tilmans, S.; Wells, G.F.; Tarpeh, W.A. Systematic Evaluation of Emerging Wastewater Nutrient Removal and Recovery Technologies to Inform Practice and Advance Resource Efficiency. *ACS EST Eng.* **2021**, *1*, 662–684. [CrossRef]

39. Gil, J.; Túa, L.; Rueda, A.; Montaño, B.; Rodríguez, M.; Prats, D. Monitoring and analysis of the energy cost of an MBR. *Desalination* **2010**, *250*, 997–1001. [CrossRef]

40. Van Dijk, L.; Roncken, G. Membrane bioreactors for wastewater treatment: The state of the art and new developments. *Water Sci. Technol.* **1997**, *35*, 35–41. [CrossRef]

41. Mohammadi, M.; Mohammadi, P. Developing single-substrate steady-state model for biohydrogen production in continuous anaerobic activated sludge-rotating biological contactor system: Novel insights on the process. *Int. J. Energy Res.* **2022**, *46*, 2041–2050. [CrossRef]

MDPI
St. Alban-Anlage 66
4052 Basel
Switzerland
Tel. +41 61 683 77 34
Fax +41 61 302 89 18
www.mdpi.com

Energies Editorial Office
E-mail: energies@mdpi.com
www.mdpi.com/journal/energies